Advances in Neurobiology

Volume 42

Series Editor

Alexej Verkhratsky, University of Manchester
Manchester, UK

Advances in Neurobiology covers basic research in neurobiology and neurochemistry. It provides in-depth, book-length treatment of some of the most important topics in neuroscience including molecular and pharmacological aspects. The main audiences of the series are basic science researchers and graduate students as well as clinicians including neuroscientists (neurobiologists and neurochemists) and neurologists. *Advances in Neurobiology* is indexed in PubMed, Google Scholar, and the Thompson Reuters Book Citation Index.

Linda J. Noble-Haeusslein • David M. Schnyer
Editors

Traumatic Brain Injuries

Navigating the Unique Terrain of the Injured Young and Aged Brains

Springer

Editors
Linda J. Noble-Haeusslein
Departments of Neurology and Psychology,
Dell Medical School and the College
of Liberal Arts
The University of Texas at Austin
Austin, TX, USA

David M. Schnyer
Department of Psychology
The University of Texas at Austin
Austin, TX, USA

ISSN 2190-5215 ISSN 2190-5223 (electronic)
Advances in Neurobiology
ISBN 978-3-031-69831-6 ISBN 978-3-031-69832-3 (eBook)
https://doi.org/10.1007/978-3-031-69832-3

This Springer imprint is published by the registered company Springer Nature Switzerland AG
The registered company address is: Gewerbestrasse 11, 6330 Cham, Switzerland

Preface

This book focuses on two understudied groups that are most vulnerable to traumatic brain injuries; namely, the aged and the young. Traumatic brain injuries occur across all walks of life, and age at time of injury and the nature and severity of the injury are critical determinants of recovery. However, these variables contribute to the heterogeneity of the injury, which challenges the medical and surgical fields in providing optimal care across these diverse groups. These chapters address preclinical studies that offer the opportunity to control these variables, and, as such, serve as a critical pipeline for discovery and translation to the clinic. Importantly, catalysts for this translational research are inspired by findings in the clinical arena.

Austin, TX, USA

Linda J. Noble-Haeusslein
David M. Schnyer

Introduction

This book tackles the complex subject of traumatic brain injuries (TBIs) at two ends of the aging spectrum: namely, the young and the aged brains. These areas of investigation have garnered increased interest from research communities over the past decade. Based upon searches conducted from 2013 to 2023, using PubMed, the National Library of Medicine, the overall number of publications using the general search term "traumatic brain injury," increased by 42%. Of the 6810 publications in 2023, 10% and 11% of these publications were attributed to the search terms "traumatic brain injury" and either "young" or "aged," respectively.

Here we highlight key experts who have contributed to this extraordinary effort to advance our understanding of both injury and recovery in the challenging context of unique neural environments, defined by age at the time of injury. This book is divided into two major sections that address clinical (Chaps. 1, 2, 3, 4, 5 and 6) and preclinical perspectives (Chaps. 7, 8, 9, 10, 11, 12, and 13). A clear picture emerges from this collective effort that speaks to the importance of broad-based, in-depth clinical findings that serve as a platform for optimizing care as well as contributing to the refinement and validation of relevant animal models.

Clinical Perspectives

Pediatric Traumatic Brain Injuries Clinically based findings offer insight into how a traumatic brain injury adversely alters the structure and function of the brain, and the impact that these changes have on overall quality of life. In the context of pediatric traumatic brain injuries, a key challenge is that the term "pediatric" refers to a broad developmental age range spanning infancy to adolescence. Thus, a multidisciplinary approach is essential in optimizing recovery, across ages that reflect different stages of brain development as well as severity of injury. In the context of acute management of severe pediatric traumatic brain injuries, a key objective is to limit secondary damage.

Wang and Cronin (Chaps. 1 and 2, respectively) address the "climate" of the pediatric intensive care unit where medical/surgical management is guided by close monitoring of cerebral perfusion pressure, intracranial cranial pressure, temperature, EEG, and mechanical ventilation. There is a comprehensive overview of triage for the early management of patients who have sustained mild to severe closed head injuries. These chapters capture the need for multi-disciplinary teams, which is particularly relevant when encountering cases of non-accidental trauma that is all too common in children. Despite this standard of care, there remain conflicting data regarding the acute management of brain-injured children. Thus, the field continues to evolve with the objective of optimization of care in the early management of brain-injured children.

Dennis, Keleher, and Bartnik-Olson (Chap. 3) address the intersection between neuroimaging and assessment of functional outcomes in brain-injured children. The authors offer an extensive review of imaging modalities and consider those changes in the structure of the brain that are predictive of functional outcome. Such an approach underscores the promise of using advanced brain imaging as a prognostic tool to better understand recovery of function. With future possibilities of increased sample sizes and longitudinal studies, there is an opportunity to develop patient-tailored interventions for this vulnerable population.

Traumatic Brain Injuries in the Aged Management of traumatic brain injuries in the elderly poses unique challenges including pre-existing conditions that impede the recovery process. Gheim, Broderick, and Stein (Chap. 4) offer an in-depth overview of dementia and address the increased risk for acceleration toward dementia that may occur with the elderly who sustain traumatic brain injuries. While early age traumatic brain injuries have been linked to dementia, injury to the aged brain has also been associated with late onset depression, which may interact with dementia risk to reveal a rather muddled picture of an individual's functional outcome following traumatic brain injury.

Gheim, Broderick, and Stein (Chap. 5) have also addressed how a late life traumatic brain injury alters patterns of sleep. While the elderly often experience changes in patterns of sleep, a traumatic brain injury may also alter this pattern, giving rise to significant sleep-wake disorders that can impact recovery and regain of function. The authors address potential mechanisms underlying these disorders, including both direct injury to the circuitry that modulate sleep as well as non-selective disruption and the putative involvement of the glymphatic system.

McGee and Schnyer (Chap. 6) focus on an understudied issue; namely, how earlier life traumatic brain injuries alter the trajectory of aging in an otherwise normal population. Considerable focus has been on the long-term repercussions of repetitive injury in early adulthood, most associated with contact sports. However, few studies have examined the long-term repercussions of a single traumatic brain injury earlier in life. This chapter reviews what is known and sets forth a framework to better understand this important issue going forward.

Preclinical Perspectives

While preclinical models of pediatric traumatic brain injuries have been developed in a variety of species, including piglets and rabbits, a majority of this research has been conducted in rodents where there has been an opportunity to study the severity and nature of the injury, and the early and long-term consequences of that injury on behavior. This collective effort has revealed the vulnerability of the developing brain to injury, the mechanisms underlying secondary damage and behavioral recovery, and strategies to support recovery.

Pediatric Traumatic Brain Injuries Moschonas, Annas, Zamudio-Flores, Jaravis, Lajud, Bondi, and Klein (Chap. 7) provide a thorough overview of four classic pediatric rodent models; namely, controlled cortical impact, weight drop, fluid percussion, and abusive head trauma. These models, which produce distinct features consistent with clinical findings, are in line with the known clinical heterogeneity of traumatic brain injuries. Thus, these models serve as the basis for understanding not only the kinetics of the injury that likely differ across injury types but also responsiveness to therapeutic interventions.

The research conducted by Gorthy, Balleste, Placeres-Uray, and Atkins (Chap. 8) is based upon a key clinical finding; namely, that some children, who sustain mild traumatic brain injuries, show remarkable recovery whereas others experience prolonged symptoms. These investigators present compelling evidence that early life stress is a modifier of recovery after an early age brain injury. Such findings offer insight into the vulnerability of the young brain to adverse childhood events, a finding which has served to catalyze a broader clinical assessment of pre-injury health conditions in brain-injured children.

There is a growing body of literature that recognizes age at time of injury and sex as a biological variable as determinants of recovery after a traumatic brain injury. Semple and Mychasiuk (Chap. 9) highlight the maturational trajectory of the broadly distributed social brain network where early-age brain injuries may disrupt this network, resulting in the emergence of anti-social behaviors. Equally important, there are well-defined sex differences in social behaviors in both rodents and humans. This variable likely plays a key role in defining unique signatures of social behaviors after an early age brain injury.

Traumatic brain injuries result in the engagement of injury severity-dependent inflammation. Leonard, Ladner, Harris, Jager, and Theus (Chap. 10) address this topic, focusing on neuroimmune interactions in the context of both brain development and aging. These authors describe the complexity of these interactions, where aging alters the profile of the neuroinflammatory response. The latter is exemplified in immune profiles that vary between brain-injured young children and adults. These differences are attributed to an immature immune response with limited ability to counter the impact of a brain injury. As such, developmental trajectories are disrupted within an environment of neurodegeneration. These events adversely impact maturation-dependent behaviors and are implicated in post-traumatic epilepsy. It is noteworthy that the aging population likewise has a unique immune profile that is characterized by immune senescence. These collective observations speak to brain injury-related, age-dependent unique immune signatures that disrupt brain development and impair recovery.

Few studies have examined the cerebellum after a traumatic brain injury. Donovan, Lozano, Rao, Rodriguez, and Noble-Haeusslein (Chap. 11) highlight the remote vulnerability of the cerebellum in diverse adult models of traumatic brain injuries to the rodent brain. Recent clinical studies have likewise demonstrated the vulnerability of the cerebellum in children and adolescents after accidental traumatic brain injuries (refer to Chap. 3). This review addresses the pathobiology of the cerebellum, focusing on diaschisis, mechanotransduction, and inflammatory networks and the social consequences of cerebellar disruption in brain-injured children.

Traumatic Injuries in the Aged and Aging Brains The aged population is prone to fall-related traumatic brain injuries and even mild brain injuries may contribute to cognitive decline. Both clinical findings and preclinical models reveal characteristics of the aged brain that likely give rise to its vulnerability. The emerging premise, based upon preclinical findings, is that the pathophysiology of a traumatic brain injury and potential for recovery varies across the aging spectrum, with heightened vulnerability in the young and the aged. In these chapters, the authors address those factors that influence outcomes in a model brain injury in aged animals.

In their extensive review, Cheng, Anwer, Fan, Cheung, Zhang, and Wellington (Chap. 12) identify multiple pathways that may worsen recovery, based upon preclinical models of aging, including age-related mitochondrial dysfunction, oxidative stress, and a pro-inflammatory environment. These authors identify challenges in preclinically modeling where the intent is to generate a consistent injury severity across different age groups where there are changes in the mechanical properties of the brain as well as brain volume. They also support blood biomarkers which may guide clinical diagnosis and practice,

Race, Moschonas, Klein, and Bondi (Chap. 13) address age-related differences in affective, cognitive, and neurobehavioral deficits and putative factors that contribute to these differences. While age-related co-morbidities, including diabetes, hypertension, and premorbid dementia, likely contribute to deficits, few preclinical models incorporate these variables into the experimental design. The authors also consider the biological factors that influence post-injury recovery including oxidative stress and mitochondrial dysfunction, dysregulated inflammation, and the more limited capacity for neuroplasticity.

Contents

Part I
Clinical Perspectives of Traumatic Brain Injuries

Chapter 1
Intracranial Pressure and Its Related Parameters in the Management of Severe Pediatric Traumatic Brain Injury

Vincent Y. Wang

Abstract There are a number of challenges in the management of acute traumatic brain injuries in children. Beyond their relatively broad age range, which spans neonates to late adolescence, these children may likewise present with coexisting injuries. Thus, their management often necessitates a multidisciplinary team, who coordinate medical/surgical management during their hospitalization in the intensive care unit, as well as specialists in pediatric neurology and rehabilitation during postoperative recovery. Here we address standard of care for acute management, based upon established guidelines and focusing on intracranial pressure, cerebral perfusion pressure, and autoregulation. We also consider the controversies related to monitoring intracranial pressure and methods for sedation and treatment.

Keywords Intracranial pressure · Intensive care unit · Cerebral perfusion pressure · Decompressive craniectomy

Abbreviations

ABP	Arterial blood pressure
AIS	Abbreviated Injury Score
ADAPT trial	Approaches and Decisions on Acute Pediatric TBI Trial
BEST-TRIP trial	Benchmark Evidence from South American Trials: Treatment of Intracranial Pressure
CA	Cerebral autoregulation
CBF	Cerebral blood flow
CPP	Cerebral perfusion pressure
CSF	Cerebrospinal fluid

V. Y. Wang (✉)
Department of Neurosurgery, Dell Medical School, University of Texas at Austin, Austin, TX, USA
e-mail: vywang@ascension.org

L. J. Noble-Haeusslein, D. M. Schnyer (eds.), *Traumatic Brain Injuries*, Advances in Neurobiology 42, https://doi.org/10.1007/978-3-031-69832-3_1

DECRA trial	Decompressive Craniectomy trial
DC	Decompressive craniectomy
EVD	External ventricular drain
GOS	Glasgow Outcome Scale
GOS-E	Glasgow Outcome Scale extended
GOS-EP	Glasgow Outcome Scale Extended Pediatric
HTS	Hypertonic saline
ICP	Intracranial pressure
ICU	Intensive care unit
NIRS	Near-infrared spectroscopy
NMDA receptor	N-methyl-D-aspartate receptor
PRx	Pressure reactivity index
RESCUE-ICP	Randomised Evaluation of Surgery with Craniectomy for Uncontrollable Elevation of Intracranial Pressure
TBI	Traumatic brain injury (Note: TBI-TRC is the name of the database)
TQUIP	Trauma Quality Improvement Program
TCD	Transcranial Doppler

1.1 Introduction

Traumatic brain injuries (TBIs) are a leading cause of death and disability in the pediatric population. According to the Centers for Disease Control and Prevention, there were over 640,000 emergency room visits due to pediatric TBIs in 2013, resulting in 18,000 hospitalizations and 1500 deaths [1]. Among survivors of moderate or severe pediatric TBIs, 61% have significant disability [1]. Various anatomical factors predispose pediatric patients to more severe TBIs. Young children have a higher head to body ratio, weaker neck muscles, and smaller facial bones, resulting in a higher probability of head injury in children compared to adults [2].

Management of pediatric TBIs poses unique challenges to the clinical providers. Pediatric patients range from neonate to adolescents, and their physiologic parameters vary widely among different age groups. Many of these patients have other coexisting injuries, and management requires a multidisciplinary team of pediatric intensivists, pediatric trauma surgeons, and pediatric neurosurgeons. Other specialists such as pediatric neurologists and pediatric physical medicine and rehabilitation physicians are also integral members of the team. In Chap. 2, the authors provide a comprehensive overview of the evaluation and management of a pediatric TBI patient, from prehospital care to rehabilitation. In this chapter, we will focus our discussion on intracranial pressure (ICP) and its related parameters, including cerebral perfusion pressure (CPP) and autoregulation. The chapter will discuss the current standard of care of ICP and CPP, based on recommendations from the Guidelines for the Management of Pediatric Severe Traumatic Brain Injury, Third Edition: Update of the Brain Trauma Foundation Guidelines (known as the Guideline below). We will also review literature that is published after the 2019 guideline including studies from the ADAPT trial (Approaches and Decisions on Acute Pediatric TBI Trial) as well as selected topics that are not covered by the Guideline.

1.2 Intracranial Pressure Monitoring and Management

A primary goal in the management of traumatic brain injuries, whether adults or children, is to minimize secondary injury to the brain. Some of the interventions include general intensive care unit (ICU) care such as blood pressure support and ventilation support to avoid hypotension and hypoxia. Specific to TBI, ICP monitoring and treatment have been the pillars of the management of TBI patients and have been recommended in all the guidelines since the first edition of the adult and pediatric guidelines. However, there remains significant variation in the practice of ICP monitoring and management nationally and internationally [3, 4].

1.2.1 ICP: To Monitor or Not

The current Guideline recommends ICP monitoring to improve overall outcome [5]. A number of retrospective studies provide evidence for ICP monitoring in pediatric TBI patients. Alkhoury and Kyriakides studied the National Trauma Data Bank and found that the use of ICP monitoring was associated with a lower mortality rate compared to no ICP monitoring [6]. Similarly, Aliali et al., using data from the Trauma Quality Improvement Program (TQIP) from 2010 to 2012, also reported that the use of ICP monitoring was associated with lower mortality at both the patient and hospital levels, although this study was not used as part of the evidence in the Guideline [7]. In another retrospective study, Bennett et al. found that hospitals with a higher rate of ICP monitoring had better patient outcomes with lower rate of mortality or severe disability using data from the Pediatric Health Information System [8]. However, in a subsequent study, Bennett et al. reported no significant difference in this data set, in terms of mortality and functional survival, using propensity-weighted effectiveness analysis [9]. Two additional studies have come out since the publication of the 2019 guideline on ICP monitoring. Delaplain et al. analyzed another cohort from the TQIP database and found that ICP monitoring was an independent risk factor for inpatient mortality [10]. However, Shibahashi et al., using a propensity score matching analysis with data from a national inpatient database from Japan, found that ICP monitoring is associated with lower inpatient mortality rate [11].

1.2.2 ICP and Controversies

The value of ICP monitoring remains controversial. Most of the available data reflect prospectively collected data from various trauma or inpatient databases. One common issue in these various datasets is that the rate of ICP monitor placement is relatively low. Oftentimes, the monitored patients may have more severe injury. While there are many potential explanations for the conflicting data, one possibility is a selection bias of placement of ICP monitors in only the more severely injured patients, based on the treating physicians' assessment. While statistical tools, such as multivariate analysis or propensity weighted analysis, may help with the analysis of the data in relationship to the severity of injury, many of the available datasets such as TQIP may not capture all the clinical and radiographic data that identify the group of patients who will benefit from ICP monitoring. A new randomized controlled trial, similar to the adult BEST-TRIP trial, is currently ongoing and may shed more light on the issue [12]. Also, studies like TRACK-TBI or CENTER-TBI, which prospectively collect a more detailed clinical, radiographic, and biomarker dataset with high-density data, may also provide the basis to identify the patient group who will benefit from ICP monitoring.

1.2.3 ICP Monitoring Technology

There are a number of different technologies available for ICP monitoring, including both invasive and noninvasive technologies [13]. Invasive approaches that rely on an intraparenchymal ICP monitor and external ventricular drain (EVD) remain the mainstay of ICP monitoring in clinical practice. EVD has the added advantage of being able to drain cerebrospinal fluid (CSF) as a therapeutic intervention to lower ICP. CSF drainage via EVD is recommended as a way to manage increased ICP in the 2019 Guideline [5]. It shall be noted that this recommendation was based on limited data from three small-scale studies, with a combined 56 pediatric patients in total [5]. This prompted a large-scale study under the ADAPT trial. In this large international observational comparative effectiveness study, 1000 children with severe TBI with ICP monitoring were enrolled [14]. Thirty-one percent had an EVD placed with CSF diversion while 69% had an intraparenchymal ICP monitor without CSF diversion [14]. Using propensity weighted analysis of 98 pairs of subjects, there was no difference in the six-month outcome based on Glasgow outcome scale-extended for pediatrics (GOS-EP) [14]. However, the ICP was lower in the CSF diversion group [14]. It shall be noted that the current study did not report the therapeutic intensity of the two groups. Therefore, it is not clear whether the intraparenchymal monitor group may have received more pharmacological ICP lowering treatments like hyperosmolar therapy or sedatives/paralytics.

1.2.4 Threshold for Intervention

The primary goal for ICP management is to prevent intracranial hypertension, but what constitutes intracranial hypertension is a matter of debate. In adults, normal ICP is 0–5 mmHg in the supine position and −5–0 mmHg in the upright position [15]. There is less data for normal ICP in children and with more conflicting data [15]. While some studies suggest that normal ICP range is lower in children compared to adults, this is not substantiated in other studies [15]. Despite these differences, the threshold of 20mmgHg has been used for decades as the threshold for treatment of intracranial hypertension, based on retrospective analysis of outcome of patients with different ICP thresholds [16]. In the last edition of the adult guideline, the threshold for treatment of elevated ICP has been changed to 22 mmHg, based on a single study [16]. The current pediatric Guideline recommended a target threshold of less than 20 mmHg for ICP [5]. Several retrospective studies have evaluated the different thresholds in the pediatric population in relationship to outcome. Miller Ferguson et al. studied the outcome of patients with three different thresholds (<14 mmHg, <20 mmHg, <30 mmHg) and found that there was no significant difference in terms of their Glasgow Outcome Score (GOS) [17]. In fact, the model with 20 mmHg had the best fit, although it did not reach statistical difference significance [17]. Mehta et al. analyzed a subset of the Miller Ferguson cohorts, focusing on patients younger than 2 years old, and they did not find a difference in outcome by GOS comparing threshold of 15 mmHg versus 20 mmHg [18]. Adelson et al. did a subanalysis of patients in a hypothermia trial and found that patients with good outcomes had a significantly lower ICP than patients with poor outcomes [19]. It shall be noted that in the Miller Ferguson/Mehta cohort, the threshold for treatment for elevated ICP was 20 mmHg while in the Adelson study, there were different age-specific thresholds (greater than or equal to 15 mmHg for 0–24 months old, greater than or equal to 18 mmHg for 25–96 months old and greater than or equal to 20 mmHg for 96–156 months old) [17, 18, 20]. A number of other studies using 20 mmHg as threshold for intracranial hypertension showed that intracranial hypertension was associated with poorer outcome or increased mortality [5]. In these studies, patients with ICP higher than 20 mmHg may have poorer outcomes and therefore infer that controlling ICP to a level below 20 mmHg may improve outcome. However, there is no direct evidence to support this position. If extensive effort is needed to get the ICP below 20 mmHg, the outcome can also be adversely affected. For example, if a significant amount of sedatives or hyperosmolar therapy is needed to control the ICP, the side effects of these interventions will need to be considered. It is also important to emphasize that a pressure of 20 mmHg is the goal that should be targeted for interventions. However, this does not necessarily represent the threshold for initiations of interventions. It is possible that initiating ICP-control intervention at a lower threshold may achieve better ICP control for a patient.

1.2.5 ICP Elevation Treatment

ICP elevations are treated with multiple different interventions. General ICU care such as maintaining the head of bed above 30° or avoiding a tight cervical collar is important as part of the ICP management. The sections below discussed specific ICP lowering interventions.

Sedation and analgesic medications are part of routine ICU care of trauma patients. Since pain and agitation will increase ICP, appropriate use of sedative and analgesic medications is also important for control of ICP as these medications can prevent pain and agitation from elevating the ICP. The Guideline, however, specifically allows the choice and dosing of sedatives, analgesics, and neuromuscular blocking agents to the treating physicians due to the lack of outcome data [5]. The only recommendation from the Guideline is that bolus administration of midazolam and/or fentanyl should be avoided for control of ICP given the risk of hypotension [5].

One agent that may be worth additional discussion is ketamine. Ketamine is a noncompetitive *N*-methyl-D-aspartate receptor (NMDA receptor) antagonist [21]. It has been used as an anesthetic agent over the last 50 years. Unlike other anesthetic agents, ketamine does not induce vasodilation and does not have a negative chronotropic effect on the heart. Therefore it does not induce hypotension and other side effects on the cardiovascular system, making it an ideal agent in patients with hypotension [21]. There had been some initial concern that ketamine would increase CBF and ICP, and hence it should not be used widely in TBI patients. However, based upon subsequent studies, there is no overall evidence to suggest that ketamine administration results in increased ICP in a consistent manner [21]. Indeed, a number of recent studies suggest that ketamine may be a potential therapeutic agent for treatment of elevated ICP in both pediatrics and adults. Bar-Joseph et al. used ketamine either as a sedation agent to prevent further ICP during distressing interventions or as an ICP-lowering agent in 30 pediatric patients with elevated ICP (>18 mmHg) [22]. ICP was lower after ketamine administration with both indications, with an overall 30% decrease in ICP and an average increase of 4 mmHg of CPP [22]. In a more recent study with 33 pediatric patients, Laws et al. found that when ketamine was used as a standard sedation agent, there was little effect on ICP or CPP [23]. However, when bolus ketamine was given for treatment of elevated ICP, there was a significant decrease in ICP with an elevation of CPP [23]. These preliminary findings suggest that bolus ketamine may be a potential therapeutic agent for elevated ICP in pediatric TBI. In addition to its effect on ICP and CPP, ketamine also may be helpful in the treatment of spread depolarization, a complication seen in severe TBI patients.

CBF is known to be inversely related to the $PaCO_2$ level in blood. Therefore, hyperventilation that lowers $PaCO_2$ will lower both CBF and hence ICP. While hyperventilation was used in the past for ICP elevation treatment, it has been falling out of favor due to the potential ischemic effect of lowering CBF. The current Guideline recommends against prophylactic severe hyperventilation to a $PaCO_2$ less than 30 mmHg during the initial 48 h [5]. If hyperventilation is to be used, advanced neuromonitoring may be needed to evaluate potential cerebral ischemia [5].

1.2.6 Hyperosmolar Therapy

Hyperosmolar therapy has been used for over 100 years and remains the workhorse for the management of elevated ICP [24]. Although many agents have been used in the past, mannitol and hypertonic saline are the two main agents currently used in the ICU. Mannitol was first introduced in the 1960s for the treatment of elevated ICP and has been the agent of choice until the last 10–20 years. The use of hypertonic saline (HTS) has been gaining popularity for treatment of elevated ICP and cerebral edema and has become the recommended agent in guidelines published by Neurocritical Care Society and also the current pediatric TBI Guideline [5, 25].

Mannitol is a sugar molecule and is generally prepared as a 20% solution in the case of ICP treatment. Mannitol decreases ICP through two different mechanisms. Immediately after administration in adult TBI patients, mannitol decreases viscosity of blood, which will, in turn, increase CBF [26]. In a brain with intact autoregulation, vasoconstriction will serve to maintain a constant CBF [26]. The decreased cerebral blood volume hence would decrease ICP, and this effect has been shown to achieve its greatest effect within the first 10 min, with a rebound in cerebral blood volume at around 75 min [26]. Importantly, this effect on ICP is dependent on intact autoregulation mechanism [26, 27]. Mannitol also serves as an osmotic diuretic, whereby water will move from brain parenchyma to blood [24]. Thus, it may lower blood volume, resulting in the development of hypotension.

Hypertonic saline (HTS) has similar rheological and osmotic effects and thus leads to a reduction in ICP. An important advantage of HTS is that unlike mannitol, it does not cause hypotension (in fact it can sometimes increase blood pressure), making it easier to use in trauma patients [24]. HTS may also have other advantages due to its immunomodulatory, neurochemical, and vasoregulatory effects [24].

Over the past 10–20 years, multiple studies have focused on the use of HTS for the treatment of elevated ICP. However, there has been significant variation in terms of the concentration of the HTS or the manner of administration (bolus versus continuous, targeting specific sodium goal or targeting ICP goal). Thus, it remains debatable whether HTS is necessarily superior to mannitol as a ICP reduction agent in adults and children. In a Cochrane review from 2020, Chen et al. found no convincing evidence that HTS is superior to mannitol in the treatment of acute TBI [28]. Similarly, in a recent meta-analysis by Bernhardt et al., HTS was evaluated as a treatment for uncontrollable ICP. The findings in support of HTS did not reach statistical significance [29].

The current Guideline has multiple recommendations regarding the use of HTS for treatment of intracranial hypertension. There is a level II recommendation for the. The use of 3% HTS bolus for treatment of intracranial hypertension [5], a level III recommendation for continuous infusion of HTS in patients with intracranial hypertension [5]. There is also a level III recommendation for the use of 23.4% bolus HTS for treatment of refractory ICP elevation [5]. Notably, there is no recommendation for use of mannitol, due to the lack of available studies that met inclusion criteria for the guideline establishment [5].

It should be noted that the evidence used by the Guideline to support the above recommendation is actually quite limited. In the case of the recommendation for the use of 3% HTS bolus, the Guideline was based on two studies with a total of 34 patients [5]. Shein et al. conducted a study involving 16 patients that compared an HTS bolus to fentanyl or pentobarbital bolus. The HTS bolus was not only more effective for treatment of elevated ICP; it also increased CPP [30]. Complementary to this research, Fisher et al. [31] found that HTS bolus administration was associated with a lower ICP than normal saline administration and also required fewer other treatments. Evidence used for the recommendation of continuous infusion likewise reflects a paucity of research. It is based on one prospective study and one retrospective study, and the use of 23.4% recommendation was based on one retrospective study [5].

Additional studies have since been published after the Guideline. Wellard et al. [32] evaluated 30 patients who were given HTS bolus for high ICP and noted that 56% of them responded with a decrease in ICP. In this study, responders would have a lower ICP and decreased pressure reactivity index (PRx) while the nonresponders would have a higher CP and increased PRx [32]. Importantly, when HTS was given when ICP was not elevated, it resulted in higher ICP and increased PRx, suggesting that HTS should only be used to treat but not to prevent intracranial hypertension [32].

Perhaps the most important study was the ADAPT trial that was published in 2022. In this study, 518 patients received either hypertonic saline or mannitol [33]. Overall, administration of HTS was associated with a slightly better decrease in ICP than mannitol (−1.03 mmHg vs −0.2 mm Hg) [33]. In an adjusted model, HTS administration was shown to be more effective than mannitol for ICP reduction when treating episodes of ICP >25 mmHg, but there was no significant difference in situations of ICP>20 mmHg or >30 mmHg [33]. There are several important observations in this study. This large-scale study provided additional data to suggest that HTS may be slightly superior than mannitol for treatment of elevated ICP, in line with other published data [29]. However, the difference was modest at best. Also, it should be noted that the average decrease in ICP was quite small in response to each dose of treatment, only 1 mmHg. However, 78% of the doses were given when the ICP was below 20 mmHg [33]. This would suggest that the effect of bolus HTS is most effective when given when ICP is significantly elevated [33]. A caveat to this study was that clinical outcomes were not reported. Thus, an important issue that remains to be resolved is whether treatment of ICP with hyperosmolar therapy actually leads to better outcomes. While elevated ICP is known to be associated with worse outcome and that hyperosmolar therapy has the potential of lowering ICP, there has not been any solid data that confirm an improved outcome in response to hyperosmolar therapy [29]. This is partly why the current adult guideline by the Brain Trauma Foundation does not have a specific recommendation about the use of hyperosmolar therapy. The use of hyperosmolar therapy, therefore, should be dictated by the overall clinical picture of the individual patient.

Decompressive craniectomy (DC) has been used for over 100 years to reduce brain swelling [34]. However, there has been much controversy about the utility of this procedure in improving outcome. While it has been generally accepted that DC can reduce intracranial pressure effectively, there was often a concern whether performing DC in TBI patients will reduce mortality but result in patients in severely disabled or persistent vegetative state. In the late 2000s, evidence emerged that DC for malignant middle cerebral artery stroke significantly improved the functional outcome of patients compared to medical management. This partly reignited the interest in DC for TBI as well. The DECRA trial was published in 2011, and its result showed that while DC would decrease ICP, it was associated with a worse outcome [35]. However, there were a number of criticisms of the DECRA trial such as the surgical technique used, cohort characteristics of patients, etc. and that there was reservation about the generalization of the data from this trial. RESCUE-ICP, another randomized controlled trial, was published in 2016 and showed that secondary DC for refractory intracranial hypertension resulted in lower mortality but higher proportion of patients in persistent vegetative state, lower, and upper severe disability category [36].

Such large randomized controlled trial data are not available in pediatric TBI. The current suggested recommendation from the Guideline is a level III recommendation that DC be used to treat neurologic deterioration, herniation, or intracranial hypertension refractory to medical management [5]. In general, data from various treatment series, retrospective studies, and one randomized controlled trial suggest that ICP decreases after DC, and fewer interventions are needed for ICP elevation during the postoperative period [5, 37]. In terms of outcome, the data are more conflicting. In a prospective study, Cho et al. reported a significantly improvement in patients who had elevated ICP and underwent DC [38] whereas a retrospective study by Josan et al. showed better outcomes in DC group than the non-DC group [39].

While other studies may not show a significant difference in outcome after DC compared to medical treatment, these data do not necessarily refute the utility of DC. For example, Taylor et al. did not report a statistically significant difference in outcomes between the medical treatment group and DC. Such findings might be partly due to differences in the statistical methods that were applied to each group. In this particular study, 54% of the DC group (13 patients) had a favorable GOS score (median time 5.9 months) versus 14% in the medical treatment group (14 patients), and this did not reach statistical significant difference [40]. It is possible that with a larger sample size, a statistically significant difference may be found. Several retrospective studies compared a DC cohort with a medical treated cohort. Mhanna et al. showed that there was no difference in terms of survival rate and median GOS score between the DC group and a matched non-DC group [41]. However, 42% of the patients in the DC group achieved a GOS score of 5, compared to none in the non-DC group [41]. Another point to note in this study was that despite decompressive craniectomy, many patients in the DC group still had significantly elevated ICP, which was in contrast to most of the other studies that showed that DC patients were more likely to have ICP controlled [41]. This may suggest that the DC group in this study may have sustained more severe injuries. Similarly, in a study by Thomale et al., the authors found no difference in outcome between the conservatively treated group and the DC group [42]. However, in 92% of the conservatively managed patients, ICP was controlled with first tier therapy [42]. In contrast, 11 out of the 14 DC patients' ICP were not controlled with first tier therapy, suggesting that the ICP burden was significantly higher in the DC group (the other three patients in the DC group had craniectomy during surgical treatment of mass lesions) [42]. Lastly, Bruns et al. retrospectively analyzed patients who underwent DC and compared their outcome to patients who did not undergo DC using a prospective trauma database with patients from Germany, Austria, and Switzerland [43]. The authors found that the DC group had a significantly higher mortality rate and poor outcome compared to the non-DC group [43]. The observation was likely due to the significantly worse injuries sustained in the DC group. The authors also analyzed the data based on AIS head and neck score [43]. In the AIS head and neck 3 or 4 group, the non-DC group had better outcome compared to the DC group, while for the AIS head and neck 5 group, the DC group had better outcome scores [43]. Several interesting factors could be noted in the data. The average Glasgow Coma Scale score was significantly lower in the DC group compared to the non-DC group in the AIS head/neck 3 and 4 cohort, suggesting that the DC patients had a more severe head injury [43]. It is also somewhat unclear why an AIS head and neck 3 or 4 patient will be undergoing DC, as these patients are by definition lost consciousness less than 24 h [43]. This will not fit the profile of the typical patient with refractory ICP. It is possible that many of these patients underwent craniectomy for other pathologies, such as depressed or open skull fractures. Such patients would have a different clinical profile than a patient with refractory ICP elevation. It should be noted that in the AIS head and neck 5 cohort, the DC patients did have better outcomes [43]. This will be more in line with other studies.

In summary, although high-quality data from a randomized controlled trial or large prospective database is lacking for DC in pediatric TBI, data available so far suggest that DC is an effective treatment for refractory ICP and is likely associated with improved outcome compared to medical management alone. An interesting observation is that in the pediatric population, DC survivors often have a GOS of 4 or even 5 [44, 45]. This is in contrast to adult TBI patients, where survivors generally have a lower GOS/GOS-E score [36]. This suggests that the benefit of DC is potentially even greater in the pediatric population.

1.3 Cerebral Perfusion Pressure

Cerebral perfusion pressure (CPP) is defined by mean arterial pressure (MAP) minus the mean ICP. It is thought to be the major gradient driving cerebral blood flow (CBF). In the normal physiological state, CBF is maintained by autoregulation over a wide range of CPP values. However, autoregulation is often impaired in the injured brain. Therefore, maintaining an adequate CPP is important to minimize risk of ischemia to the brain [46].

While it is not clear whether there is significant age variation in ICP, CPP variation is highly likely among different age groups since blood pressure is significantly lower in the younger age group. Having an age-specific threshold may be important, but getting a sufficiently large enough dataset to establish age-specific threshold is difficult. It is difficult to combine data from different studies due to a number of factors. Difference reference point for zero reference for ICP or MAP can lead to substantial measurement difference [5]. The variation in treatment protocols and the different statistical methods used also limit the ability to compare results from different studies [5].

In general, a higher mean CPP is associated with better outcome in various studies [5, 19, 47]. 40 mmHg has been considered as the minimal threshold for CPP in the pediatric population [5]. Downard et al. analyzed 118 patients' GOS scores in relationship to mean CPP divided into deciles from 40 to above 70 and found no significant difference in GOS among the different deciles [48]. Similarly, Miller Ferguson et al. analyzed various CPP thresholds (<40, <45, <50, <55, and <60 mmHg) in relationship to outcome by GOS and found no significant difference among the different thresholds [17]. Other studies had largely confirmed that pediatric patients with CPP below 40 mmHg were more likely to have poor outcomes. Vavilala et al. studied the discharge GOS of 236 patients and found that CPP above 40 mmHg was associated with good discharge GOS [49]. Similarly, Figaji et al. found that patients with unfavorable outcome (measured by GOS) have more episodes with CPP below 40 mmHg [50]. There was however no difference in terms of the number of episodes of CPP below 50 mmHg between the favorable and unfavorable outcome cohorts [50].

Several studies have attempted to define age-specific threshold for CPP. Allen et al. analyzed the TBI-TRC database of New York state for age-specific CPP threshold with 317 pediatric patients [46]. The investigators predefined a high and a low threshold CPP for different age groups as 40 mmHg/30 mmHg for 0–5 years old, 50 mmHg/35 mmHg for 6–11 years old, and 60 mmHg/50 mmHg for 12 years old and older [46]. CPP events were categorized as CCP-H (above both high and low threshold), CCP-B (between high and low threshold), and CPP-L (below low threshold) [46]. In the 0–11 age group, there was a significant decrease in survival with exposure to CPP-B and CPP-L events [46]. There was also a decrease in survival in all age groups with exposure to CPP-L events [46]. The data suggested that the CPP threshold should be age specific, with a CPP threshold of above 40 mmHg for 0–5 years old and above 50 mmHg for 6–17 years old [46]. In a much smaller series (n = 36), Kapapa et al. found that patients with CPP value`below an age-specific threshold (>40 mmHg up to 1 month, >45 mmHg 2 month to 1 years old, >50 mmHg 1–7 years old, 55–60 above 7 years old) even for just single episode had a significantly worse outcome [51].

Taken together, these data suggest CPP be maintained at a minimum of 40 mmHg as recommended by the Guideline [5]. It is recommended that a CPP target between

40 and 50 mmHg should be used to ensure the 40 mmHg threshold is not breached and that there may be age-specific thresholds with infants at lower end and older children at the upper end of the range [5].

1.4 ICU Management Algorithm

The 2019 guideline includes an ICU management algorithm for ICP and CPP (Fig. 1.1) [52]. This particular algorithm outlines a general approach to a severe pediatric TBI patient. Baseline clinical care includes evaluation by CT scan, insertion of an ICP monitor, adequate sedation and analgesia, mechanical ventilation, normothermia, adequate intravascular volume, maintenance of hemoglobin above 7 g/dL, treatment for any coagulopathy, positioning of the head of bed at 30°, and initiation of appropriate antiepileptic prophylaxis (see Fig. 1.1) [52]. If there are signs of herniation, a defined pathway for treatment is initiated. This includes hyperventilation, hyperosmolar therapy, and CSF drainage and possibly repeat CT (Fig. 1.1) [52]. If indicated, surgical intervention may be warranted.

In other patients, ICP monitoring and treatment pathways should be followed. The protocol here generally recommends treatment if ICP is above 20 mmHg for over 5 min. The alg takes a stepwise approach, using CSF drainage, hyperosmolar therapy, additional sedation, neuromuscular blockade, additional hyperosmolar

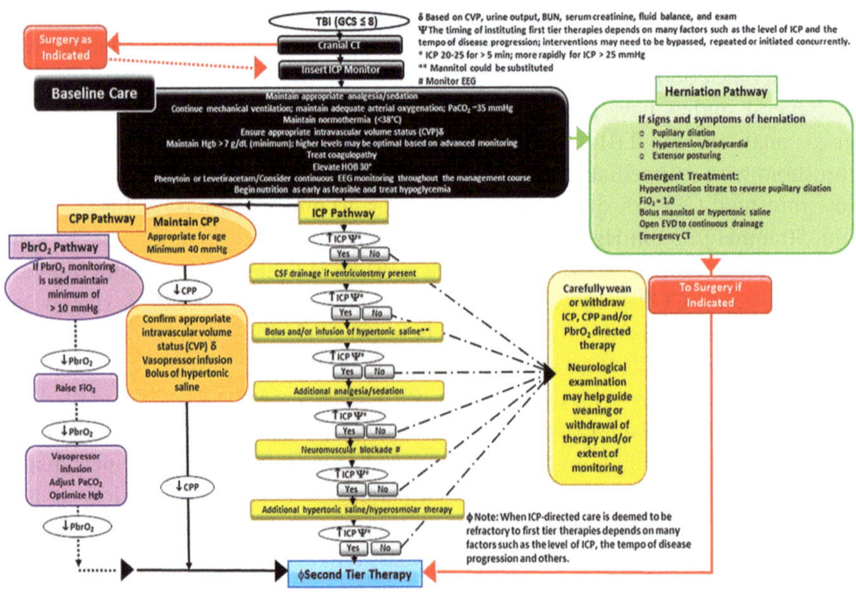

Fig. 1.1 ICU Management Algorithm for ICP and CPP. Reprinted with Permission. Kochanek et al., Pediatr Crit Care Med. 2019;20 (3):269–79

therapy, and subsequently second-tier therapy for ICP treatment [52]. A CPP pathway is also described, with the use of appropriate vasopressors and maintenance of intravascular volume status to maintain the appropriate CPP [52].

While this algorithm presents a basic framework for treatment of ICP and CPP, much still relies on clinical judgments by treating physicians. For example, the ICP pathway takes a stepwise approach to treating elevated ICP. However, in practice, it is not unusual to use a combinatorial approach especially in case of an ICP crisis. Also, there is no clear definition as to when to transition to the next stage of therapy or to bypass this step altogether. Lastly, CPP treatments sometimes can also lead to elevated ICP and how to prioritize ICP and CPP treatments are based on clinical judgments [33].

1.5 Autoregulation and Pressure Reactivity Index

A major premise behind the rationale for treating elevated ICP and low CPP is maintenance of CBF. In a healthy young brain, CBF is maintained at a relatively constant state through the mechanism of cerebral autoregulation (CA). CA is complex and involves responses to metabolites, blood pressure, serum CO_2 levels, and hypoxia [53]. Among all the factors, blood pressure in CA has been studied the most. Historically, CA can be measured by increasing and decreasing blood pressure and then measuring the changes in CBF directly. However, such methods are cumbersome and do not provide continuous data on CA. Moreover, significant changes in blood pressure may also expose the injured brain tissue to increased ICP or ischemia. Therefore, the current methodologies involve the use of changes of CBF in response to the spontaneous fluctuations in arterial blood pressure (ABP). Three methods have been used clinically for assessment of CA. The most widely used is the pressure reactivity index (PRx). PRx is calculated based on the correlation coefficient between the slow waves (0.005–0.05 Hz) of ABP and ICP [54]. While plotted against CPP, a U-shaped curve is produced and the lowest point represents the optimal CPP [54]. Another way to assess CA is with transcranial doppler (TCD). TCD measures mean blood flow velocity; the correlation coefficient between mean blood flow velocity and CPP is known as Mx [54]. Lastly, near-infrared spectroscopy (NIRS) is a noninvasive method that measures regional oxygen level. Data from NIRS can be used to calculate the cerebral oximetry index (Cox).

Among the three methodologies, PRx has been studied the most often [53]. In addition to adult studies, a number of studies have also been done in the pediatric population. Abecasis et al. studied three indexes, PRx, Mx, and Cox. Of these indexes, PRx was most robust and correlated with outcome. In 2023, two relatively large clinical series have shown that impaired CA as measured by PRx value was associated with poor outcomes [55, 56]. In fact, in the Smith et al.'s study, PRx was noted to be more predictive of outcome than ICP and CPP [55]. There is currently a large ongoing prospective study to collect data from multiple sites on CA in pediatric TBI [57].

It should be noted that most of the current data only show an association between impaired CA (i.e., elevated PRx) and poor outcome. There is currently little data to show that particularly targeting the PRx value will lead to improved outcome. One interesting parameter is the optimal CPP, which is the CPP level that the cerebral blood vessels react best to maintain a constant CBF. Optimal CPP is individualized and in fact will vary over time for each patient. Optimal CPP calculation may be particularly attractive in pediatrics, given the significant age variation in CPP threshold. Optimal CPP targeting may allow for an individualized treatment strategy.

1.6 Conclusion

Modern ICU management of severe TBIs aims to reduce secondary injury. Cerebral swelling and ischemic changes due to CBF impairment are two potential mechanisms that can cause secondary injury. Yet, the optimal strategies to monitor and treat these issues remain unclear in the field of TBI. This is especially an issue in pediatric TBIs. As physiological conditions such as blood pressure and ICP change rapidly during growth, a single target may not be appropriate for the pediatric aged groups that range from neonates to young adults. However, the development of age-appropriate targets and guidelines will require a tremendous amount of data. International collaborations with large prospective databases, similar to the adult age-focused TRACK-TBI (Transforming Research and Clinical Knowledge in TBI) and CENTER-TBI models will be important for development and shaping of these guidelines.

Acknowledgments None.

Disclosures None.

References

1. Schuchat A. The management of traumatic brain injury in Children: an executive summary. In: Houry D, Baldwin G, editors. https://www.cdc.gov/traumatic-brain-injury/media/pdfs/TBI-ReporttoCongress-508.pdf
2. Aslan N, Yıldızdaş D. Severe pediatric traumatic brain injury treatment approaches of pediatric intensivists in Turkey PICUs: national survey results. Turk Arch Pediatr. 2022;57(4):421–7.
3. Dean NP, Boslaugh S, Adelson PD, Pineda JA, Leonard JR. Physician agreement with evidence-based recommendations for the treatment of severe traumatic brain injury in children. J Neurosurg. 2007;107(5 Suppl):387–91.
4. Morris KP, Forsyth RJ, Parslow RC, Tasker RC, Hawley CA, Group UPTBIS, et al. Intracranial pressure complicating severe traumatic brain injury in children: monitoring and management. Intensive Care Med. 2006;32(10):1606–12.

5. Kochanek PM, Tasker RC, Carney N, Totten AM, Adelson PD, Selden NR, et al. Guidelines for the management of pediatric severe traumatic brain injury, third edition: update of the brain trauma foundation guidelines. Pediatr Crit Care Med. 2019;20(3S Suppl 1):S1–S82.
6. Alkhoury F, Kyriakides TC. Intracranial pressure monitoring in children with severe traumatic brain injury: national trauma data bank-based review of outcomes. JAMA Surg. 2014;149(6):544–8.
7. Alali AS, Gomez D, Sathya C, Burd RS, Mainprize TG, Moulton R, et al. Intracranial pressure monitoring among children with severe traumatic brain injury. J Neurosurg Pediatr. 2015;16(5):523–32.
8. Bennett TD, Riva-Cambrin J, Keenan HT, Korgenski EK, Bratton SL. Variation in intracranial pressure monitoring and outcomes in pediatric traumatic brain injury. Arch Pediatr Adolesc Med. 2012;166(7):641–7.
9. Bennett TD, DeWitt PE, Greene TH, Srivastava R, Riva-Cambrin J, Nance ML, et al. Functional outcome after intracranial pressure monitoring for children with severe traumatic brain injury. JAMA Pediatr. 2017;171(10):965–71.
10. Delaplain PT, Grigorian A, Lekawa M, Mallicote M, Joe V, Schubl SD, et al. Intracranial pressure monitoring associated with increased mortality in pediatric brain injuries. Pediatr Surg Int. 2020;36(3):391–8.
11. Shibahashi K, Ohbe H, Matsui H, Yasunaga H. Intracranial pressure monitoring in children with severe traumatic brain injury: a propensity score matching analysis using a nationwide inpatient database in Japan. J Neurosurg. 2023;139(6):1514–22.
12. Chesnut R, Temkin N, Pridgeon J, Sulzbacher S, Lujan S, Videtta W, et al. Protocol for a randomized trial comparing intracranial pressure monitor-based management of severe pediatric traumatic brain injury with management based on imaging and clinical examination without intracranial pressure monitoring. J Neurosurg. 2024;94(1):72–9.
13. Agrawal S, Abecasis F, Jalloh I. Neuromonitoring in children with traumatic brain injury. Neurocrit Care. 2024;40(1):147–58.
14. Bell MJ, Rosario BL, Kochanek PM, Adelson PD, Morris KP, Au AK, et al. Comparative effectiveness of diversion of cerebrospinal fluid for children with severe traumatic brain injury. JAMA Netw Open. 2022;5(7):e2220969.
15. Pedersen SH, Lilja-Cyron A, Astrand R, Juhler M. Monitoring and measurement of intracranial pressure in pediatric head trauma. Front Neurol. 2019;10:1376.
16. Carney N, Totten AM, O'Reilly C, Ullman JS, Hawryluk GW, Bell MJ, et al. Guidelines for the management of severe traumatic brain injury, fourth edition. Neurosurgery. 2017;80(1):6–15.
17. Miller Ferguson N, Shein SL, Kochanek PM, Luther J, Wisniewski SR, Clark RS, et al. Intracranial hypertension and cerebral hypoperfusion in children with severe traumatic brain injury: thresholds and burden in accidental and abusive insults. Pediatr Crit Care Med. 2016;17(5):444–50.
18. Mehta A, Kochanek PM, Tyler-Kabara E, Adelson PD, Wisniewski SR, Berger RP, et al. Relationship of intracranial pressure and cerebral perfusion pressure with outcome in young children after severe traumatic brain injury. Dev Neurosci. 2010;32(5–6):413–9.
19. Adelson PD, Ragheb J, Kanev P, Brockmeyer D, Beers SR, Brown SD, et al. Phase II clinical trial of moderate hypothermia after severe traumatic brain injury in children. Neurosurgery. 2005;56(4):740–54; discussion –54.
20. Adelson PD, Bratton SL, Carney NA, Chesnut RM, du Coudray HE, Goldstein B, et al. Guidelines for the acute medical management of severe traumatic brain injury in infants, children, and adolescents. Chapter 19. The role of anti-seizure prophylaxis following severe pediatric traumatic brain injury. Pediatr Crit Care Med. 2003;4(3 Suppl):S72–5.
21. Gregers MCT, Mikkelsen S, Lindvig KP, Brøchner AC. Ketamine as an anesthetic for patients with acute brain injury: a systematic review. Neurocrit Care. 2020;33(1):273–82.
22. Bar-Joseph G, Guilburd Y, Tamir A, Guilburd JN. Effectiveness of ketamine in decreasing intracranial pressure in children with intracranial hypertension. J Neurosurg Pediatr. 2009;4(1):40–6.

23. Laws JC, Vance EH, Betters KA, Anderson JJ, Fleishman S, Bonfield CM, et al. Acute effects of ketamine on intracranial pressure in children with severe traumatic brain injury. Crit Care Med. 2023;51(5):563–72.
24. Boone MD, Oren-Grinberg A, Robinson TM, Chen CC, Kasper EM. Mannitol or hypertonic saline in the setting of traumatic brain injury: what have we learned? Surg Neurol Int. 2015;6:177.
25. Cook AM, Morgan Jones G, Hawryluk GWJ, Mailloux P, McLaughlin D, Papangelou A, et al. Guidelines for the acute treatment of cerebral edema in neurocritical care patients. Neurocrit Care. 2020;32(3):647–66.
26. Muizelaar JP, Wei EP, Kontos HA, Becker DP. Mannitol causes compensatory cerebral vasoconstriction and vasodilation in response to blood viscosity changes. J Neurosurg. 1983;59(5):822–8.
27. Muizelaar JP, Lutz HA, Becker DP. Effect of mannitol on ICP and CBF and correlation with pressure autoregulation in severely head-injured patients. J Neurosurg. 1984;61(4):700–6.
28. Chen H, Song Z, Dennis JA. Hypertonic saline versus other intracranial pressure-lowering agents for people with acute traumatic brain injury. Cochrane Database Syst Rev. 2020;1(1):CD010904.
29. Bernhardt K, McClune W, Rowland MJ, Shah A. Hypertonic saline versus other intracranial-pressure-lowering agents for patients with acute traumatic brain injury: a systematic review and meta-analysis. Neurocrit Care. 2024;40(2):769–84.
30. Shein SL, Ferguson NM, Kochanek PM, Bayir H, Clark RS, Fink EL, et al. Effectiveness of pharmacological therapies for intracranial hypertension in children with severe traumatic brain injury--results from an automated data collection system time-synched to drug administration. Pediatr Crit Care Med. 2016;17(3):236–45.
31. Fisher B, Thomas D, Peterson B. Hypertonic saline lowers raised intracranial pressure in children after head trauma. J Neurosurg Anesthesiol. 1992;4(1):4–10.
32. Wellard J, Kuwabara M, Adelson PD, Appavu B. Physiologic characteristics of hyperosmolar therapy after pediatric traumatic brain injury. Front Neurol. 2021;12:662089.
33. Kochanek PM, Adelson PD, Rosario BL, Hutchison J, Miller Ferguson N, Ferrazzano P, et al. Comparison of intracranial pressure measurements before and after hypertonic saline or mannitol treatment in children with severe traumatic brain injury. JAMA Netw Open. 2022;5(3):e220891.
34. Schirmer CM, Ackil AA, Malek AM. Decompressive Craniectomy. Neurocrit Care. 2008;8(3):456–70.
35. Cooper DJ, Rosenfeld JV, Murray L, Arabi YM, Davies AR, D'Urso P, et al. Decompressive craniectomy in diffuse traumatic brain injury. N Engl J Med. 2011;364(16):1493–502.
36. Hutchinson PJ, Kolias AG, Timofeev IS, Corteen EA, Czosnyka M, Timothy J, et al. Trial of decompressive craniectomy for traumatic intracranial hypertension. N Engl J Med. 2016;375(12):1119–30.
37. Ardissino M, Tang A, Muttoni E, Tsang K. Decompressive craniectomy in paediatric traumatic brain injury: a systematic review of current evidence. Childs Nerv Syst. 2019;35(2):209–16.
38. Cho DY, Wang YC, Chi CS. Decompressive craniotomy for acute shaken/impact baby syndrome. Pediatr Neurosurg. 1995;23(4):192–8.
39. Josan VA, Sgouros S. Early decompressive craniectomy may be effective in the treatment of refractory intracranial hypertension after traumatic brain injury. Childs Nerv Syst. 2006;22(10):1268–74.
40. Taylor A, Butt W, Rosenfeld J, Shann F, Ditchfield M, Lewis E, et al. A randomized trial of very early decompressive craniectomy in children with traumatic brain injury and sustained intracranial hypertension. Childs Nerv Syst. 2001;17(3):154–62.
41. Mhanna MJ, Mallah WE, Verrees M, Shah R, Super DM. Outcome of children with severe traumatic brain injury who are treated with decompressive craniectomy. J Neurosurg Pediatr. 2015;16(5):508–14.

42. Thomale UW, Graetz D, Vajkoczy P, Sarrafzadeh AS. Severe traumatic brain injury in children--a single center experience regarding therapy and long-term outcome. Childs Nerv Syst. 2010;26(11):1563–73.
43. Bruns N, Kamp O, Lange K, Lefering R, Felderhoff-Müser U, Dudda M, et al. Functional short-term outcomes and mortality in children with severe traumatic brain injury: comparing decompressive craniectomy and medical management. J Neurotrauma. 2022;39(13–14):944–53.
44. Ballestero MFM, Furlanetti LL, Augusto LP, Chaves PHC, Santos MV, de Oliveira RS. Decompressive craniectomy for severe traumatic brain injury in children: analysis of long-term neuropsychological impairment and review of the literature. Childs Nerv Syst. 2019;35(9):1507–15.
45. Elsawaf Y, Anetsberger S, Luzzi S, Elbabaa SK. Early decompressive craniectomy as management for severe traumatic brain injury in the pediatric population: a comprehensive literature review. World Neurosurg. 2020;138:9–18.
46. Allen BB, Chiu YL, Gerber LM, Ghajar J, Greenfield JP. Age-specific cerebral perfusion pressure thresholds and survival in children and adolescents with severe traumatic brain injury*. Pediatr Crit Care Med. 2014;15(1):62–70.
47. Narotam PK, Burjonrappa SC, Raynor SC, Rao M, Taylon C. Cerebral oxygenation in major pediatric trauma: its relevance to trauma severity and outcome. J Pediatr Surg. 2006;41(3):505–13.
48. Downard C, Hulka F, Mullins RJ, Piatt J, Chesnut R, Quint P, et al. Relationship of cerebral perfusion pressure and survival in pediatric brain-injured patients. J Trauma. 2000;49(4):654–8; discussion 8–9.
49. Vavilala MS, Kernic MA, Wang J, Kannan N, Mink RB, Wainwright MS, et al. Acute care clinical indicators associated with discharge outcomes in children with severe traumatic brain injury. Crit Care Med. 2014;42(10):2258–66.
50. Figaji AA, Zwane E, Thompson C, Fieggen AG, Argent AC, Le Roux PD, et al. Brain tissue oxygen tension monitoring in pediatric severe traumatic brain injury. Part 1: relationship with outcome. Childs Nerv Syst. 2009;25(10):1325–33.
51. Kapapa T, König K, Pfister U, Sasse M, Woischneck D, Heissler H, et al. Head trauma in children, part 2: course and discharge with outcome. J Child Neurol. 2010;25(3):274–83.
52. Kochanek PM, Tasker RC, Bell MJ, Adelson PD, Carney N, Vavilala MS, et al. Management of pediatric severe traumatic brain injury: 2019 consensus and guidelines-based algorithm for first and second tier therapies. Pediatr Crit Care Med. 2019;20(3):269–79.
53. Donnelly JE, Young AMH, Brady K. Autoregulation in paediatric TBI-current evidence and implications for treatment. Childs Nerv Syst. 2017;33(10):1735–44.
54. Abecasis F, Dias C, Zakrzewska A, Oliveira V, Czosnyka M. Monitoring cerebrovascular reactivity in pediatric traumatic brain injury: comparison of three methods. Childs Nerv Syst. 2021;37(10):3057–65.
55. Smith CA, Rohlwink UK, Mauff K, Thango NS, Hina TS, Salie S, et al. Cerebrovascular pressure reactivity has a strong and independent association with outcome in children with severe traumatic brain injury. Crit Care Med. 2023;51(5):573–83.
56. Velle F, Lewén A, Howells T, Hånell A, Nilsson P, Enblad P. Cerebral pressure autoregulation and optimal cerebral perfusion pressure during neurocritical care of children with traumatic brain injury. J Neurosurg Pediatr. 2023;31(5):503–13.
57. Agrawal S, Placek MM, White D, Daubney E, Cabeleira M, Smielewski P, et al. Studying Trends of Auto-Regulation in Severe Head Injury in Paediatrics (STARSHIP): protocol to study cerebral autoregulation in a prospective multicentre observational research database study. BMJ Open. 2023;13(3):e071800.

Chapter 2
Evaluation and Management of Traumatic Brain Injuries in the Pediatric Intensive Care Unit

Michael Cronin

Abstract In this overview, intended for a multidisciplinary readership, we address the challenges in early management of children who have sustained mild, moderate, or severe traumatic brain injuries. Pediatric traumatic brain injuries (pTBIs) present unique diagnostic and management challenges as compared with adults. Proper management requires careful interpretation of data and strong clinical judgment. Children with injuries due to nonaccidental trauma present unique diagnostic and management challenges. In the pediatric intensive care unit, care is provided as part of a multidisciplinary, collaborative team, with medical and surgical interventions tailored to injury severity. Intensive care focuses on managing cerebral perfusion, temperature, mechanical ventilation, and continuous EEG monitoring to mitigate secondary brain injury. Long-term recovery emphasizes multidisciplinary rehabilitation and support to address physical, cognitive, and emotional challenges.

Keywords Prehospital care · Therapeutics · Diagnostics · Intensive care · Nonaccidental trauma · Rehabilitation

Abbreviations

CBF	Cerebral blood flow
CPP	Cerebral perfusion pressure
CSF	Cerebrospinal fluid
CT	Computed tomography
EVD	External ventricular drain
ICP	Intracranial pressure

M. Cronin (✉)
Department of Pediatrics, Dell Medical School, University of Texas at Austin, Austin, TX, USA

Dell Children's Medical Center of Central Texas, Austin, TX, USA
e-mail: mtc2659@eid.utexas.edu

L. J. Noble-Haeusslein, D. M. Schnyer (eds.), *Traumatic Brain Injuries*, Advances in Neurobiology 42, https://doi.org/10.1007/978-3-031-69832-3_2

NAT	Nonaccidental trauma
$PaCO_2$	Arterial carbon dioxide
PaO_2	Arterial oxygen
PICU	Pediatric intensive care unit
pTBIs	Pediatric traumatic brain injuries
TXA	Antifibrinolytic agent—tranexamic acid

2.1 Introduction

Pediatric traumatic brain injuries (pTBIs) often present differently as compared with adults, due, in part, to the evolving nature of the developing brain in response to trauma [1]. Recognizing the signs and symptoms of a pTBI and conducting a thorough clinical assessment are important for effective management. Pediatric patients may lack the verbal skills to accurately describe their symptoms. Cognitive and behavioral changes can be subtle and might be mistaken for typical developmental behaviors or transient emotional responses [2]. Children may also be less cooperative with a physical examination, allowing more covert signs to go undetected. Thus, wherever possible, it is important to pay close attention to information from parents and other caregivers, especially if there are new unexplained changes in the child's behavior. Children are also at higher risk of suffering from unintentional injuries, which may be missed on the initial history and examination [3, 4]. Caregivers may not be entirely forthcoming upon presentation to a healthcare facility, and it is important for clinicians to have a high index of suspicion, especially when exam findings do not correlate well with the stated history. Infants younger than 6 months of age and children who present with facial injuries, seizures, decreased mental status, abnormal respiratory patterns, vomiting, and irritability are all risk factors for nonaccidental traumatic brain injury, and clinicians should consider if neuroimaging is warranted [5]. It is important to identify these patients early because their risk of death is much higher than in children who present with known trauma [3, 5, 6].

2.2 Management

2.2.1 *Prehospital Care, Stabilization, and Resuscitation*

Prompt recognition and management of a pTBI by emergency medical services are critical to improve outcomes. Immediate priorities include ensuring a patent airway, adequate ventilation to prevent hypoxia, and close monitoring for signs of decompensation vital signs due to children's limited physiologic reserve. Pediatric patients have a smaller total lung capacity and anatomy which is more at risk for developing upper airway obstruction, which increases the risk of severe hypoxia. Pediatric airway anatomy may also pose challenges for clinicians who have limited experience intubating younger patients. Establishing intravenous access is important to mitigate the risks of hypotension and may be challenging especially in smaller patients. The cervical spine is immobilized, taking care to avoid hyperflexion due to a child's proportionally larger head size, compared to that of the adult. Rapidly transporting the child to a specialized trauma center, while minimizing delays from unnecessary interventions and facilitating prearrival communication, is vital [1].

Although prompt stabilization and transport to definitive care should be the priority of first responders, early medical interventions may have benefit in specific situations. The CRASH-3 (Clinical Randomisation of an Antifibrinolytic in Significant Head Injury) trial, published in 2019, aimed to understand the effects of the antifibrinolytic agent tranexamic acid (TXA) in patients with TBIs. The study's results indicated a modest reduction in head injury-related mortality without an increase in adverse events or disability when TXA was administered within 3 h of injury. This early therapeutic window is crucial as the benefits of TXA diminished significantly when administered TXA was administered at later time points [7]. These results have paved the way for the incorporation of TXA into the management protocols for pTBIs, underlining the importance of rapid assessment and intervention in the early phases postinjury to maximize the therapeutic benefits [8]. The results of this trial continue to be evaluated and debated in the clinical community, as patients with the most severe injuries did not appear to benefit from TXA, and further study is needed to refine our understanding of TXA's role in management of pTBIs [9, 10].

2.2.2 *Initial Diagnostic and Therapeutic Interventions*

Upon arrival in the emergency department, the initial assessment prioritizes the airway, breathing, and circulation. Any immediately life-threatening conditions are stabilized prior to moving on to the neurological assessment. The medical team determines the child's level of consciousness, often utilizing a standardized tool such as the Pediatric Glasgow Coma Scale [2, 11]. Children with moderate to severe pTBIs often require intubation for airway protection, which is performed by an experienced provider in order to avoid exacerbating the underlying injury [1].

Imaging with computer tomography (CT) may not always be necessary for patients with mild TBIs, and guidelines exist to ensure unnecessary exposure to radiation [12]. The decision for imaging hinges on several factors: the age of the child, the nature of the injury (e.g., high-impact events), and the presence of symptoms such as persistent vomiting or a severe headache. Patients with mTBIs and no other injuries are often able to be discharged home after a period of observation. Parents and other caregivers are provided with clear guidelines for postinjury care, emphasizing the importance of a phased return to school and physical activities and the necessity of follow-up appointments to monitor the child's progress and recovery [12].

Moderate to severe pTBIs should receive urgent neuroimaging, typically CT scans, to discern the type, location, and extent of brain injury and guide potential surgical interventions. A collaborative approach involving neurosurgery, pediatric intensive care, and trauma teams is initiated to develop a comprehensive and targeted care plan. The decision to pursue surgical intervention with a severe pTBI is multifaceted, influenced by clinical, radiological, and physiological factors [1]. Epidural or subdural hematomas, large contusions, or depressed skull fractures may necessitate urgent surgical decompression. The size, location, and effect on surrounding structures play a role in the urgency and type of surgery required. Even in the presence of potential surgical lesions on CT, the child's clinical state is vital in determining the best course of action. Rapid deterioration in consciousness, new neurological deficits, or signs of brain herniation may necessitate emergent surgical intervention.

2.2.3 Intracranial Pressure (ICP) Monitoring

In general, placement of some form of intracranial pressure (ICP) monitor is recommended for severe pTBI [defined as Glasgow Coma Scale (GCS) 3–8] after initial resuscitation and confirmation of an abnormal CT scan. The proper management of secondary brain injury, resulting from factors including cerebral swelling or progression of hemorrhage, often necessitates continuous ICP monitoring to guide therapeutic interventions. Elevated ICP, if not promptly addressed, may lead to brain herniation and irreversible damage or death [1, 11].

The two primary ICP monitoring methodologies are external ventricular drains (EVDs) and intraparenchymal monitors. For placement of an EVD, a catheter is introduced into the lateral ventricle, generally under stereotactic or freehand guidance informed by anatomical landmarks. This drain is connected to a transducer that is capable of measuring ICP either continuously or intermittently. An EVD is considered the gold-standard metric for ICP measurement. The EVD also has the capability of facilitating therapeutic drainage of cerebrospinal spinal fluid (CSF) in the event of acute increases in ICP. There are limitations to an EVD. These include risk of hemorrhage during placement, especially if accessing the ventricular space proves challenging (as in the case of severe cerebral edema). As the EVD is also an open system, interfacing with the external environment, it is a potential source of infection. Obstruction of the catheter by debris or clots can impair both measurement of ICP and drainage capacities, and there is risk that overdrainage of CSF can induce secondary complications such as subdural hematomas [13].

Intraparenchymal monitors are probes directly positioned within the cerebral parenchyma, equipped with fiberoptic tip sensors and strain gauge sensors that measure ICP. These probes are closed systems that have reduced infectious risks compared with EVDs, and the less invasive nature of these probes typically results in a diminished hemorrhagic risk upon insertion. They are particularly beneficial in scenarios where ventricular anatomy is distorted or compressed, rendering ventricular access challenging. The major limitation of intraparenchymal monitors is that they are incapable of CSF drainage. Fiberoptic systems may develop sensor drift over prolonged durations, which necessitate recalibration, removal, or replacement.

Though the invasive footprint is minimal, there remains a risk of parenchymal injury or minor hemorrhagic events upon insertion [14].

2.2.4 *Additional Neurosurgical Interventions*

Decompressive hemicraniectomy (DHC) is a surgical intervention employed in the management of severe pTBIs, where there is a refractory elevated intracranial pressure (ICP) despite maximal medical therapy. The procedure entails the removal of a large portion of the skull thereby affording space for the swollen brain tissue to expand without causing herniation or compromising cerebral perfusion. The goal is to salvage the uninjured portion of the brain by allowing the damaged tissue to swell outside of the bounds of the skull. In the pediatric population, the immature brain's plasticity and adaptability often provide a theoretical advantage for better postoperative neurological recovery [15]. The decision to undertake DHC must weigh the potential benefits of ICP reduction and improved cerebral perfusion against surgical risks, which include infection, subdural effusion, hygroma formation, and seizures. Additionally, the timing, indications, and postoperative management, including the reconstruction of eventual skull, require meticulous planning and a multidisciplinary approach [1].

The use of decompressive craniectomy in pTBIs is controversial, and the overall efficacy, long-term outcomes, and potential complications of this intervention have been the subject of rigorous debate in the medical community [16]. One of the most cited sources of controversy emanates from the DECRA (Decompressive Craniectomy in Patients with Severe Traumatic Brain Injury) trial, which indicated that while early bifrontotemporoparietal decompressive craniectomy decreased ICP and the length of stay in the intensive care unit, it did not improve outcomes. In fact, patients in the craniectomy group had worse functional outcomes at 6 months compared to the standard-care group [15]. The RESCUEicp (Randomized Evaluation of Surgery with Craniectomy for Uncontrollable Elevation of ICP) trial, published in Lancet in 2016, highlighted that while DC reduced mortality at 6 months postinjury, many survivors were left with severe physical and cognitive impairments [17]. Not all patients with TBI and elevated ICP may benefit uniformly from surgical decompression, and the decision requires nuanced clinical judgment [1, 16].

2.2.5 *Management in the Pediatric Intensive Care Unit (PICU)*

After initial stabilization, the care in the PICU focuses on mitigating secondary brain injury and optimizing conditions that favor recovery. The overall strategy is to provide supportive care to ensure adequate cerebral oxygen supply and minimize metabolic demand where possible. This involves the management of factors that can exacerbate brain injury such as hypoxia, hypotension, and increased ICP. Adequate oxygenation and ventilation are ensured, and mechanical ventilation may be necessary for airway protection and controlled ventilation. Hyperosmolar therapy with hypertonic saline or mannitol can be used to manage increased ICP, and careful fluid management is crucial [1, 18].

Cerebral perfusion pressure (CPP) and its appropriate management are critical aspects in the care of pTBIs. CPP is defined as the difference between mean arterial pressure and intracranial pressure, and it represents the pressure gradient driving cerebral blood flow (CBF). Maintaining adequate CPP is essential for ensuring sufficient blood flow to the brain, thereby providing necessary oxygen and nutrients to the injured brain tissue [1, 19, 20].

While studies of brain-injured adults have found that maintaining a CPP greater than 50 mmHg has been associated with improved outcomes [21], the optimal CPP for pediatric patients remains uncertain. According to the most recent guidelines from the Brain Trauma Foundation for pediatric TBI [1], the optimal target is suggested to be between 40 and 50 mmHg. This range, considered to provide adequate CBF while minimizing the risk of exacerbating cerebral edema and ICP elevation, is extrapolated from the adult trauma data and the results of other trials examining pTBI [1]. Individualized CPP targets, considering factors like age and the severity of the injury, are also emphasized in the guidelines. A recent retrospective analysis of pediatric TBI patients suggested that CPP targets, even higher than recommended by the most recent guidelines, may be associated with improved outcomes [19].

2.2.6 Cerebral Autoregulation

Cerebral autoregulation refers to the brain's ability to maintain stable CBF despite fluctuations in systemic blood pressure. This autoregulatory mechanism is vital for ensuring consistent and adequate blood supply to the brain, which is crucial for delivering essential nutrients and oxygen to the brain tissue. In the context of pTBIs, cerebral autoregulation can be compromised, leading to significant clinical implications. When autoregulation is impaired, CBF becomes directly dependent on systemic blood pressure. This dependency means that any changes in blood pressure will directly impact CBF. Hypotension can lead to cerebral ischemia, exacerbating secondary brain injury, while hypertension can contribute to increased intracranial pressure and cerebral edema, further compounding the injury [20].

Maintaining CPP becomes particularly crucial in the setting of impaired cerebral autoregulation in pTBIs. The recommended CPP target of 40–50 mmHg is designed to ensure adequate CPP and minimize secondary brain injury while avoiding the deleterious effects of excessive blood pressure augmentation. Unfortunately, impairment and recovery of the brain's autoregulatory capacity is a dynamic process that evolves over several days or longer following a severe brain injury. At present, there are no clear guidelines for how best to account for the changes in cerebral autoregulation following pTBI. For this reason, continuously monitoring and managing ICP and CPP and adjusting treatment strategies to the individual patient's clinical status are essential. The utilization of multimodal monitoring, including ICP, CPP, and other advanced neuromonitoring techniques, may aid in tailoring the management approach to optimize CBF and minimize secondary injury [1].

2.2.7 Targeted Temperature Management

Temperature management for children with TBIs is of great importance. The developing brain in children is particularly sensitive to temperature fluctuations, with deviations from normothermia having the potential to exacerbate secondary injury [22]. Elevated brain temperatures can accelerate metabolic demands, exacerbate cerebral edema, and amplify the release of excitotoxic neurotransmitters, all of which can further compromise an already injured brain. A patient recovering from a severe trauma is at a heightened inflammatory state and especially predisposed to develop fever, even in the absence of infection. Noninvasive, whole-body temperature management systems may be helpful at quickly achieving and maintaining normothermia and are frequently used in patients recovering from both TBI and cardiac arrest [1, 22]. The role of therapeutic hypothermia (i.e., targeting lower than normal body temperatures) is controversial [23]. While some evidence suggests that hypothermia may decrease metabolic demand, curb inflammatory responses, and mitigate the progression of secondary injury, it has also been associated with complications including coagulopathies, infections, and arrhythmia [24]. Current guidelines do not recommend therapeutic hypothermia for most patients with TBI, but they acknowledge that it may be considered in select cases [1].

2.2.8 Ventilator Management

Careful attention to ventilator management is also an important aspect of care for the pediatric patient recovering from severe TBI. The relationship between ventilation and CBF is predominantly modulated through changes in arterial carbon dioxide ($PaCO_2$) and arterial oxygen (PaO_2) levels. Hypoventilation produces elevated $PaCO_2$, which leads to cerebral vasodilation that may increase cerebral blood volume, thereby exacerbating cerebral edema and elevating (ICP). On the other hand, reduced $PaCO_2$, a consequence of hyperventilation, induces cerebral vasoconstriction, which decreases cerebral blood volume and can transiently reduce ICP. Excessive vasoconstriction from aggressive hyperventilation can compromise cerebral perfusion, potentially leading to a secondary ischemic injury. In a similar fashion, decreased PaO_2 levels can also lead to vasodilation. For pediatric TBI patients, whose recovering brains already have heightened vulnerability, these ventilatory-driven vascular changes can be especially consequential. The selection of ventilator settings must be delicately balanced to ensure optimal cerebral oxygenation and perfusion [1, 18].

Continuous electroencephalography (cEEG) monitoring. Continuous EEG monitoring in the PICU provides real-time insight into the electrical activity of the brain. EEG enables clinicians to promptly identify and address seizures, many of which may be subclinical and not discernible through clinical examination alone. In pTBI, such subclinical seizures or nonconvulsive status epilepticus can exacerbate secondary brain injury by increasing cerebral metabolic demands, potentially leading to further damage [25]. The early detection of these abnormalities with cEEG guides antiepileptic interventions, potentially mitigating the associated risks and improving outcomes.

Another potential advantage of cEEG is the ability to assist in gauging the depth of sedation and in some settings assessing cerebral perfusion and overall brain function. By providing a continuous stream of data on cortical function, cEEG informs clinicians in tailoring sedative and analgesic regimens, ensuring that patients are neither under- nor oversedated. Certain cEEG patterns can also hint at compromised CBF, prompting timely interventions to restore adequate CPP. As our understanding of brain injury deepens, the applications of cEEG in the PICU continue to expand, emphasizing its role as a critical component of comprehensive neuromonitoring in the management of TBIs.

2.2.9 Nonaccidental Trauma

Nonaccidental trauma (NAT), also sometimes referred to as abusive head trauma, presents challenges unique to pTBIs. These are injuries that are inflicted on the child, often by family members or other caregivers [3]. Early diagnosis and intervention for NAT are critical, yet the subtle and varied clinical presentations may compromise timely identification. Children experiencing TBI from NAT may present with a range of nonspecific symptoms, such as irritability, vomiting, altered consciousness, or even cardiac arrest, making the diagnosis more challenging. The absence of a reliable history or explanation for the injury further complicates the clinical picture. Children suffering injury from NAT may not receive immediate medical attention, leading to a delay in diagnosis and treatment, which can exacerbate the injury and decrease the likelihood of recovery [5].

The injuries sustained from abuse, often seen in infants and toddlers, tend to be more severe and have different biomechanical properties [3]. Children, especially infants and toddlers, are especially vulnerable to severe brain injury from abuse due to their developing neuroanatomy, large head-to-body proportions, and the lack of protective muscle mass, especially in the neck and shoulders [3]. The force applied in NAT is typically repetitive and involves both acceleration and deceleration mechanisms, leading to distinct injury patterns. Diffuse axonal injury, subdural hematomas, and retinal hemorrhages are commonly observed in NAT, and these patterns of injury can often alert healthcare providers to the possibility of abuse. The long-term outcomes for these children often involve significant neurologic and developmental impairment, reinforcing the importance of ongoing rehabilitation, support, and monitoring [26].

Another distinctive aspect of NAT is the significant likelihood of additional noncranial injuries, such as fractures, internal organ injuries, or evidence of previous injuries, indicative of ongoing abuse [27]. These findings often necessitate a broader and more detailed medical evaluation, including skeletal surveys and a thorough ophthalmologic examination.

The management of TBI resulting from NAT requires a multidisciplinary approach that not only focuses on the immediate clinical needs but also addresses the psychosocial aspects [28]. The healthcare team must work closely with child protection services and law enforcement to ensure the child's safety and the prevention of further harm. Healthcare providers play a crucial role in recognizing potential signs of abuse and are mandated to report suspected cases to appropriate authorities.

2.2.10 Rehabilitation and Follow-Up

Management in the PICU is only one small, albeit crucial, part of a child's recovery from traumatic brain injury. A great deal of the recovery after pTBI occurs when children are under the care of an experienced physical medicine and rehabilitation team [29]. Follow-up care after pTBI requires ongoing monitoring and assessment to identify and address emerging challenges and to ensure the provision of necessary support and services, such as physical, occupational, and speech therapy. Regular medical evaluations, neurological assessment, cognitive testing, and psychosocial evaluation facilitate the timely identification of issues and allow for adjustments in the management plan. Close coordination with educational and community resources is essential to support the child's academic and social needs, ensuring appropriate accommodations and interventions are in place to promote successful reintegration and continued development [30]. Recognizing and managing the potential long-term complications of pTBI, such as learning difficulties, behavioral issues, and emotional challenges, underscores the importance of a robust follow-up framework.

2.2.11 Outcomes Following a pTBI

Outcomes after pTBI are highly variable and are influenced by several factors including the severity of the injury, age at the time of injury, and the timeliness and quality of medical intervention [31, 32]. Postinjury challenges may range from physical impairments, such as motor dysfunction and sensory deficits, to cognitive and behavioral issues, including attention deficits, memory problems, and emotional dysregulation [2, 32]. Even a mild pTBI may have lasting effects on a child's cognitive and academic function, potentially leading to learning difficulties and impacting overall educational attainment [33]. The varied and often persistent nature of these deficits highlights the critical need for long-term, multidisciplinary management to optimize functional recovery and quality of life.

Despite these potential challenges, many brain-injured children demonstrate remarkable resilience and adaptability, particularly with the provision of appropriate and timely interventions and support [31]. Early and ongoing rehabilitative interventions play a significant role in enhancing recovery, helping to improve functional outcomes, and supporting children in achieving their maximum potential postinjury [30–32]. Collaboration between healthcare providers, families, and educational institutions is crucial in ensuring that children receive the necessary support and accommodations, fostering positive outcomes and aiding in the mitigation of long-term complications and challenges.

Acknowledgments None

Disclosures None

References

1. Kochanek PM, Tasker RC, Carney N, et al. Guidelines for the management of pediatric severe traumatic brain injury, third edition: update of the Brain Trauma Foundation guidelines. Executive Summary Pediatr Crit Care Med. 2019;20(3):280–9. https://doi.org/10.1097/PCC.0000000000001736.
2. Coulter IC, Forsyth RJ. Paediatric traumatic brain injury. Curr Opin Pediatr. 2019;31(6):769–74. https://doi.org/10.1097/MOP.0000000000000820.
3. Christian CW, Block R, the Committee on Child Abuse and Neglect. Abusive head trauma in infants and children. Pediatrics. 2009;123(5):1409–11. https://doi.org/10.1542/peds.2009-0408.
4. Barlow KM, Minns RA. Annual incidence of shaken impact syndrome in young children. Lancet. 2000;356(9241):1571–2. https://doi.org/10.1016/S0140-6736(00)03130-5.
5. Jenny C. Analysis of missed cases of abusive head trauma. JAMA. 1999;281(7):621. https://doi.org/10.1001/jama.281.7.621.
6. Adamsbaum C, Méjean N, Merzoug V, Rey-Salmon C. How to explore and report children with suspected non-accidental trauma. Pediatr Radiol. 2010;40(6):932–8. https://doi.org/10.1007/s00247-010-1591-0.
7. Effects of tranexamic acid on death, disability, vascular occlusive events and other morbidities in patients with acute traumatic brain injury (CRASH-3): a randomised, placebo-controlled trial. Lancet. Published online October 2019. https://doi.org/10.1016/S0140-6736(19)32233-0.
8. Hawryluk GWJ, Lulla A, Bell R, et al. Guidelines for prehospital management of traumatic brain injury 3rd edition: executive summary. Neurosurgery. Published online September 26, 2023. https://doi.org/10.1227/neu.0000000000002672.
9. Maas AIR, Steyerberg EW, Citerio G. Tranexamic acid in traumatic brain injury: systematic review and meta-analysis trumps a large clinical trial? Intensive Care Med. 2021;47(1):74–6. https://doi.org/10.1007/s00134-020-06305-x.
10. Taccone FS, Citerio G, Stocchetti N. Is tranexamic acid going to CRASH the management of traumatic brain injury? Intensive Care Med. 2020;46(6):1261–3. https://doi.org/10.1007/s00134-019-05879-5.

11. Borgialli DA, Mahajan P, Hoyle JD, et al. Performance of the pediatric glasgow coma scale score in the evaluation of children with blunt head trauma. Walthall J, ed. Acad Emerg Med. 2016;23(8):878–84. https://doi.org/10.1111/acem.13014
12. Kuppermann N, Holmes JF, Dayan PS, et al. Identification of children at very low risk of clinically-important brain injuries after head trauma: a prospective cohort study. Lancet. 2009;374(9696):1160–70. https://doi.org/10.1016/S0140-6736(09)61558-0.
13. Chung DY, Olson DM, John S, et al. Evidence-based management of external ventricular drains. Curr Neurol Neurosci Rep. 2019;19(12):94. https://doi.org/10.1007/s11910-019-1009-9.
14. Hawthorne C, Piper I. Monitoring of intracranial pressure in patients with traumatic brain injury. Front Neurol. 2014;5:121. https://doi.org/10.3389/fneur.2014.00121.
15. Cooper DJ, Rosenfeld JV, Murray L, et al. Decompressive craniectomy in diffuse traumatic brain injury. N Engl J Med. 2011;364(16):1493–502. https://doi.org/10.1056/NEJMoa1102077.
16. Hawryluk GWJ, Rubiano AM, Totten AM, et al. Guidelines for the management of severe traumatic brain injury: 2020 update of the decompressive craniectomy recommendations. Neurosurgery. 2020;87(3):427–34. https://doi.org/10.1093/neuros/nyaa278.
17. Hutchinson PJ, Kolias AG, Timofeev IS, et al. Trial of decompressive craniectomy for traumatic intracranial hypertension. N Engl J Med. 2016;375(12):1119–30. https://doi.org/10.1056/NEJMoa1605215.
18. Taran S, Cho SM, Stevens RD. Mechanical ventilation in patients with traumatic brain injury: is it so different? Neurocrit Care. 2023;38(1):178–91. https://doi.org/10.1007/s12028-022-01593-1.
19. Woods KS, Horvat CM, Kantawala S, et al. Intracranial and cerebral perfusion pressure thresholds associated with inhospital mortality across pediatric neurocritical care. Pediatr Crit Care Med. Published online 2020. https://doi.org/10.1097/PCC.0000000000002618.
20. Lele AV, Vavilala MS. Cerebral autoregulation-guided management of adult and pediatric traumatic brain injury. J Neurosurg Anesthesiol. 2023;35(4):354–60. https://doi.org/10.1097/ANA.0000000000000933.
21. Carney N, Totten AM, O'Reilly C, et al. Guidelines for the management of severe traumatic brain injury, fourth edition. Neurosurgery. 2017;80(1):6–15. https://doi.org/10.1227/NEU.0000000000001432.
22. Newmyer R, Mendelson J, Pang D, Fink EL. Targeted temperature management in pediatric central nervous system disease. Curr Treat Options Pediatr. 2015;1(1):38–47. https://doi.org/10.1007/s40746-014-0008-y.
23. Tasker RC, Vonberg FW, Ulano ED, Akhondi-Asl A. Updating evidence for using hypothermia in pediatric severe traumatic brain injury: conventional and bayesian meta-analytic perspectives*. Pediatr Crit Care Med. 2017;18(4):355–62. https://doi.org/10.1097/PCC.0000000000001098.
24. Dietrich WD, Bramlett HM. Therapeutic hypothermia and targeted temperature management in traumatic brain injury: clinical challenges for successful translation. Brain Res. 2016;1640:94–103. https://doi.org/10.1016/j.brainres.2015.12.034.
25. Lee H, Mizrahi MA, Hartings JA, et al. Continuous electroencephalography after moderate to severe traumatic brain injury. Crit Care Med. 2019;47(4):574–82. https://doi.org/10.1097/CCM.0000000000003639.
26. Jackson JE, Beres AL, Theodorou CM, Ugiliweneza B, Boakye M, Nuño M. Long-term impact of abusive head trauma in young children: outcomes at 5 and 11 years old. J Pediatr Surg. 2021;56(12):2318–25. https://doi.org/10.1016/j.jpedsurg.2021.02.019.
27. Smith EB, Lee JK, Vavilala MS, Lee SA. Pediatric traumatic brain injury and associated topics. Anesthesiol Clin. 2019;37(1):119–34. https://doi.org/10.1016/j.anclin.2018.10.002.
28. Iqbal O'Meara A, Sequeira J, Miller Ferguson N. Advances and future directions of diagnosis and management of pediatric abusive head trauma: a review of the literature. Front Neurol. 2020;11:118. https://doi.org/10.3389/fneur.2020.00118.
29. Jonsson CA, Catroppa C, Godfrey C, Smedler AC, Anderson V. Cognitive recovery and development after traumatic brain injury in childhood: a person-oriented, longitudinal study. J Neurotrauma. 2013;30(2):76–83. https://doi.org/10.1089/neu.2012.2592.

30. Wade SL, Zhang N, Yeates KO, Stancin T, Taylor HG. Social environmental moderators of long-term functional outcomes of early childhood brain injury. JAMA Pediatr. 2016;170(4):343. https://doi.org/10.1001/jamapediatrics.2015.4485.
31. Babikian T, Asarnow R. Neurocognitive outcomes and recovery after pediatric TBI: meta-analytic review of the literature. Neuropsychology. 2009;23(3):283–96. https://doi.org/10.1037/a0015268.
32. Petranovich CL, Smith-Paine J, Wade SL, et al. From early childhood to adolescence: lessons about traumatic brain injury from the Ohio Head Injury Outcomes study. J Head Trauma Rehabil. 2020;35(3):226–39. https://doi.org/10.1097/HTR.0000000000000555.
33. Kamins J, Giza CC. Concussion—mild traumatic brain injury. Neurosurg Clin N Am. 2016;27(4):441–52. https://doi.org/10.1016/j.nec.2016.05.005.

Chapter 3
Neuroimaging Correlates of Functional Outcome Following Pediatric TBI

Emily L. Dennis, Finian Keleher, and Brenda Bartnik-Olson

Abstract Neuroimaging plays an important role in assessing the consequences of TBI across the postinjury period. While identifying alterations to the brain is important, associating those changes to functional, cognitive, and behavioral outcomes is an essential step to establishing the value of advanced neuroimaging for pediatric TBI. Here we highlight research that has revealed links between advanced neuroimaging and outcome after TBI and point to opportunities where neuroimaging could expand our ability to prognosticate and potentially uncover opportunities to intervene.

Keywords Pediatric traumatic brain injury · Neuroimaging · Functional outcome · Cognitive outcome · Behavioral outcome

Abbreviations

AD	Axial diffusivity
ADHD	Attention deficit disorder
ALFF	Amplitude of low-frequency fluctuations
ASL	Arterial spin labeling
CBF	Cerebral blood flow
Cho	Choline
Cr	Creatine
CT	Computed tomography

E. L. Dennis · F. Keleher
Department of Neurology, University of Utah School of Medicine, Salt Lake City, UT, USA
e-mail: emily.dennis@hsc.utah.edu; finian.keleher@hsc.utah.edu

B. Bartnik-Olson (✉)
Department of Radiology, School of Medicine, Loma Linda University Medical Center, Loma Linda, CA, USA
e-mail: bbartnik@llu.edu

L. J. Noble-Haeusslein, D. M. Schnyer (eds.), *Traumatic Brain Injuries*, Advances in Neurobiology 42, https://doi.org/10.1007/978-3-031-69832-3_3

DMN	Default mode network
dMRI	Diffusion magnetic resonance imaging
DWI	Diffusion-weighted imaging
EEG	Electroencephalography
ENIGMA	Enhancing Neuro Imaging Genetics Through Meta-Analysis
FA	Fractional anisotropy
fALFF	Fractional amplitude of low-frequency fluctuations
FCD	Functional connectivity density
FDG	Fluorodeoxyglucose
FLAIR	Fluid-attenuated inversion recovery
fMRI	Functional magnetic resonance imaging
fNIRS	Functional near-infrared spectroscopy
GCS	Glasgow Coma Scale
Gln	Glutamine
Glu	Glutamate
Glx	Glutamate + glutamine
GOS	Glasgow outcome scale
GOS-E	Glasgow outcome scale—extended
ICA	Independent component analysis
IPL	Inferior parietal lobe
ISO	Isotropic volume fraction
IQ	Intelligence quotient
LGI	Local gyrification index
Lac	Lactate
MD	Mean diffusivity
MEG	Magnetoencephalography
mI	Myoinositol
MRI	Magnetic resonance imaging
MRS	Magnetic resonance spectroscopy
MRSI	Magnetic resonance spectroscopic imaging
mTBI	Mild TBI
NAA	N-acetyl-aspartate
NS	Neurite density
ODI	Orientation distribution index
OCD	Obsessive-compulsive disorder
PCC	Posterior cingulate cortex
PCPCS	Pediatric cerebral performance category scale
PET	Positron emission tomography
RD	Radial diffusivity
RDI	Restricted diffusion imaging
ReHo	Regional homogeneity
ROI	Regions of interest
rsfMRI	Resting-state functional magnetic resonance imaging
RSN	Resting-state networks
SVS	Single voxel spectroscopy

SUV	Standardized uptake value
SWI	Susceptibility weighted imaging
T1w-MRI	T1-weighted magnetic resonance imaging
T2w-MRI	T2-weighted magnetic resonance imaging
TAI	Traumatic axonal injury
TBI	Traumatic brain injury
TBM	Tensor-based morphometry
VBM	Voxel-base morphometry
VBR	Ventricle to brain ratio
vmPFC	Ventromedial prefrontal cortex
WMH	White matter hyperintensities

3.1 Introduction

Traumatic brain injury (TBI) is among the most frequent pediatric neurological disorders, contributing significantly to childhood morbidity and mortality. Although a significant portion of children with TBI regain ambulatory and self-care skills, many have long-term neuropsychological and behavioral deficits, including impaired intellectual abilities and reduced academic performance (attention, memory, learning, problem-solving, processing speed, language, visual perception, and visual motor skills). Neuroimaging can better characterize the nature and evolution of injury and the underlying mechanisms that lead to progressive neurodegeneration, recovery, or subsequent plasticity. While acute neuroimaging, such as computed tomography (CT), is critical for assessing damage and determining what immediate interventions may be necessary, such as placement of a shunt to relieve increased intracranial pressure, more advanced neuroimaging modalities, such as magnetic resonance imaging (MRI), are more often used in research settings. Neuroimaging biomarkers can provide important information about the underlying brain damage associated with TBI and can help predict clinical outcomes and guide treatment strategies. Previous reviews of neuroimaging in pediatric TBI have covered neuroimaging in pediatric mild-severe TBI in general [1–4], specific modalities such as diffusion MRI [5], and longitudinal studies [6]. Here we focus on neuroimaging studies that specifically investigate the relationship between imaging findings and outcome following moderate-severe TBI. Studies that also included mild TBI (mTBI) are reviewed, but those that solely studied mTBI are not.

In the following sections, we briefly describe the neuroimaging modalities used to study structural, functional, and metabolic changes following pediatric mild-severe TBI. Each section includes a basic description of the imaging technique and analysis methods, followed by a discussion of the overall findings. A summary of the studies can be found in Tables 3.1, 3.2 and 3.3 and includes the following details: population characteristics of patient and comparison groups, injury severity, postinjury interval between injury and imaging, and analysis methods. As the postinjury period is quite dynamic, the interval between injury and neuroimaging is critical for

Table 3.1 Summary of studies assessing structural changes using T1-weighted MR imaging and their relationship to outcome

Study	Population characteristics	Age at injury	Severity	Time since injury	Outcome category	Analysis method
Levin et al. (1989) [15]	**TBI: n = 21 Comparison: none**		**Mild-severe**			**Lesions**
Mendelsohn et al. (1992) [16]	TBI: n = 55 (39M, 16F) Comparison: none	5–17 years	Mild-severe	3 months–years	Functional	Lesions
Mendelsohn et al. (1992) [18]	**TBI: n = 34 Comparison: none**		**Moderate-severe**		**Functional**	**Lesions**
Levin et al. (1993) [196]	TBI: n = 76 Comparison: n = 57 (50M, 66F)	6–16 years	Moderate-severe	≥ 3 months	Cognitive—memory, fluency, inhibition	Lesions
Levin et al. (1994) [32]	TBI: n = 62 Comparison: n = 54 (79M, 39F)	5–16 years	Moderate - severe	Unclear	Cognitive—verbal memory	Lesions
Berryhill et al. (1995) [24]	**TBI: n = 76 Comparison: n = 57**	**5–15 years**	**Severe**	**≥ 3 months**	**Functional**	**Lesions**
Levin et al. (1996)[a] [33]	TBI: n = 77 Comparison: n = 104 (108M, 73F)	5–16 years	Mild-severe	3 and 12 months	Cognitive—memory	Lesions

Bowen et al. (1997)[a] [25]	TBI: n = 22 (9M, 13F) Comparison: none	6–16 years	Moderate-severe	Within 4 weeks	Cognitive—memory, motor	VBR
Koelfen et al. (1997) [28]	TBI: n = 59 Comparison: n = 59 (82M, 36F)	6–15 years	Mild-severe	1–5 years	Functional; cognitive—intelligence, language, motor, arithmetic	Lesions
Levin et al. (1997) [22]	**TBI: n = 22 Comparison: none**			**3 months–3 years**	**Functional, behavior**	**Lesions**
Levin et al. (1997)[a] [30]	**TBI: n = 151 Comparison: n = 189**			**3 years**	**Cognitive—executive function**	**Lesions**
Barker et al. (1999) [26]	TBI: n = 38 Comparison: n = 36 (74M, 0F)	16–30 years	Not reported	≥ 6 weeks	Cognitive—memory, intelligence, processing speed	VBR
Benavidez et al. (1999)[197]	TBI: n = 51 Comparison: n = 16 (41M, 26F)	8–16 years	Mild-severe	≥ 3 months	Cognitive—verbal listening	Area and lesions
Herskovits et al. (1999)[a] [59]	TBI: n = 76 Comparison: none (M/F not reported)	4–19 years	Moderate-severe	3 months	Psychiatric—ADHD	Lesions

(continued)

Table 3.1 (continued)

Study	Population characteristics	Age at injury	Severity	Time since injury	Outcome category	Analysis method
Gerring et al. (2000)[a] [60]	**TBI: n = 99 Comparison: none**	**4–19 years**	**Moderate-severe**	**3 months**	**Psychiatric—ADHD**	**Lesions**
Verger et al., 2001 [39]	**TBI: n = 19 Comparison: n = 19**		**Moderate-severe**	**≥ 6 years**	**Cognitive—processing speed, visuospatial, memory**	**Volume/area**
Herskovits et al. (2002)[a] [68]	TBI: n = 94 Comparison: none (53M/41F)	4–19 years	Moderate-severe	3 months	Psychiatric—PTSD	Lesions
Prasad et al. (2002)[a] [19]	TBI: n = 60 Comparison: none (28M/32F)	1 month–6 years	Moderate-severe	Within 7 days	Functional	Lesions
Levin et al. (2004)[a]	**TBI: n = 144 Comparison: none**		**Mild-severe**	**3 months**	**Cognitive—working memory**	**Lesions**
Levin et al. (2004)[a] [21]	TBI: n = 80 Comparison: none (60M/20F)	5–18 years	Mild-severe	≥ 3 months	Behavior	Lesions
Max et al. (2004) [65]	**TBI: n = 94 Comparison: n = 22**	**5–14 years**			**Psychiatric—ADHD**	**Lesions**
Vasa et al. (2004)[a] [67]	TBI: n = 95 Comparison: none (54M/41F)	4–19 years	Severe	3 months	Psychiatric—ADHD	Lesions

Salorio et al. (2005)[a] [29]	TBI: n = 76 Comparison: none (46M/30F)	5–16 years	Moderate-severe	3 months	Psychiatric—PTSD	Lesions
Serra-Grabulosa et al. (2005) [35]	TBI: n = 16 Comparison: n = 16 (28M/4F)	17.8 years (average)	Severe	9.7 years (average)	Cognitive—verbal and visual learning	Volume
Bonnier et al. (2007) [20]	TBI: n = 50 Comparison: none (M/F not reported)	< 4 years	Severe	≥ 3 months	Functional	Lesions
Braga et al. (2007) [27]	TBI: n = 23 Comparison: none (13M/10F)	7–13 years	Severe	≥ 1 year	Cognitive—intelligence	Lesions
Catroppa et al. (2008) [31]	TBI: n = 53 Comparison: n = 13 (43M/27F)	7–12 years	Mild-severe	5 years	Cognitive—memory	Lesions
Grados et al. (2008)[a] [64]	TBI: n = 80 Comparison: none (39M/33F)	6–18 years	Severe	3 months	Psychiatric—OCD	Lesions

(continued)

Table 3.1 (continued)

Study	Population characteristics	Age at injury	Severity	Time since injury	Outcome category	Analysis method
Merkley et al. (2008) [38]	TBI: n = 16 (8M/8F) Comparison: n = 16	9–16 years	Moderate-severe	3.1 years (average)	Cognitive—memory	Cortical thickness
Slawik et al. (2009) [43]	TBI: n = 33 Comparison: none (20M/13F)	8–13 years	Severe	3–5 years	Cognitive—intelligence	Volume (VBM)
McCauley et al. (2010) [37]	TBI: n = 40 Comparison: n = 41 (56M/25F)	7–17 years	Moderate-severe	3 months	Cognitive—memory	Cortical thickness
Wu et al. (2010)[a] [79]	TBI: n = 23 Comparison: n = 25 (33M/15F)	7–17 years	cMild, moderate, severe	3 and 18 months	Cognitive—processing speed	Volume
Max et al. (2011)[a] [63]	TBI: n = 177 Comparison: none (125M/52F)	5–14 years	Mild-severe	3 months	Psychiatric	Lesions
Max et al. (2012)[a] [62]	TBI: n = 177 Comparison: none (125M/52F)	5–14 years	Mild-severe	3 months	Psychiatric	Lesions
Anderson et al. (2013)[a] [48]	TBI: n = 93 Comparison: n = 44 (92M/44F)	5–15 years	Mild-severe	8 weeks	Cognitive—social	Lesions

Bigler et al. (2013) [56]	TBI: n = 12 Comparison: none (9M/3F)	8–12 years	cMild	1–5 years	Cognitive—social (TOM)	Lesions, qualitative atrophy
Max et al. (2013)[a] [61]	TBI: n = 70 Comparison: none (47M/23F)	5–14 years	Mild	3 and 6 months	Psychiatric—novel disorders	Lesions
Yeates et al. (2013) [58]	TBI: n = 55 Comparison: n = 32 (53M/34F)	8–13 years	cMild, moderate, severe	≥ 4 years	Social/classroom acceptance	Volume (VBM)
Levan et al. (2015) [51]	TBI: n = 23 Comparison: none (15M/8F)	8–17 years	Mild-severe	1–8 years	Cognitive—social, cognitive proficiency	Cortical thickness
Dennis et al. (2016)[a] [14]	TBI: n = 36 Comparison: n = 45 (49M/22F)	8–18 years	Moderate-severe	1–6 months and 13–19 months	Cognitive—processing speed, working memory, verbal memory, switching	Volume (TBM)
Levan et al. (2016)[a] [57]	TBI: n = 23 Comparison: none (15M/8F)	7–16 years	Mild-severe	1–8 years	Cognitive—executive function, social	Cortical thickness

(continued)

Table 3.1 (continued)

Study	Population characteristics	Age at injury	Severity	Time since injury	Outcome category	Analysis method
Ryan et al. (2016)[a] [46]	TBI: n = 93 Comparison: n = 34 (91M/46F)	5–15 years	Mild-severe	6 weeks and 24 months	Cognitive—fatigue	Volume
Ryan et al. (2016)[a] [54]	TBI: n = 98 Comparison: n = 33 (92M/39F)	5–15 years	Mild-severe	2–8 weeks and 6 and 24 months	Cognitive—social (TOM)	Volume
Ryan et al. (2017)[a] [54]	TBI: n = 103 Comparison: n = 34 (100M/37F)	Not reported	Mild-severe	6 weeks	Cognitive—social (TOM)	Volume
Dennis et al. (2017)[a] [5]	TBI: n = 21 Comparison: n = 26 (31M/16F)	8–18 years	cMild, moderate, severe	2–5 and 13–19 months	Cognitive—processing speed, working memory, verbal learning, set shifting	Volume (TBM)
Catharine et al. (2019)[a] [45]	TBI: n = 16 Comparison: none (9M/7F)	Not reported	Moderate-severe	1–5 years	Cognitive—executive function	Volume
Crasta et al. (2019) [47]	TBI: n = 14 Comparison: none (8M/6F)	11–18 years	Moderate-severe	1–14 years	Motor	Volume
Hoskinson et al. (2019) [53]	TBI: n = 45 Comparison: n = 43 (52M/35F)	8–13 years	cMild, moderate, severe	1–5 years	Cognitive—social (TOM)	Volume

Ferrazzano et al. (2021) [23]	TBI: n = 24 Comparison: none (10M/14F)	9–18 years	Severe	1– 2 years	Functional, cognitive—intelligence	VBR, CC area
King et al. (2021)[a] [41]	TBI: n = 83 Comparison: n = 33 (74M/42F)	5–16 years	Mild-severe	1–88 days	Cognitive—executive function	Structural covariance
Ryan et al. (2021)[a] [42]	**TBI: n = 112 Comparison: n = 43 (100M/55F)**	**5–16 years**	**Mild-severe**	**6 weeks**	**Cognitive—executive function, behavior**	**Volume**
Wilde et al. (2021) [52]	TBI: n = 17 Comparison: n = 17 (23M/11F)	6–15 years	Mild-severe	6–15 years	Cognitive—social competence	Gyrification index
Anderson et al. (2022)[a] [49]	TBI: n = 107 Comparison: none (74M/33F)	5–15 years	Mild-severe	2–8 weeks and 6, 12, and 24 months	Cognitive—social competence	FS volume
Greer et al. (2022) [44]	TBI: n = 32 Comparison: n = 30 (55M/7F)	8–13 years	cMild, moderate, severe	Within 1 year	Cognitive—executive function	Thickness/deformation

(continued)

Table 3.1 (continued)

Study	Population characteristics	Age at injury	Severity	Time since injury	Outcome category	Analysis method
Khan et al. (2022)[a] [50]	TBI: n = 103 Comparison: n = 43 (95M/51F)	5–15 years	Mild-severe	2–8 weeks	Cognitive—social	"Neuropathology on CT or MRI"
Ryan et al. (2023)[a] [66]	TBI: n = 112 Comparison: n = 47 (100M/55F)	5–16 years	Mild-severe	6 weeks	Psychiatric—ADHD	Volume

[a]Denotes longitudinal study design. *cMild* complicated mild TBI, *TOM* theory of mind, *VBR* volume to brain ratio. Bold rows indicate papers for which we could not access the full text

Table 3.2 Summary of studies measuring T2, diffusion, or susceptibility changes and their relationship to outcome

Study	Population characteristics	Age at injury	Severity	Time since injury	Outcome category	Analysis method
T2-weighted MR imaging						
Power et al. (2007) [71]	TBI: n = 36 Comparison: none	6–14 years	Moderate-severe	≥ 5 years	Cognitive—attention	Lesion location and volume
Sigmund et al. (2007) [69]	TBI: n = 40 Comparison: none	1.5–18 years	Not reported	1 week	Functional	Lesion count and volume
Lipszyc et al. (2014) [70]	TBI: n = 29 Comparison: n = 50	8–17	Moderate-severe	3 months	Cognitive—inhibition	Lesion count and volume
Smitherman et al. (2016) [72]	TBI: n = 63 Comparison: none	0–17	Moderate-severe	1 week	Functional	Lesion location and volume
Ressel et al. (2020)[a] [99]	TBI: n = 21 Comparison: none 10M, 11F	7–19 years	Not reported	1–144 days	Cognitive	Lesion scoring
Wright et al. (2020) [198]	TBI: n = 60 Comparison: none 39M, 21F	8–13 years	cMild, moderate, severe	1–4 years	Cognitive—processing speed	Volume
Janas et al. (2022) [199][a]	TBI: n = 55 Comparison: none	1 month–19 years	Moderate-severe	< 30 days	Functional	DAI lesion grade
Diffusion-weighted MR imaging						
Wozniak et al. (2007) [105]	TBI: n = 14 Comparison: n = 14 11M, 17F	10–18 years	Mild-moderate	6–12 months	Behavior, cognitive—processing speed, working memory, executive functioning, motor speed	ROI
Levin et al. (2008) [103]	TBI: n = 32 Comparison: n = 36 52M, 18F	7–17 years	Moderate-severe	3 months	Functional, cognitive—processing speed	Tractography
Kurowski et al. (2009) [107]	TBI: n = 14 Comparison: n = 12 13M, 8F	6–9 years	cMild, moderate, and severe	≥ 1 year	Cognitive—executive functioning, attention	ROI

(continued)

Table 3.2 (continued)

Study	Population characteristics	Age at injury	Severity	Time since injury	Outcome category	Analysis method
Wu et al. (2010)[awrig]	TBI: n = 23 Comparison: n = 25 33M, 15F	7–17 years	Moderate-severe	3 and 18 months	Cognitive—processing speed	ROI
Wilde et al. (2011) [97]	TBI: n = 40 Comparison: n = 42 52M, 27F	7–17 years	Moderate-severe	days	Cognitive—working memory	ROI
Juranek et al. (2012) [101]	TBI: n = 21 Comparison: n = 16 28M, 13F	6–16 years	Moderate-severe	≥ 3 months	Psychiatric—anxiety	ROI
Max et al. (2012) [102]	TBI: n = 44 Comparison: n = 44 62M, 26F	7–17 years	cMild, moderate, severe	3 months	Psychiatric—novel disorders	ROI
Dennis et al. (2015) [200]	TBI: n = 32 Comparison: n = 31 40M, 23F	8–19 years	cMild, moderate, severe	1–5 and 13–19 months	Cognitive—processing speed, working memory, verbal learning, set shifting	Tractography
Dennis et al. (2015b) [201]	TBI: n = 32 Comparison: n = 31 40M, 23F	8–19 years	cMild, moderate, severe	1–5 months	Cognitive—processing speed, working memory, verbal learning, set shifting	Tractography
Faber et al. (2016) [108]	TBI: n = 21 Comparison: n = 16 28M, 8F	10–18 years	Moderate-severe	≥ 5 years	Cognitive—executive functioning, attention	Tractography
Genc et al. (2017)[a] [98]	TBI: n = 21 Comparison: n = 16 70M, 38F	5–14 years	Mild-severe	1–4 weeks and 3–45 months	Cognitive—processing speed	TBSS
Ressel et al. (2017) [96]	TBI: n = 14 Comparison: n = 15 15M, 14F	1–19 years	Not reported	1–100 days	Functional—functional independence	ROI
Volpe et al. (2017) [109]	TBI: n = 25 Comparison: none	6–15 years	Mild-severe	2–9 years	Cognitive—processing speed, reasoning, working memory, IQ	ROI

(continued)

Table 3.2 (continued)

Study	Population characteristics	Age at injury	Severity	Time since injury	Outcome category	Analysis method
Yuan et al. (2017)[a] [105]	TBI: n = 17 Comparison: n = 11 13M, 15F	9–18 years	cMild, moderate, severe	≥ 1 year	Behavior, cognitive—executive functioning, attention	Graph theory
Königs et al. (2018) [106]	TBI: n = 37 Comparison: n = 27 35M, 29F	8–14 years	cMild, moderate, severe	9 months–6 years	Behavior, cognitive—intelligence, working memory, verbal memory, attention	TBSS
Ressel et al. (2020)[a] [99]	TBI: n = 21 Comparison: none 10M, 11F	7–19 years	Not reported	1–144 days	Cognitive	ROI
Dennis et al. (2021) [7]	TBI: n = 244 Comparison: n = 263 320M, 187F	5–20 years	cMild, moderate, severe	1 week–years	Behavior, cognitive—executive functioning	ROI
Mayer et al. (2022)[a] [100]	TBI: n = 204 Comparison: n = 173 221M, 156F	8–18 years	Mild	1–11 days and 4 months	Cognitive—verbal memory, executive functioning	Voxelwise
Susceptibility-weighted imaging						
Tong et al. (2004) [80]	TBI: n = 40 Comparison: none M/F not reported	2–18 years	cMild, moderate, severe	7 ± days	Functional	Lesion count, lesion volume, location
Sigmund et al. (2007) [69]	TBI: n = 40 Comparison: none M/F not reported	2–18 years	cMild, moderate, severe	7 ± days	Functional	Lesion count, lesion volume
Babikian et al. (2005) [83]	TBI: n = 18 Comparison: none 12M, 6F	2–18 years	cMild, moderate, severe	1–3 years	Cognitive—intelligence, processing speed, executive function	Lesion count, lesion volume, location

(continued)

Table 3.2 (continued)

Study	Population characteristics	Age at injury	Severity	Time since injury	Outcome category	Analysis method
Beauchamp et al. (2013) [81]	TBI: n = 106 Comparison: n = 43 70M, 36F	5–16 years	Mild, cMild, moderate, severe	2–8 weeks	Behavior, social, cognitive – intelligence, IQ, processing speed	Lesion count, lesion volume, location
Ryan et al. (2015)[a] [84]	TBI: n = 112 Comparison: n = 43 76M, 36F	5–15 years	Mild, cMild, moderate, severe	2–8 weeks	Behavior, social, cognitive—language competence, attention	Lesion count, lesion load, location
Ryan et al. (2015)[a] [85]	TBI: n = 104 Comparison: n = 43 74M, 30F	6–15 years	Mild, cMild, moderate, severe	2–8 weeks	Cognitive—social (TOM)	Lesion count, lesion load, location
Ryan et al. (2015)[a] [90]	TBI: n = 112 Comparison: n = 43 76M, 36F	5–15 years	Mild, cMild, moderate, severe	2–8 weeks	Cognitive—social (TOM)	Lesion count, lesion load, location
Ryan et al. (2016) [202][a]	TBI: n = 78 Comparison: n = 40 51M, 27F	6–15 years	Mild, moderate, severe	2–8 weeks	Cognitive—intelligence, social	Lesion count, lesion volume, lesion load, location
Anderson et al. (2017) [203][a]	TBI: n = 74 Comparison: n = 39 69M, 44F	5–15 years	Mild-severe	1 month	Cognitive—social	Lesions
Vander Linden et al. (2019) [89]	TBI: n = 18 Comparison: n = 18 11M, 7F	11–17 years	Moderate-severe	1–5 years	Cognitive—working memory, processing speed, spatial planning, executive function	Lesion count, distribution, anatomical grading system
Ryan et al. (2023) [66]	TBI: n = 42 Comparison: none 32M, 10F	5–15 years	Mild, cMild, moderate, severe	2–8 weeks	Functional—fatigue	Lesion count, location
Baker et al. (2022) [82]	TBI: n = 42 Comparison: none 26M, 17F	0.5–17 years	Severe	4 months	Functional	Lesion location

[a]Denotes a longitudinal study design. *cMild* complicated mild, *TOM* theory of mind, *DAI* diffuse axonal injury, *ROI*, region of interest, *TBSS*, tract-based spatial statistics

Table 3.3 Summary of studies assessing postinjury metabolism (MR spectroscopy, perfusion imaging, positron emission tomography) and brain function (functional MR imaging) and their relationship to outcome

Study	Population characteristics	Age at injury	Severity	Time since injury	Outcome category	Analysis method
MR spectroscopy						
Ashwal et al. (2000) [162]	TBI: n = 53 M/F not reported Comparison: none	1–180 months	Mild-severe	2–31 days	Functional	*Method*: SVS—short echo time metabolite ratios. *Regions*: midoccipital gray matter
Brenner et al. (2003) [171]	TBI: n = 22 M/F not reported Comparison: none	3–19 years	Mild-severe	0–22 days	Functional, cognitive—intelligence, attention/executive function, learning/memory	*Method*: SVS short echo time metabolite ratios. *Regions*: midoccipital gray matter echo
Parry et al. (2004) [167]	TBI: n = 22 M/F not reported Comparison: n = 30	20 months–13 years	Moderate-severe	3 months–12 years	Cognitive—intelligence, memory, executive function/verbal memory	*Method*: SVS long echo time and short echo time metabolite levels. *Regions*: right frontal white matter, hippocampus, midcorpus callosum
Ashwal et al. (2004) [170]	TBI: n = 38 M/F not reported Comparison: none	1.6–17 years	Moderate-severe	1–17 days	Functional	*Method*: SVS short echo time metabolite levels and ratios. *Regions*: midoccipital gray matter and parietal white matter
Ashwal et al. (2004) [172]	TBI: n = 38 M/F not reported Comparison: none	2–17 years	Moderate-severe	1–17 days	Functional	*Method*: SVS short echo time metabolite levels and ratios. *Regions*: midoccipital gray matter and parietal white matter
Hunter et al. (2005) [176]	TBI: n = 7 (2M, 5F) Comparison: n = 12	7–12 years	Mild, moderate, severe	6 weeks–3 years	Cognitive—intelligence	*Method*: MRSI short echo time metabolite levels and ratios. *Region*: centrum semiovale

(continued)

Table 3.3 (continued)

Study	Population characteristics	Age at injury	Severity	Time since injury	Outcome category	Analysis method
Holshouser et al. (2005) [173]	TBI: n = 40 M/F not reported Comparison: n = 49	2–18 years	Moderate-severe	3–12 days	Functional	*Method*: MRSI intermediate echo time metabolite ratios. *Region*: midcorpus callosum
Makoroff et al. (2005) [179]	TBI: n = 11 (6M, 5F) Comparison: none	4–26 weeks	"Inflicted/NAT grouped based on the presence of lactate"	1–19 days	Functional	*Method*: SVS short echo time metabolite levels and ratios. *Regions*: frontal white matter, midoccipital gray
Babikian et al. (2006) [164]	TBI: n = 20 (13M, 7F) Comparison: none	1–18 years	Mild, moderate, severe	1–17 days	Cognitive—intelligence, attention/working memory, verbal learning/memory	*Method*: MRSI intermediate echo time metabolite ratios. SVS short echo time metabolite levels and ratios. *Regions*: corpus callosum (MRSI), midoccipital gray and parietal white matter
Yeo et al. (2006)[a] [164]	TBI: n = 36 (30M, 6F) Comparison: 14	6–18 years	cMild, moderate, severe	4–158 days; 5, 13, and 24 weeks	Cognitive—intelligence, learning and memory, executive function	*Method*: MRSI short echo time metabolite ratios. *Region*: centrum semiovale
Walz et al. (2008) [166]	TBI: n = 10 (6M, 4F) Comparison: 10	3–6 years	cMild, moderate, severe	12 months	Cognitive—learning ability, academic achievement, pragmatic language	*Method*: SVS long echo time metabolite levels. *Regions*: mid-frontal gray matter, left frontal white matter
Babikian et al. (2010)[a] [165]	TBI: n = 10 (8M, 2F) Comparison: 15	8–18 years	Moderate-severe	3–8 months; 14.7 months	Cognitive—intelligence/processing speed, learning and memory, visual memory, motor coordination	*Method*: MRSI intermediate echo time metabolite ratios. *Region*: corpus callosum

Aaen et al. (2010) [169]	TBI: n = 44 (28M, 16F) Comparison: none	1–12 months	Mild, moderate, severe	1–30 days MRS	Functional	*Method*: MRSI intermediate echo time metabolite ratios. *Region*: corpus callosum
Babikian et al. (2018) [168]	TBI: n = 55 (20M, 11F) Comparison: 24	14–17 years	Moderate-severe	3–8 months; 16 months (average)	Cognitive—intellectual functioning, verbal memory, psychomotor skills, working memory	*Method*: volumetric EPSI short echo time metabolite ratios. *Region*: whole brain
Dennis et al. (2018) [175]	TBI: n = 29 (21M, 8F) Comparison: 33	8–18 years	Moderate-severe	2–5 months, 13–19 months	Cognitive—processing speed, working memory, verbal learning, executive function	*Method*: volumetric EPSI short echo time metabolite ratios. *Region*: whole brain
Holshouser et al. (2019)[a] [174]	TBI: n = 64 (47M, 17F) Comparison: 63	7–11 years	cMild-moderate, severe	6–17 days, 12 months.	Functional Cognitive—intelligence, attention, and memory	*Method*: MRSI intermediate echo metabolite ratios. *Regions*: midcorpus callosum through midbrain
Berger et al. (2023)[a] [153]	TBI: n = 64 (47M, 17F) Comparison: 63	5–18 years	cMild-moderate, severe	6–17 days, 12 months.	Functional, cognitive—intelligence, attention, and memory	*Method*: MRSI intermediate echo global metabolite ratios. *Regions*: midcorpus callosum through midbrain
Perfusion-weighted MR imaging						
Newsome et al. (2012) [186]	TBI: n = 7 (3M, 4F) Comparison: n = 7	13–19 years	Moderate- severe	2.4 years (average)	Social, cognitive—verbal reasoning and executive function, intelligence	*Method*: arterial spin labeling with region of interest analysis

(continued)

Table 3.3 (continued)

Study	Population characteristics	Age at injury	Severity	Time since injury	Outcome category	Analysis method
Positron emission tomography						
Worley et al. (1995) [191]	TBI: n = 22 (12M, 10F) Comparison: none	4 months–19 years	Severe	18–643 days, 5–38 months	Functional, cognitive	*Method*: regional PET score (normal, reduced)
Task-based functional MR imaging						
Karunanayake et al. (2007) [115]	TBI: n = 8 Comparison: n = 9 10M, 7F	6–9 years	Mild-severe	1–3 years	Cognitive—verbal memory, executive function	
Karamer et al. (2010) [116]	TBI: n = 7 Comparison: n = 13 9M, 11F	7–13 years	Mild-moderate	2–5 years	Cognitive—associative memory	
Tlustos et al. (2011) [117]	TBI: n = 11 Comparison: n = 11 13M, 9F	12–16 years	Mild-severe	2 years (average)	Cognitive—executive function	
Wilde et al. (2011) [97]	TBI: n = 40 Comparison: n = 41 54M, 27F	7–17 years	Moderate-severe	4.0 days	Cognitive—verbal memory	
Scheibel et al. (2011) [118]	TBI: n = 9 Comparison: n = 9 10M, 8F	13–19 years	Moderate-severe	4 years	Cognitive—social (TOM)	

Chiu et al. (2013) [119]	TBI: n = 11 Comparison: n = 8 (M/F not reported)	14–17 years	Mild-moderate	2 years (average)	Behavior—risk taking	
Morgań et al. (2013) [120]	TBI: n = 32 Comparison: n = 17 22M, 27F	12–24 years	Mild-severe	2–16 years	Speech	
Fufaeva (2014) [121]	TBI: n = 62 Comparison: none (M/F not reported)	4–17 years	Severe	Not reported	Cognitive—verbal memory, executive function	
Tlustos et al. (2015) [122]	TBI: n = 11 Comparison: n = 14 13M, 12F	16 years (average)	Mild-moderate	> 12 months	Behavior, cognitive	
Resting-state functional MR imaging						
Stephens et al. (2018) [123]	TBI: n = 11 Comparison: n = 11 6M, 5F	12–18 years	Mild-moderate	~12 months	Behavior, cognitive—executive function	Seed-based
Stephens et al. (2017) [148]	TBI: n = 17 Comparison: n = 14 22M, 9F	12–18 years	Mild-severe	1–5 years	Behavior, cognitive—impulsivity	Seed-based

(continued)

Table 3.3 (continued)

Study	Population characteristics	Age at injury	Severity	Time since injury	Outcome category	Analysis method
Tuerk et al. (2020) [149]	TBI: n = 11 Comparison: n = 11 6M, 5F	11–16 years	cMild, moderate, severe	1–6 years	Behavior, cognitive	ROI to ROI
Botchway et al. (2022) [150]	TBI: n = 11 Comparison: n = 11 6M, 5F	8–14 years	Mild-severe	1–6 years	Behavior, cognitive	ICA, graph theory
Vaughn et al. (2022) [151]	TBI: n = 11 Comparison: n = 11 6M, 5F	8–15 years	Mild-severe	7 weeks and 7 months	Psychiatric—anxiety	BVAR-connect

[a]Denotes longitudinal study design. *cMild* complicated mild, *TOM* theory of mind, *SVS* single-voxel spectroscopy, *MRSI* magnetic resonance spectroscopic imaging, *ROI* region of interest, *ICA*, independent components analysis, *BVAR* Bayesian multisubject vector autoregressive

accurate interpretation of results. The time postinjury of brain scans is listed in all tables, exactly as was reported in the original papers. To aid interpretation of the results discussed, we have clustered studies into three postinjury periods. While the exact boundaries between these windows are somewhat arbitrary, it is informed by prior literature [3, 5, 7–10]. For the purposes of this chapter, "acute" refers to imaging within 2 months of injury, "postacute" for imaging 2–6 months postinjury, and "chronic" more than 6 months postinjury.

3.2 Structural MRI

3.2.1 T1-Weighted MRI

T1-weighted MRI (T1w-MRI) has been used to study changes in brain structure for nearly 40 years [11]. T1w-MRI generates a three-dimensional anatomical image of the brain and can be used to calculate various morphological metrics. The most basic is lesion volume and location, where lesions are manually or semimanually traced on the T1w-MRI, and the number, location, and/or volume of the lesion burden is associated with outcome. The next level in complexity is measuring volume, either of the whole brain or of specific regions of interest (ROIs). Using FreeSurfer and related tools, researchers can also estimate the cortical thickness by measuring the distance between the gray matter-white matter boundary and the pial boundary, either vertex-wise across the brain or averaged within an ROI [11]. FreeSurfer can also be used to extract cortical surface area, cortical volume, and local gyrification index (LGI), a measure of how extensive cortical folding patterns are. Voxel-wise approaches include voxel-based morphometry (VBM) and tensor-based morphometry (TBM). VBM involves tissue segmentation and registration to a template, resulting in a measure of gray matter density that reflects volume [12]. TBM uses the warp information from mapping a participant image to a template image to measure regional differences in brain volume [13, 14]. Structural covariance networks involve parcellating the brain into a set of ROIs based on an atlas and calculating the correlations between the volumes of these regions.

Lesion studies are the earliest and the most commonly reported in the T1w-MRI literature, with the first paper published in 1989 [15]. The first papers to examine correlates of lesions after TBI investigated functional outcome measured by the Glasgow Outcome Scale (GOS and GOS-Extended [GOSE]) [16–18]. Lesions present within the first week postinjury are predictive of poorer functional outcome [19], with studies in the postacute and chronic periods highlighting frontal and subcortical lesions as particularly predictive of poor outcome [16, 20–22]. General atrophy measured chronically by the ventricle-to-brain ratio (VBR) is associated with poorer outcome [23], as are volumetric reductions in the prefrontal gray matter [24]. All T1w-MRI papers reviewed here may be found in Table 3.1.

Cognitive outcomes are by far the most common outcomes examined. While associations between volume changes and cognitive outcome were not detected within the first 2 months postinjury [25, 26], many investigations beyond this early time point have reported significant results. Beyond 1 year postinjury, cerebellar lesions and deep white matter lesions have been linked with poorer performance on intelligence measures [27, 28], as has corpus callosum area [23]. Studies have shown broad links between lesions and poorer verbal memory, set shifting, and problem-solving performance [14, 29, 30], with some studies specifically reporting associations with lesions in the corpus callosum [31] and frontal cortex [32, 33]. Various memory domains are frequently impacted after TBI, including working, visual, verbal, and prospective memory [34]. Increases in ventricular volume and decreases in white matter volume have also been associated with poorer verbal and visual memory performance [35, 36]. Cortical thinning has been reported across a wide postinjury range and is associated with poorer prospective and working memory performance [37, 38]. Processing speed is another domain that is often impacted after TBI [34], but the literature regarding neuroimaging correlates is more mixed, with some studies showing associations with volume [36, 39] while others have reported null results [26, 40]. Looking broadly at executive function, which could include any combination of attention, inhibition, processing speed, working memory, and set shifting, studies have shown disruption in volume and structural covariance acutely in patients with poor executive function 1–2 years later [41, 42]. These structural changes appear to persist, with loss of gray matter volume years postinjury associated with executive function deficits [43]. Several null studies have been published as well, with deficits in neuroimaging metrics but no relation reported with executive function performance [44, 45]. In a large longitudinal study, Ryan and colleagues found that decreases in the volume of a corticostriatal network are associated with higher cognitive fatigue [46]. Lastly, while null results have been reported when examining acute correlates of motor function [25], motor performance years postinjury is associated with cortical and deep white matter lesions, as well as smaller volume of the motor and premotor cortices, as well as total brain volume [47].

There have been a number of investigations into associations between alterations in brain structure and social function. While some studies have reported that lesions and volume changes within the first weeks postinjury are not associated with social cognition [48, 49], others have found acute pathology in the frontal cortex to be predictive of self-esteem months later [50]. Additionally, volume and thickness deficits and alterations in cortical gyrification in frontotemporal regions are linked with poorer performance on theory of mind and social competence measures as early as 2 weeks and as late as 15 years postinjury (Fig. 3.1) [51–57]. Additionally, social acceptance has been associated with white matter volume in the parietal and limbic lobes [58].

While psychiatric disorders can be a contributing factor for sustaining a TBI, injury can also lead to worsening symptoms and the emergence of novel psychiatric disorders. In the postacute phase, lesions in the basal ganglia were predictive of attention-deficit hyperactivity disorder (ADHD) months later [59, 60], while lesions in the frontal cortex have been associated with novel psychiatric disorders [61], including depression [62], anxiety [63], and obsessive-compulsive disorder (OCD) [64]. Null results have also been reported for lesion studies [65]. Smaller volume of regions in the default mode network (DMN, described in more detail in the rsfMRI section) including the vmPFC, PCC, IPL, and hippocampus is associated with

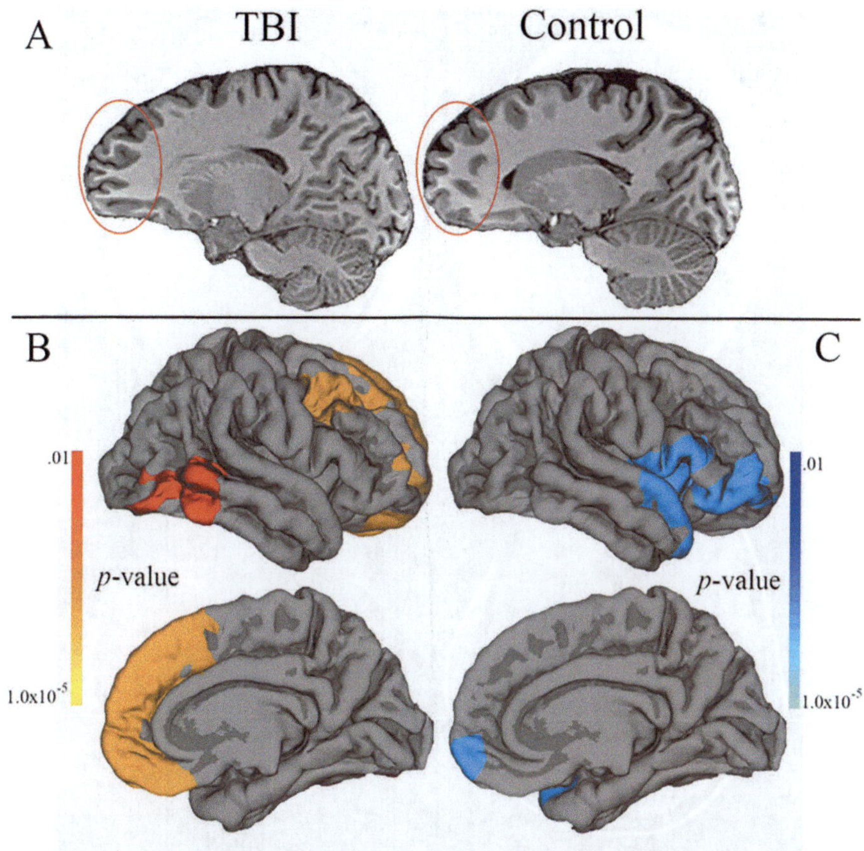

Fig. 3.1 Differences between TBI and control in local gyrification index (LGI). In panel **a**, representative participants from the TBI and control groups are shown. The TBI participant has more extensive cortical folding in the frontal cortex. Panel **B** displays the regions with significant group differences in gyrification, with color corresponding to p-value. Panel **C** displays regions where LGI was associated with parent-rated Vineland-II Socialization domain scores, again with color corresponding to p-value. (Adapted with permission from the authors from Wilde et al. (2021) [52])

increased risk of developing ADHD [66]. Conversely, lesions in frontolimbic areas are also associated with decreased anxiety [67] and reexperiencing [68]. These may represent damage that inhibits natural responses to stimuli.

3.2.2 T2-Weighted MRI

T2-weighted MRI (T2w-MRI) sequences, including fluid-attenuated inversion recovery (FLAIR), are often used to identify lesions which appear as hyperintensities. While frequently part of the standard neuroimaging battery in TBI, it is less often the focus of research studies. Lesions detected by T2w-MRI acutely were far more predictive of functional outcome than CT (Fig. 3.2) [69].

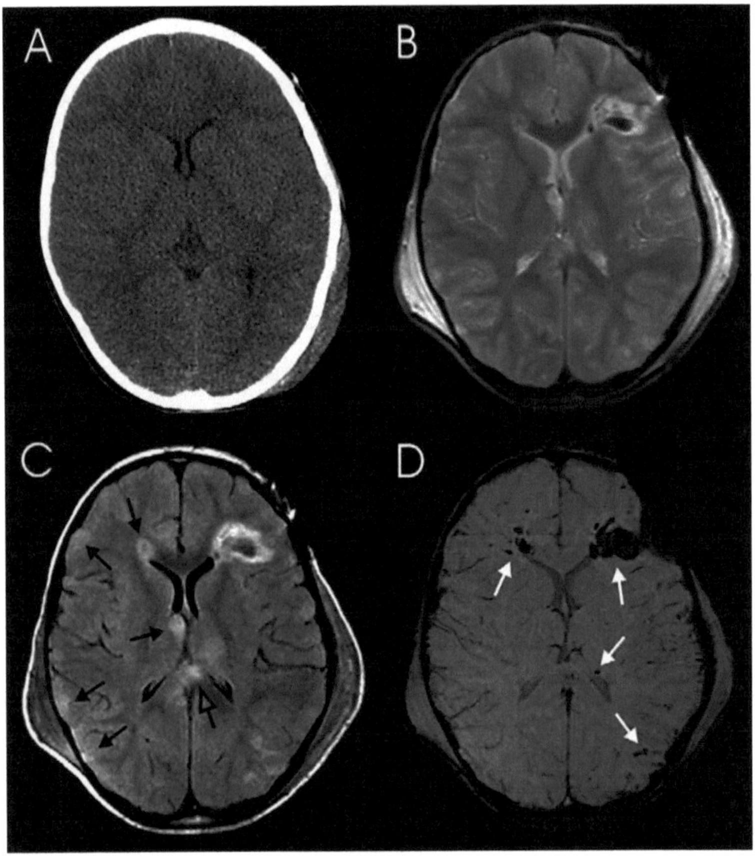

Fig. 3.2 Imaging from a 5-year-old male ejected from a car during a rollover. Initial GCS was 4 and outcome on PCPCS was moderate. (**A**) Initial computed tomography obtained at the level of the basal ganglia within 24 h was normal. The magnetic resonance imaging was taken 2 days later and showed (**B**) T2—the greatest extent of injury at the left frontal periventricular white matter, with other lesions at the right periventricular white matter, thalami, and less obviously in the splenium of the corpus callosum. (**C**) FLAIR highlights the edema lesions (black arrows) more clearly than T2, particularly the lesion in the splenium of the corpus callosum (open arrow). (**D**) Susceptibility-weighted imaging study demonstrates that the two lesions in the periventricular white matter have a hemorrhagic component not seen on computed tomography and not well seen on the other magnetic resonance imaging sequences. The lesion in splenium of the corpus callosum is also seen along with another small hemorrhagic lesion anterior to the occipital horn of the lateral ventricle (white arrows). This illustrates an important feature of magnetic resonance sequences such that T2 depicts a variety of lesion types while the FLAIR and susceptibility-weighted imaging more definitively highlight specific lesions. (Reprinted with permission from Sigmund et al. (2007) [69])

In the postacute phase, lesions in the frontal cortex in particular were associated with poorer inhibition [70]. These lesions can persist years postinjury and remain related to cognitive function, although general lesion load appears more predictive of function at this point than lesions in specific regions (i.e., frontal lobe [71]). White matter hyperintensities (WMH) are similarly predictive of. poorer outcomes measured by the GOSE, with the worst outcomes in patients with WMH in multiple regions (i.e., cortex, basal ganglia, and brainstem [72]).

3.3 Susceptibility-Weighted Imaging

Susceptibility-weighted imaging (SWI) is a high spatial resolution three-dimensional gradient echo T2* MRI technique that accentuates the paramagnetic properties of blood and is sensitive to intravascular venous deoxygenated blood as well as extravascular blood products [73, 74]. Following TBI, SWI is used to detect intraparenchymal microhemorrhages associated with traumatic axonal injury (TAI) owing to its increased sensitivity compared to CT, T1w, or T2w MRI (Fig. 3.3) [69, 73, 75–79]. All SWI papers reviewed here can be found in Table 3.2.

Early work in pediatric TBI focused on quantifying the number, volume, and distribution of hemorrhage ("SWI lesions") and its relationship to injury severity (Glasgow Coma Scale [GCS], duration of coma, intracranial pressure) and neurological outcome (GOS or Pediatric Cerebral Performance Category Scale [PCPCS]) [69, 80–82]. While there is significant variation in SWI lesion volume and number within study populations, TBI patients with poor neurological outcomes consistently show a greater number and volume of lesions [69, 80]. Moreover, a larger distribution of lesions within multiple cerebral lobes, the cerebellum, and/or deep gray matter was similarly associated with poor neurological outcome, suggesting a dose-response relationship between injury severity and SWI lesion number [69, 80, 81]. Initial exploratory analyses identified negative associations between lesion number and volume with global measures of learning and memory, with SWI lesions in deeper brain regions (basal ganglia, thalamus, brainstem) more strongly predicting poor neuropsychological functioning as early as 6 months [81] and as far out as 4 years postinjury [83].

The use of SWI to predict recovery of social cognition, pragmatic communication, and executive function has been investigated (Table 3.2). Social cognition skills were associated with global lesion volume in the early postacute (4–6 weeks) and 6 months following TBI, suggesting an underlying structural disconnection between anatomical regions involved in the social network [84]. Moreover, the presence of SWI lesions in the corpus callosum in the early postacute period was associated with deficits in pragmatic communication at 6 months, with persistent deficits at 24 months associated with the presence of lesions in frontal and temporal regions [85], supporting the hypothesis that injury to the corpus callosum, frontal, and temporal brain regions during critical developmental periods compromise structural connectivity and lead to downstream, long-term consequences [86–88]. Although there was no correlation between the number of SWI lesions and persistent executive function impairment, there was an association with the location of SWI lesions, specifically the corpus callosum and deep subcortical regions [83, 85, 89].

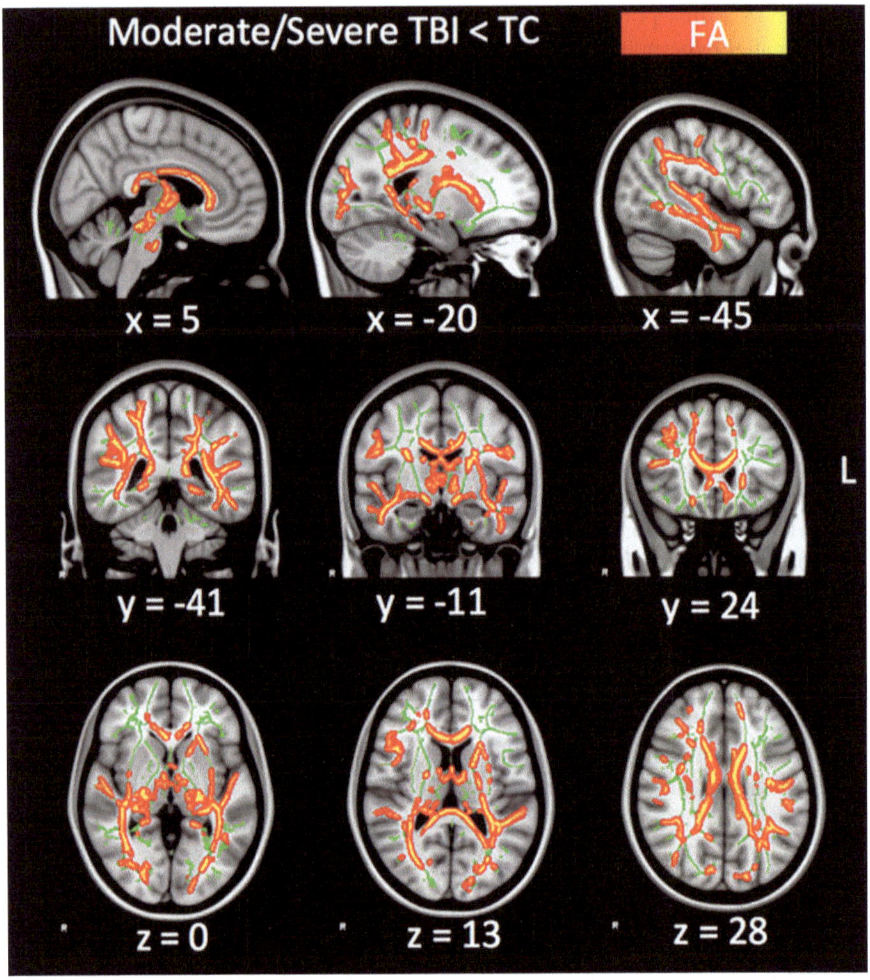

Fig. 3.3 Voxel-wise comparison of FA using threshold-free cluster enhanced correction in TBSS. Note. Results of voxel-wise group comparisons showing the parts of the whole brain skeleton (at FA > 0.2, in green) with differences in terms of FA between the moderate/severe TBI group as compared to the TC group. Lower values in the moderate/severe TBI group as compared to other groups are displayed in red-yellow. The results are overlaid on a MNI152 1 mm T1 brain in radiological convention (right = left), and for visualization purposes, regions in the whole brain skeleton with significant group differences were "thickened" toward the full width of the white matter tract. TC trauma control group, TBI traumatic brain injury, FA fractional anisotropy. (Reprinted with permission from Königs et al. [107])

The interaction between age at injury, lesion number, and outcome has also been investigated. When the injury occurred during late childhood (10–11 years of age), the number of SWI lesions was associated with a 24-month social outcome; however, when the injury was acquired later, during the adolescent period (12–15 years of age), both the extent and number of SWI lesions were related to 6- and 24-month social cognition outcomes [90, 91]. This finding suggests that the long-term social consequences of the diffuse damage may not be realized until the social skills reach maturity. Traumatic axonal injury (TAI) has important implications for the developing brain, disrupting structural brain development and interfering with immature brain connectivity. Despite the heterogeneity in number and location of SWI lesions, the time since injury, and age at injury among the studies reviewed here, SWI is highly sensitive to the presence of TAI, and there is mounting evidence to support its role in outcome prediction. Future longitudinal studies with larger sample sizes and robust clinical outcomes, with careful characterization of lesion location and distribution, will strengthen these initial findings.

3.4 Diffusion-Weighted MRI

Diffusion-weighted MRI (DWI) measures the movement of water in the brain to model white matter tracts and assess the level of organization [92]. As mentioned above, TAI can occur after TBI, particularly in more severe injuries, but TAI is not detectable in conventional imaging such as CT. DWI was first applied to TBI around 20 years ago and its use has accelerated due to its higher sensitivity to TAI [93]. Since then, the number of papers using DTI to examine outcome after TBI has increased exponentially [11]. The most commonly examined metric from DWI is fractional anisotropy (FA), a ratio which reflects the degree to which water is diffusing along the axon rather than across it, commonly interpreted as a proxy for myelination, although other factors such as inflammation can be at play. Other metrics include mean diffusivity (MD), the magnitude of diffusion; radial diffusivity (RD), diffusion perpendicular to the axon; and axial diffusivity (AD), diffusion along the length of the axon. In recent years, more advanced metrics have emerged including ODI, orientation distribution index; ND, neurite density; ISO, isotropic volume fraction or free water; and RDI, restricted diffusion imaging. Lower FA and higher MD and RD are generally interpreted as a sign of pathology such as demyelination or decreased axonal density. Higher ODI and lower ND and higher ISO can reflect less cohesive axonal bundles and lower neuronal density, and higher RDI and ISO can indicate inflammation [94–96].

In the immediate postinjury phase, bleeding and edema may lead to temporarily increased FA, but this typically resolves to decreased FA postacutely and beyond, although literature in uncomplicated mild TBI is mixed [5, 93]. Acutely, higher FA in the corticospinal tract has been associated with better functional independence [97] and lower FA/higher MD in frontoparietal systems with slower reaction time [98] and processing speed [99]. Ressel et al. (2020) found corpus callosum FA to be predictive of outcome after rehabilitation [100]. Examining the more advanced metrics from NODDI in an mTBI sample, Mayer et al. (2022) found higher ISO and ND in mTBI, along with consistent changes in more standard DTI metrics that persisted over 4 months postinjury, concurrent with persisting deficits in executive function and memory [101].

In the postacute and chronic phases, lower FA is associated with psychiatric outcomes, including increased anxiety and increased risk for novel psychiatric disorders [102, 103], as well as functional outcomes (GOSE [104]). A similar pattern of alterations is shown with more parent- and self-reported behavioral problems [7, 105–107]. DWI metrics have also been associated with performance in numerous cognitive domains, including executive functioning [7, 98, 105, 106, 108, 109], verbal and working memory [98, 105, 107, 110], attention [106–109], and processing speed [40, 104, 105, 110]. In some cases, these associations can be seen more than 5 years postinjury, reflecting the long-term implications for pediatric TBI [7, 107, 109, 111].

3.5 Functional MRI

Functional MRI (fMRI) encompasses a variety of neuroimaging methods designed to measure regional changes in brain metabolism over time. Most conventional fMRI methods use blood oxygenation level-dependent (BOLD) contrast, which results from the change in magnetic field that occurs around red blood cells when hemoglobin becomes deoxygenated. This metabolic change can result from the presentation of various stimuli or experimental tasks (termed task-based fMRI) or can occur as a result of unregulated processes in the resting brain (resting-state fMRI, rsfMRI) [112]. In task-based fMRI, regional changes in BOLD signal in response to a task are interpreted as upregulated or "activated" regions. In rsfMRI, lower-frequency changes in BOLD signal are measured to determine synchronous activation between spatially distinct regions at rest, that is, in the absence of any visual, auditory, or other stimuli.

3.5.1 Task-Based fMRI

While both methods were developed in the early 1990s, task-based fMRI quickly outpaced rsfMRI in clinical and research applications due to its employment in clinical psychiatry, treatment monitoring, presurgical planning, and cognitive neuroscience [112]. Despite such broad applications, it was not until nearly a decade later that task-based fMRI studies of TBI began to emerge. Following TBI, investigators started using fMRI to characterize functional changes after insult, investigate the neuroplasticity of injury/recovery, and examine correlations between alterations in brain dynamics and behavior [113]. However, functional imaging of children remained uncommon due to the unique challenges that occur in this population, such as participant-related complications (difficulty understanding the task/scan, following instructions, or sitting still, parental or child anxiety, etc.) [114], technical obstacles (lack of child-specific equipment, difficulty recruiting), or analytical drawbacks (such as lack of pediatric brain templates or child-appropriate processing techniques) [115]. Indeed, many early fMRI studies of pediatric TBI operated in part to demonstrate the feasibility of functional imaging in children.

A common goal of foundational fMRI research in pediatric TBI was to classify the neural substrates of well-known negative behavioral, cognitive, and psychological outcomes of TBI, such as deficits in verbal fluency/intelligence [116], memory [98, 117], and cognitive control [118]. One common finding among these early investigations was that differences in patterns of brain activation were correlated with task performance, even when no structural abnormalities were present on conventional MR. These findings demonstrated the complexity of neural alterations following TBI and the utility of fMRI as a technique to quantify these changes. Consequently, ongoing investigations began to more commonly implement fMRI to investigate more diverse tasks related to outcome after TBI, such as social attribution [119], risk-taking [120], speech [121], executive function [122], emotional processing, inhibition [123], and motor control [124]. These tasks often targeted specific disorders that occur after TBI. For example, one study demonstrated associations between speech dysarthria after TBI and increased activation in regions related to speech compared to typically developing controls [121] (Fig. 3.4). Despite the diversity of investigated tasks and brain regions, one consistent finding across studies is that participants with TBI routinely demonstrate greater activation during task administration. Brain regions that exhibited overaction included the frontal and parietal cortices and the cingulate [98, 118, 119, 121, 123]. This overactivation has been interpreted as a consequence of inefficient processing, in which greater activation is representative of compensation involving the allocation of additional cognitive and neural resources [125].

3.5.2 *Resting-State fMRI*

While rsfMRI uses the same underlying measurements of brain metabolism, it is distinct from task-based fMRI in that it measures brain activity in the absence of any task or stimuli. The use of rsfMRI has allowed the classification of various resting-state networks (RSNs) or brain regions that exhibit synchronous fluctuations in BOLD signal at rest [126]. While there is variability in the literature, studies of rsfMRI are mostly consistent in identifying the following RSNs: the Somatomotor Dorsal, Dorsal Attention, Cinguloopercular, Auditory, Default Mode, Salience, Medial Temporal Lobe, Visual, Parietal Memory, Ventral Attention, and Frontoparietal Networks. Most of these RSNs overlap with known structural or functional networks. However, other than the Default Mode Network, all of the above networks show a decrease in neuronal activity at rest, as compared to when network-relevant tasks are performed. The Default Mode Network is therefore of particular interest in rsfMRI as it is the only RSN that exhibits elevated activity at rest, which suggests that this network demonstrates a default state of neuronal activity in the brain [127].

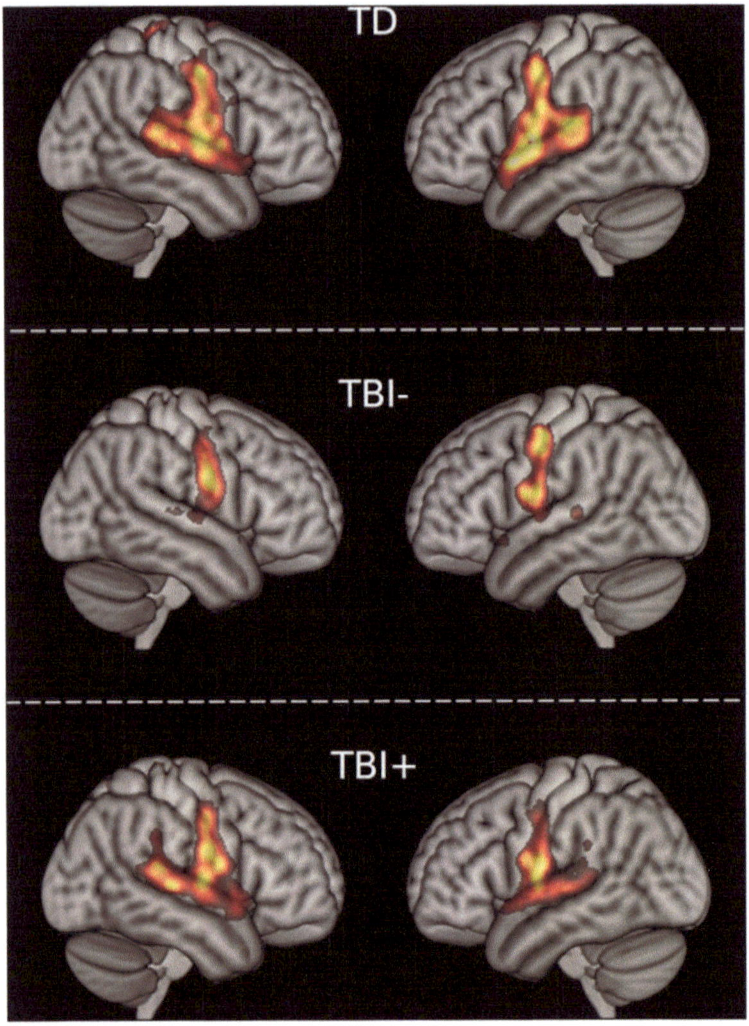

Fig. 3.4 Functional imaging comparing average activation during speaking (left) and listening (right) in typically developing children (TD), children with TBI without dysarthria (TBI-), and children with TBI with dysarthria (TBI+). All groups demonstrate bilateral activation in a typical speech network. The TBI with dysarthria group exhibited increased activation compared to the TBI without dysarthria group in the left subgyral white matter of the parietal lobe. Increased activation was also present in the TBI with dysarthria group compared to typically developing controls in the left sensorimotor cortex. Results are superimposed onto each group's mean T_1-weighted image

As the applications of rsfMRI have evolved, distinct analytic approaches have emerged to accommodate novel research aims and growing clinical demands. Functional connectivity density (FCD) was one of the first developed methods and remains a fundamental measurement of functional connectivity. FCD calculates the correlation between an individual voxel and all other voxels of the brain. Thus, it measures how globally connected a voxel is but does not determine which regions the voxel is correlated with [128]. Similar to FCD, seed-based functional connectivity approaches measure the correlation between a given region, called a seed or region of interest (ROI), with the rest of the brain. However, beyond measuring how globally connected a target region is, seed-based functional connectivity can determine other regions that exhibit synchronous activation with the target region [128]. Seed-based approaches can be used to quantify the amplitude of low-frequency fluctuations (ALFF), fractional amplitude of low-frequency fluctuations (fALFF), and measures of regional homogeneity (ReHo). The ALFF indicates the power of low-frequency fluctuations [129], while fALFF measures the ratio of low-frequency power fluctuations to the whole-brain frequency range [130].

While both ALFF and fALFF are complementary and can be used to measure differences in amplitude power between comparison groups, ALFF is calculated as the sum of amplitudes within a specific low-frequency range whereas fALFF is the fraction of the sum of amplitudes across the entire measured frequency range. ReHo represents the Kendall's coefficient of concordance between a given voxel and its 27 neighboring voxels, effectively measuring synchronization between target voxels and neighboring regions [131]. This method is limited in that it can only measure correlation between spatially adjacent voxels [131]. Apart from seed-based methods, which can only calculate connectivity based on user-defined ROIs, independent components analysis (ICA) is a data-driven approach that can measure multiple simultaneous voxel-to-voxel interactions between distinct networks in the brain, enabling the investigation of group-level as well as within-group analyses [132]. Lastly, using mathematical models from graph theory, graph theory functional connectivity analysis can measure complex attributes of connectivity, such as path length, level of modularity, centrality, nodal degree, and the clustering coefficient [133]. Beyond measuring the strength of correlation between one ROI and another, as seed-based approaches do, graph theory can be used to identify the topological properties of a target region within the whole brain [134]. While the initial uses of rsfMRI were limited, increasing attention has been drawn to its potential for research and clinical applications. Compared to other functional neuroimaging techniques, rsfMRI is advantageous in that it can be performed on patients who are incapable of participating in task-based paradigms, such as young children, sleeping/sedated patients, or patients with altered mental status [126]. Clinically, rsfMRI has been used to identify specific brain regions for presurgical planning in epilepsy [135] and brain tumor patients [136], to identify early markers of Alzheimer disease [137] and dementia [138], and to identify patients with psychiatric disorders, such as schizophrenia [139], major depressive disorder [140], autism [141], and ADHD [142]. In pediatric populations, rsfMRI has been used to investigate the development of neural networks in utero [143], infancy [144], childhood [145–147], and early adulthood [148].

Despite the advantages of rsfMRI for research in pediatric populations, the implication of rsfMRI in pediatric TBI research remains limited, with the earliest rsfMRI studies of outcome after pediatric TBI taking place as recently as 2017. Similar to task-based methods, these studies have primarily sought to establish the utility of fMRI to identify markers of dysfunction after brain injury and investigated how specific behavioral and neurocognitive impairments relate to alterations in functional connectivity. However, since rsfMRI methods preclude the concurrent administration of behavioral tasks and scan protocol, rsfMRI studies identify the correlation between measures of resting network connectivity and independent task performance. Investigations of seed-based connectivity have demonstrated distinct brain-behavior relationships in RSNs following pediatric TBI. Poorer response inhibition has been shown to be correlated with reduced functional connectivity between the motor network and caudate [149] or increased connectivity between the DMN and Brodmann area [124]. Similarly, poorer social functioning has been shown to be correlated with increased frontal-fusiform connectivity [150]. Apart from seed-based methods, one study has utilized graph theory to demonstrate relationships between atypical measures of both global and local network organization with neurocognitive functioning, working memory, memory encoding, and behavioral functioning [151]. Additionally, one study has demonstrated an association between increased connectivity in the DMN and anxiety using a novel analysis technique called Bayesian Multi-Subject Vector Autoregressive Modeling. Results from this study also demonstrated significant regional differences in positive connectivity between children with TBI and typically developing controls in the DMN (Fig. 3.5) [152].

Despite the apparent complexity of network responses to brain injury, the scope of functional MRI studies of pediatric TBI is limited. Inconsistencies in findings suggest that both negative and positive changes in connectivity, depending on the affected region/network, can enduringly impact behavioral and cognitive outcomes after pediatric TBI. Understanding the causes and consequences of these distinct network responses to pediatric TBI may be key to improving outcomes. Further, no current studies have prioritized investigating relationships between brain activation in functional MRI and other important components of pediatric TBI, such as time since injury, age, or age at injury. All past fMRI studies on pediatric TBI have been cross-sectional, and all but one [98] have only recruited participants in the chronic phase (>6 months) of injury. Future studies implicating these variables are necessary to gain insight into the processes of injury to the developing brain.

3.6 Magnetic Resonance Spectroscopy

Metabolic dysfunction is part of the pathophysiologic cascade following TBI, and proton magnetic resonance spectroscopy (MRS) has been utilized to identify metabolite changes in regions of injury and define the extent of pathology, often showing abnormalities in the injured brain even when conventional structural neuroimaging appears normal [153]. MRS is inherently quantitative as the integral of each peak in the spectrum is proportional to the concentration of that metabolite in a specified volume of interest. The most commonly measured brain metabolites include N-acetylaspartate (NAA; 2.01 ppm), an amino acid synthesized in neuronal

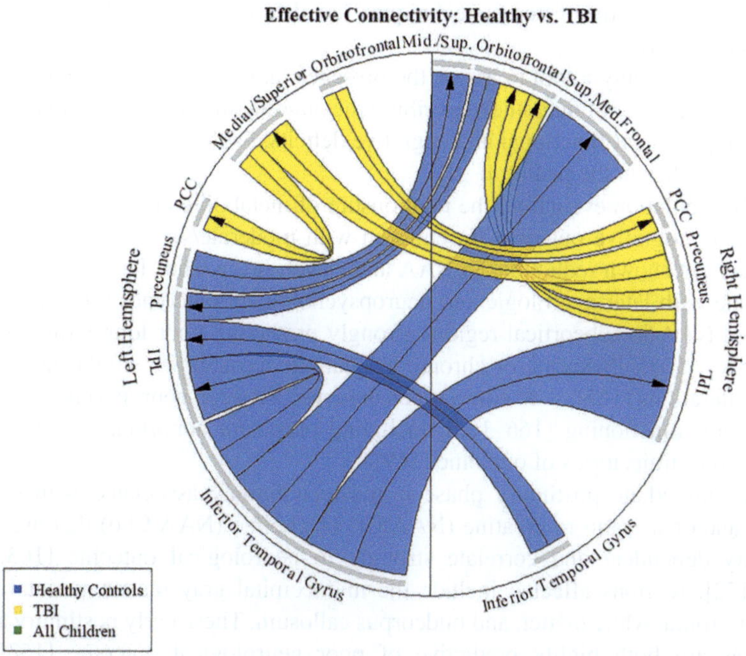

Fig. 3.5 rsfMRI: Graphical representation of positive resting-state connectivity strength within the default mode network in children with traumatic brain injury and controls. Line width represents the group-level strength of region-to-region connectivity. Blue lines depict positive connections that were present in the healthy control group. Yellow lines depict positive connections that were present in the TBI group. No positive connections were found in both the TBI and control groups

mitochondria and considered to be an indicator of neuronal energy metabolism [154]. Total creatine (Cr, 3.03 ppm) is composed of phosphocreatine and its precursor Cr, which are markers for intact brain energy metabolism. The total choline peak (Cho, 3.20 ppm) consists primarily of phosphoryl and glycerophosphoryl choline and is considered a marker for membrane synthesis or repair, inflammation, or demyelination. Glutamate (Glu), an excitatory neurotransmitter, which can be released into the extracellular space after brain injury, plays a major role in excitotoxic neuronal death [155, 156]. Its resonance overlaps with the amino acid glutamine (Gln), and the two metabolites are often reported together as Glx (2.05–2.5 ppm and 3.7 ppm). Myoinositol (mI, 3.56 ppm) is an organic osmolyte located in astrocytes that increases as a result of glial proliferation [157]. Lactate (Lac, 1.31 ppm) accumulates as a result of glycolysis, notably in conditions of decreased oxidative metabolism, and is poorly resolved in the normal brain as the threshold for detection is very low [158]. A number of localization techniques have been used to acquire MRS in pediatric TBI, including single voxel spectroscopy (SVS), multivoxel MR spectroscopic imaging (MRSI), and whole brain techniques using intermediate (135–144 ms) or short (20–60 ms) echo times. The specific advantages and disadvantages of the different techniques are detailed in a number of excellent reviews [159–162].

MRS is not only a tool to assess the presence and extent of injury but has been used as a predictor of outcome, describing the ongoing metabolic abnormalities that underlie persistent functional and cognitive deficits. Table 3.3 provides a summary of MRS studies in the pediatric

TBI population evaluating the relationship of metabolite abnormalities to functional and cognitive outcome. In children with moderate/severe TBI, acute MRS studies have shown reductions in NAA and elevations in Cho. These abnormalities correlate with later neurologic and neuropsychological outcomes [163, 164], with reduced NAA in subcortical regions strongly predicting poor long-term cognitive outcomes [164]. Postacute or chronic evaluations have shown similar neurometabolic alterations [165], with correlations observed between neuropsychological and behavioral functioning [166–168], including persistent abnormalities in subjects with poorer trajectories of outcome [169].

The immediate postinjury phase is most commonly associated with reduced NAA and/or its ratio to creatine (NAA/Cr) or choline (NAA/Cho) that are injury severity dependent and correlate strongly to neurological outcome [163, 168, 170–172]. Regions affected include the midoccipital gray matter, parietal white matter, frontal white matter, and midcorpus callosum. These early postinjury abnormalities are both highly predictive of poor neurological outcome [153, 163, 170–173] with ratios from white matter regions (parietooccipital, frontal, anterior corpus callosum) more predictive of neurological outcome than cortical gray matter ratios [153]. Additionally, the early postinjury NAA and NAA ratios correlate with long-term cognitive outcome measures including IQ, memory, attention/processing speed, executive skills, visual perception, language, motor skills, and academic achievement [164, 165, 172, 174, 175], with reduced NAA and NAA/Cr in subcortical gray matter (basal ganglia and thalami) regions strongly predicting poor long-term cognitive outcomes [164, 175]. Postacute or chronic studies have shown similar neurometabolic alterations in white matter and the corpus callosum, with correlations observed between NAA and NAA ratios and injury severity (initial GCS) [167], long-term memory, attention, and visual processing [165–169, 176]. In some cases, the association between reduced NAA/Cho and intellectual function was found 7–14 years after injury [177].

Elevated Cho/Cr has also been noted in the early postacute period, notably in the mid-occipital gray matter of infants and children with poor neurological outcome [163, 172]. Using multivoxel MRSI of the frontal and parietooccipital gray and white matter and the midcorpus callosum, an injury severity-dependent increase in Cho/Cr was also found [153, 165]. Similarly, occipital gray matter and anterior frontal white matter Cho/Cr were found to correlate with memory, [172] language, and visuomotor skill performance [165]. Increasing levels mI and Glx in the midoccipital gray and parietal white matter are associated with poor neurological outcome [171, 173] and IQ [164]. The detection of Lac is often confined to early postinjury periods of acute metabolic dysfunction associated with glutamate excitotoxicity, disordered mitochondrial metabolism, or secondary ischemic insults [178, 179].

The presence of lactate in both gray and white matter regions has been reported in patients with the worst outcomes [163, 180] and was the single most predictive MRS variable of neurological, intellectual, and neuropsychological outcomes [163, 172].

Although longitudinal studies in the pediatric population are limited, all consistently report early reductions in NAA or NAA/Cr ratios and elevated Cho or Cho/Cr ratios across multiple white matter, cortical gray matter, and subcortical (basal ganglia, thalami) gray matter and white (corpus callosum) matter regions [165, 169, 175], or whole brain gray and white matter ratios [174], consistent with neuronal injury and associated inflammation and changes in membrane metabolism. Similar to findings from immediate postinjury cross-sectional studies, early NAA ratios from the lobar gray and white matter regions and subcortical gray matter regions [175] or whole brain white matter NAA/Cr [174] were strongly correlated with a 12-month neurological outcome and performance on memory, IQ, and attention tasks. Acute NAA/Cr ratios from the basal ganglia and thalami had the highest percent accuracy for predicting memory, IQ, and a 12-month neurological outcome [175]. Moreover, longitudinal studies all demonstrate that in patients with good outcomes, most of the early regional metabolite changes return to normal levels in the chronic phase of injury (6–12 months) suggesting recovery of mitochondrial function and/or neuronal repair, whereas persistently low NAA in patients with poorer outcome suggests neuronal loss [165, 169, 174, 175]. When combined with DWI, white matter tracts with low FA and elevated MD and RD also showed low NAA (Fig. 3.6) and improved cognitive outcome prediction [176]. Persistently elevated Cho, suggesting ongoing remyelination, cellular proliferation, and membrane repair/turnover have only been reported in a few studies, with corpus callosum and lobar white matter Cho showing correlations to spatial skills, academic skills (word reading, spelling) [169], and pragmatic language [167].

Despite the sensitivity of MRS to detect injury and predict outcome, the overall number of studies is limited, especially longitudinal studies. Any inconsistency in findings is likely amplified by the small number of studies but also the significant variation in sample sizes, the duration of time between injury and MRS, regions studied, and the use of different acquisition methods (Table 3.3). These variations also present numerous challenges to harmonizing existing data [181]. Longitudinal study designs, with harmonized acquisition and analysis methods [181], will only improve our understanding of the relationship of brain metabolism to outcome.

3.7 Perfusion MR Imaging

Cerebral perfusion alterations following TBI are of particular interest as changes in cerebrovascular function have been linked to functional deficits following TBI [182]. One measure of vascular function is cerebral blood flow (CBF) which can be quantified noninvasively with arterial spin labeling (ASL) perfusion MRI that uses magnetically labeled arterial blood water as an endogenous contrast tracer [183,

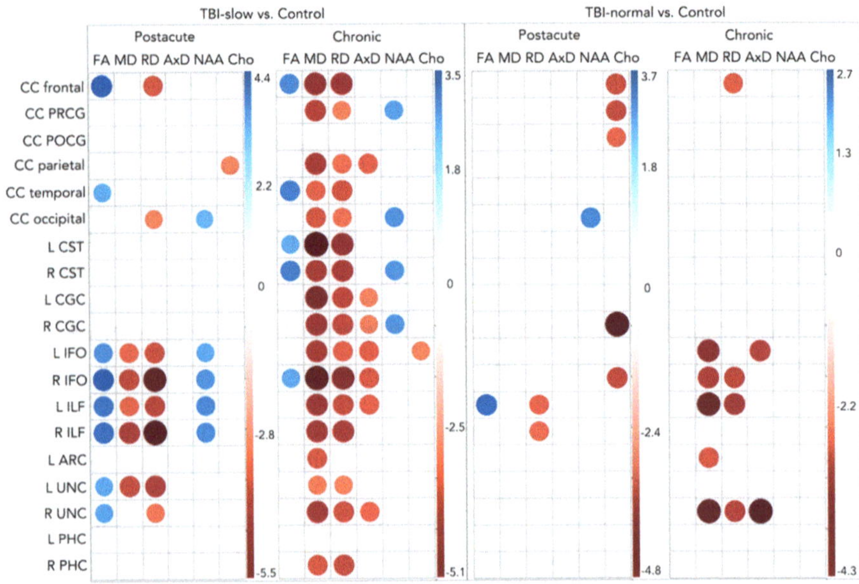

Fig. 3.6 Postacute and chronic differences between TBI-slow and control (left panels) and between TBI-normal and control (right panels). The color corresponds to the T statistic from each regression, as indicated in the color bars for each panel, with blue indicating higher values in controls and red higher values in TBI. The size of the circle corresponds to the magnitude of the T-stat. Only significant T-stats are included (after correction for multiple comparison across tracts, measures, and comparisons, FDR ($q < 0.05$). (Reprinted with permission from Dennis et al. [176])

184]. The relationship of CBF to clinical and neuropsychological outcome remains poorly understood in the pediatric moderate to severe population. In healthy children, there is an age-dependent increase in CBF from 2 to 7 years [185] followed by regional variations in the normative trajectory with CBF decreasing after puberty in males but increasing post puberty in females, [186] which likely have an important impact on functional outcome [182]. To our knowledge, only a single study has directly examined the relationship between CBF and outcome in pediatric moderate-severe TBI (Table 3.3). In the study by Newsome et al., they found that the positive relationship between resting prefrontal CBF and prefrontal activation during a social cognition task seen in typically developing adolescents was not seen in the TBI group (Fig. 3.7) [187]. Moreover, the TBI group demonstrated a positive relationship between prefrontal CBF and posterior regions of activation, including the bilateral middle temporal gyri, precuneus, and hippocampi, which suggests that following TBI, prefrontal CBF may be associated with rerouting of neuronal resources available for task completion [187]. Although preliminary, the findings highlight the impact of altered CBF on activation patterns in the brain following pediatric TBI and suggest a link between CBF deficits and long-term cognitive impairments. Future studies that examine the evolution of CBF changes across acute, subacute, and chronic postinjury periods are important to improve our understanding of the key neurovascular events that underlie both good and poor outcomes.

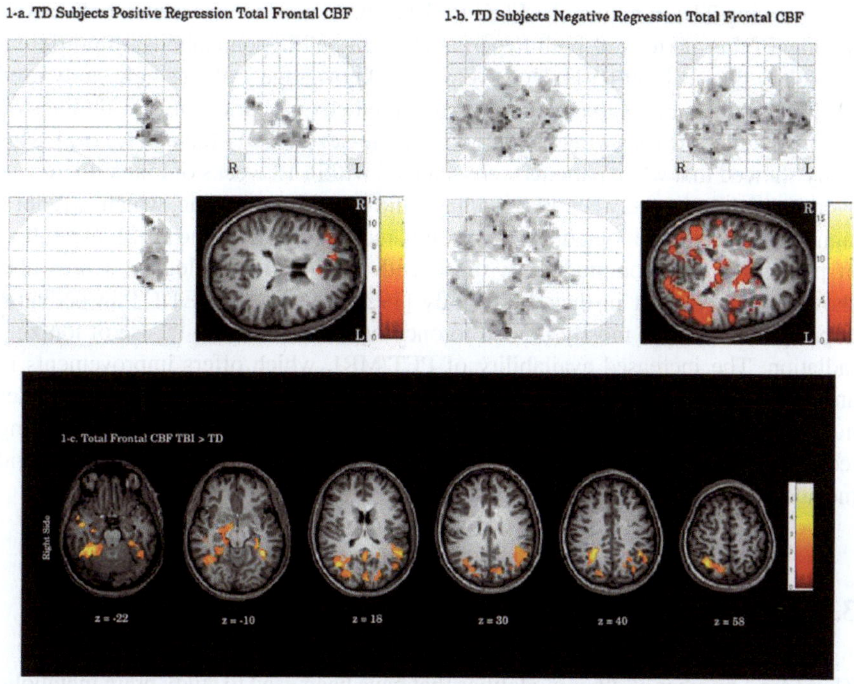

Fig. 3.7 FMRI activation during a social cognition task of making judgements about the self from the perspective of another person in typically-developing (TD) adolescents and in adolescents with moderate to severe traumatic brain injury (TBI), regressed on cerebral blood flow (CBF) values of the frontal lobes. Glass brain views are presented to illustrate the extent to which frontal CBF is related to frontal activation during normal development. (**a**) Positive regression of activation on CBF values in the TD group. (**b**) Negative regression of activation on CBF values in the TD group. (**c**) Regions where relation between CBF and activation are greater in the TBI group than in the TD group. Cluster level $p < 0.05$, corrected. The right side of the brain is presented on the viewer's left. Color scale indicates t-test values. (Reprinted with permission from Newsome et al. [187])

3.8 Positron Emission Tomography

Positron emission tomography using 2-deoxy-2-18 fluorodeoxyglucose (^{18}F-FDG-PET) has been used for over three decades to study brain glucose metabolism. ^{18}F-FDG acts as a competitive agonist of blood glucose, crossing the blood-brain barrier, where it is taken up by neurons, phosphorylated to hexokinase and FDG-6-phosphate where it accumulates within cells proportionally to the synaptic activity during the study. In static PET, volumetric concentrations of radioactivity (standardized uptake value, SUV) are mapped to CT or MRI images resulting in semiquantitative maps where regions of high radioactivity denote areas of high glucose metabolic activity. In contrast, dynamic PET acquisitions measure radioactivity on a continuous basis resulting in concentration versus time profiles which allows for kinetic modeling of glucose metabolism (rate of uptake and/or decay) [188].

Following TBI in adults, ^{18}F-FDG PET has been used to assess metabolic activity, identifying periods of both hypermetabolism and hypometabolism in different postinjury phases [189–191]. However, to our knowledge, only a single study used ^{18}F-FDG PET to assess the relationship and predictive accuracy of PET findings on cognitive outcome following pediatric moderate-severe TBI (Table 3.3) [192]. This study showed that when patients were scanned within 12 weeks of injury, there was a strong correlation between PET findings and cognitive outcome, which was not a significant predictor of outcome when compared to CT/MRI alone [192]. In a subset of patients studied longitudinally, those with improved neurological scores also showed improvement in their PET study [192]. The use of PET/CT in pediatric populations has been mostly limited to oncology studies due to its use of ionizing radiation. The increased availability of PET/MRI, which offers improvements in anatomical correlation and a significant reduction in ionizing radiation, creates numerous potential future opportunities to leverage this powerful neuroimaging technique to better understand the relationship between glucose and/or amino acid metabolism and outcome following TBI.

3.9 Other Neuroimaging Modalities

There are a number of other modalities that have been used to study brain metabolic and functional changes following TBI. Electroencephalography (EEG) was among the first clinical neurodiagnostic tools to show abnormal brain electrical activity following TBI [193]. Similarly, magnetoencephalography (MEG) measures the magnetic flux on the surface of the brain associated with underlying neuronal electrical currents [194]. Single photon emission tomography (SPECT), which measures gamma emissions from radioactive Technetium (Tc) 99 exametazime and functional near-infrared spectroscopy (fNIRS), which measures the light absorbance in superficial venous oxyhemoglobin and deoxyhemoglobin, measure neuronal activity as a function neurovascular coupling [194]. These modalities were not covered in detail as we did not identify any papers in the pediatric moderate-severe TBI population that included outcome measures. However, these modalities have the potential to provide greater depth of insight into the neurovascular consequences of injury in the developing brain.

3.10 Conclusion

Here we highlight research that has revealed links between advanced neuroimaging and outcome after TBI. While identifying alterations in brain structure or function after injury is important, further tying these changes in the brain to outcomes that are more functionally relevant for the patient, such as cognitive performance or psychological symptoms, is arguably the most important step in establishing the value of advanced neuroimaging for TBI. While there has been a tremendous amount of work done in this area, there is still considerable variance in outcome that remains unexplained. Neuroimaging has great potential to enhance our ability to prognosticate, potentially revealing compensatory processes in patients with better outcomes or subtle, insidious damage in those with worse outcomes. Early detection of risk may reveal additional opportunities to intervene.

There are a number of improvements that are necessary to move neuroimaging in TBI toward the goal of patient-tailored interventions. Larger samples are needed to identify reliable effects and to be able to parse demographic and clinical factors that may interact to impact outcome (e.g., sex). Longitudinal studies are necessary to better understand trajectories, as change after injury is dynamic and can continue decades postinjury. Additionally, longitudinal studies are particularly important in a developing population, and when multimodality study designs are combined with larger sample sizes and/or multisite collaborative efforts, future studies have the potential to identify clinically relevant phenotypes that improve outcome prediction and identify new therapeutic targets. New tools to harmonize existing data are also important, as they will enable researchers to combine datasets to achieve larger sample sizes. The ENIGMA Brain Injury consortium, among others, is actively working in this area [195].

Acknowledgments This research was supported by NIH R61 NS120249 (ELD) and R01 NS122184 (ELD and BBO).

Disclosures None.

References

1. Ashwal S, Holshouser BA, Tong KA. Use of advanced neuroimaging techniques in the evaluation of pediatric traumatic brain injury. Dev Neurosci. 2006;28:309–26.
2. Ashwal S, Tong KA, Ghosh N, Bartnik-Olson B, Holshouser BA. Application of advanced neuroimaging modalities in pediatric traumatic brain injury. J Child Neurol. 2014;29:1704–17.
3. Dennis EL, Babikian T, Giza CC, Thompson PM, Asarnow RF. Neuroimaging of the injured pediatric brain: methods and new lessons. Neuroscientist. 2018;24:652–70.
4. Hunter JV, Wilde EA, Tong KA, Holshouser BA. Emerging imaging tools for use with traumatic brain injury research. J Neurotrauma. 2012;29:654–71.
5. Dennis EL, Babikian T, Giza CC, Thompson PM, Asarnow RF. Diffusion MRI in pediatric brain injury. Childs Nerv Syst. 2017;33:1683–92.
6. Lindsey HM, Wilde EA, Caeyenberghs K, Dennis EL. Longitudinal neuroimaging in Pediatric traumatic brain injury: current state and consideration of factors that influence recovery. Front Neurol. 2019;10:1296.
7. Dennis EL, Caeyenberghs K, Hoskinson KR, Merkley TL, Suskauer SJ, Asarnow RF, et al. White Matter disruption in pediatric traumatic brain injury: results from ENIGMA Pediatric moderate to severe traumatic brain injury. Neurology. 2021; https://doi.org/10.1212/WNL.0000000000012222.

8. Imms P, Clemente A, Cook M, D'Souza W, Wilson PH, Jones DK, et al. The structural connectome in traumatic brain injury: a meta-analysis of graph metrics. Neurosci Biobehav Rev. 2019;99:128–37.
9. Schmidt AT, Hanten GR, Li X, Vasquez AC, Wilde EA, Chapman SB, et al. Decision making after pediatric traumatic brain injury: trajectory of recovery and relationship to age and gender. Int J Dev Neurosci. 2012;30:225–30.
10. Jaffe KM, Polissar NL, Fay GC, Liao S. Recovery trends over three years following pediatric traumatic brain injury. Arch Phys Med Rehabil. 1995;76:17–26.
11. Dennis EL, Keleher F, Tate DF, Wilde EA. The role of neuroimaging in evolving TBI research and clinical practice. medRxiv. 2023. https://doi.org/10.1101/2023.02.24.23286258.
12. Ashburner J, Friston KJ. Voxel based morphometry. In: Squire LR, editor. Encyclopedia of neuroscience. Oxford: Academic; 2009. p. 471–7.
13. Hua X, Leow AD, Levitt JG, Caplan R, Thompson PM, Toga AW. Detecting brain growth patterns in normal children using tensor-based morphometry. Hum Brain Mapp. 2009;30:209–19.
14. Dennis EL, Faskowitz J, Rashid F, Babikian T, Mink R, Babbitt C, et al. Diverging volumetric trajectories following pediatric traumatic brain injury. Neuroimage Clin. 2017;15:125–35.
15. Levin HS, Amparo EG, Eisenberg HM, Miner ME, High WM Jr, Ewing-Cobbs L, et al. Magnetic resonance imaging after closed head injury in children. Neurosurgery. 1989;24:223–7.
16. Mendelsohn D, Levin HS, Bruce D, Lilly M, Harward H, Culhane KA, et al. Late MRI after head injury in children: relationship to clinical features and outcome. Childs Nerv Syst. 1992;8:445–52.
17. Jennett B, Bond M. Glasgow Outcome Scale Assessment of outcome after severe brain damage: a practical scale. Lancet. 1975;480:484.
18. Mendelsohn DB, Levin HS, Harward H, Bruce D. Corpus callosum lesions after closed head injury in children: MRI, clinical features and outcome. Neuroradiology. 1992;34:384–8.
19. Prasad MR, Ewing-Cobbs L, Swank PR, Kramer L. Predictors of outcome following traumatic brain injury in young children. Pediatr Neurosurg. 2002;36:64–74.
20. Bonnier C, Marique P, Van Hout A, Potelle D. Neurodevelopmental outcome after severe traumatic brain injury in very young children: role for subcortical lesions. J Child Neurol. 2007;22:519–29.
21. Levin HS, Zhang L, Dennis M, Ewing-Cobbs L, Schachar R, Max J, et al. Psychosocial outcome of TBI in children with unilateral frontal lesions. J Int Neuropsychol Soc. 2004;10:305–16.
22. Levin HS, Mendelsohn D, Lilly MA, Yeakley J, Song J, Scheibel RS, et al. Magnetic resonance imaging in relation to functional outcome of pediatric closed head injury: a test of the Ommaya-Gennarelli model. Neurosurgery. 1997;40:432–40; discussion 440–1.
23. Ferrazzano P, Yeske B, Mumford J, Kirk G, Bigler ED, Bowen K, et al. Brain magnetic resonance imaging volumetric measures of functional outcome after severe traumatic brain injury in adolescents. J Neurotrauma. 2021;38:1799–808.
24. Berryhill P, Lilly MA, Levin HS, Hillman GR, Mendelsohn D, Brunder DG, et al. Frontal lobe changes after severe diffuse closed head injury in children: a volumetric study of magnetic resonance imaging. Neurosurgery. 1995;37:392–9; discussion 399–400.
25. Bowen JM, Clark E, Bigler ED, Gardner M, Nilsson D, Gooch J, et al. Childhood traumatic brain injury: neuropsychological status at the time of hospital discharge. Dev Med Child Neurol. 1997;39:17–25.
26. Barker LH, Bigler ED, Johnson SC, Anderson CV, Russo AA, Boineau B, et al. Polysubstance abuse and traumatic brain injury: quantitative magnetic resonance imaging and neuropsychological outcome in older adolescents and young adults. J Int Neuropsychol Soc. 1999;5:593–608.
27. Braga LW, Souza LN, Najjar YJ, Dellatolas G. Magnetic resonance imaging (MRI) findings and neuropsychological sequelae in children after severe traumatic brain injury: the role of cerebellar lesion. J Child Neurol. 2007;22:1084–9.

28. Koelfen W, Freund M, Dinter D, Schmidt B, Koenig S, Schultze C. Long-term follow up of children with head injuries-classified as 'good recovery' using the Glasgow Outcome Scale: neurological, neuropsychological and magnetic resonance imaging results. Eur J Pediatr. 1997;156:230–5.
29. Salorio CF, Slomine BS, Grados MA, Vasa RA, Christensen JR, Gerring JP. Neuroanatomic correlates of CVLT-C performance following pediatric traumatic brain injury. J Int Neuropsychol Soc. 2005;11:686–96.
30. Levin HS, Song J, Scheibel RS, Fletcher JM, Harward H, Lilly M, et al. Concept formation and problem-solving following closed head injury in children. J Int Neuropsychol Soc. 1997;3:598–607.
31. Catroppa C, Anderson V, Ditchfield M, Coleman L. Using magnetic resonance imaging to predict new learning outcome at 5 years after childhood traumatic brain injury. J Child Neurol. 2008;23:486–96.
32. Levin HS, Culhane KA, Fletcher JM, Mendelsohn DB, Lilly MA, Harward H, et al. Dissociation between delayed alternation and memory after pediatric head injury: relationship to MRI findings. J Child Neurol. 1994;9:81–9.
33. Levin HS, Fletcher JM, Kusnerik L, Kufera JA, Lilly MA, Duffy FF, et al. Semantic memory following pediatric head injury: relationship to age, severity of injury, and MRI. Cortex. 1996;32:461–78.
34. Babikian T, Asarnow R. Neurocognitive outcomes and recovery after pediatric TBI: meta-analytic review of the literature. Neuropsychology. 2009;23:283–96.
35. Serra-Grabulosa JM, Junqué C, Verger K, Salgado-Pineda P, Mañeru C, Mercader JM. Cerebral correlates of declarative memory dysfunctions in early traumatic brain injury. J Neurol Neurosurg Psychiatry. 2005;76:129–31.
36. Dennis EL, Hua X, Villalon-Reina J, Moran LM, Kernan C, Babikian T, et al. Tensor-based morphometry reveals volumetric deficits in moderate/severe pediatric traumatic brain injury. J Neurotrauma. 2016. https://doi.org/10.1089/neu.2015.4012.
37. McCauley SR, Wilde EA, Merkley TL, Schnelle KP, Bigler ED, Hunter JV, et al. Patterns of cortical thinning in relation to event-based prospective memory performance three months after moderate to severe traumatic brain injury in children. Dev Neuropsychol. 2010;35:318–32.
38. Merkley TL, Bigler ED, Wilde EA, McCauley SR, Hunter JV, Levin HS. Diffuse changes in cortical thickness in pediatric moderate-to-severe traumatic brain injury. J Neurotrauma. 2008;25:1343–5.
39. Verger K, Junque C, Levin HS, Jurado MA, Perez-Gomez M, Bartres-Faz D, et al. Correlation of atrophy measures on MRI with neuropsychological sequelae in children and adolescents with traumatic brain injury. Brain Inj. 2001;15:211–21.
40. Wu TC, Wilde EA, Bigler ED, Li X, Merkley TL, Yallampalli R, et al. Longitudinal changes in the corpus callosum following pediatric traumatic brain injury. Dev Neurosci. 2010;32:361–73.
41. King DJ, Seri S, Catroppa C, Anderson VA, Wood AG. Structural-covariance networks identify topology-based cortical-thickness changes in children with persistent executive function impairments after traumatic brain injury. Neuroimage. 2021;244:118612.
42. Ryan NP, Catroppa C, Hughes N, Painter FL, Hearps S, Beauchamp MH, et al. Executive function mediates the prospective association between neurostructural differences within the central executive network and anti-social behavior after childhood traumatic brain injury. J Child Psychol Psychiatry. 2021;62:1150–61.
43. Slawik H, Salmond CH, Taylor-Tavares JV, Williams GB, Sahakian BJ, Tasker RC. Frontal cerebral vulnerability and executive deficits from raised intracranial pressure in child traumatic brain injury. J Neurotrauma. 2009;26:1891–903.
44. Greer KM, Snyder A, Junge C, Reading M, Jarvis S, Squires C, et al. Surface-based abnormalities of the executive frontostriatial circuit in pediatric TBI. Neuroimage Clin. 2022;35:103136.

45. Catharine VL, Helena V, Ellen D, Guy V, Karel D, Karen C. Exploration of gray matter correlates of cognitive training benefit in adolescents with chronic traumatic brain injury. Neuroimage Clin. 2019;23:101827.
46. Ryan NP, Beauchamp MH, Beare R, Coleman L, Ditchfield M, Kean M, et al. Uncovering cortico-striatal correlates of cognitive fatigue in pediatric acquired brain disorder: evidence from traumatic brain injury. Cortex. 2016;83:222–30.
47. Crasta JE, Sibel J, Slomine BS, Mahone EM, Mostofsky SH, Suskauer SJ. Subtle motor signs in children with chronic traumatic brain injury. Am J Phys Med Rehabil. 2019;98:737–44.
48. Anderson V, Beauchamp MH, Yeates KO, Crossley L, Hearps SJC, Catroppa C. Social competence at 6 months following childhood traumatic brain injury. J Int Neuropsychol Soc. 2013;19:539–50.
49. Anderson V, Hearps SJC, Catroppa C, Beauchamp MH, Ryan NP. What predicts persisting social impairment following pediatric traumatic brain injury: contribution of a biopsychosocial approach. Psychol Med. 2023;53:3568–79.
50. Khan N, Ryan NP, Crossley L, Hearps S, Catroppa C, Anderson V. Global and domain-specific self-esteem after pediatric traumatic brain injury: contribution of injury characteristics and parent mental health. Neuropsychol Rehabil. 2022;32:1835–53.
51. Levan A, Baxter L, Kirwan CB, Black G, Gale SD. Right frontal pole cortical thickness and social competence in children with chronic traumatic brain injury: cognitive proficiency as a mediator. J Head Trauma Rehabil. 2015;30:E24–31.
52. Wilde EA, Merkley TL, Lindsey HM, Bigler ED, Hunter JV, Ewing-Cobbs L, et al. Developmental alterations in cortical organization and socialization in adolescents who sustained a traumatic brain injury in early childhood. J Neurotrauma. 2021;38:133–43.
53. Hoskinson KR, Bigler ED, Abildskov TJ, Dennis M, Taylor HG, Rubin K, et al. The mentalizing network and theory of mind mediate adjustment after childhood traumatic brain injury. Soc Cogn Affect Neurosci. 2019;14:1285–95.
54. Ryan NP, Catroppa C, Beare R, Silk TJ, Hearps SJ, Beauchamp MH, et al. Uncovering the neuroanatomical correlates of cognitive, affective and conative theory of mind in paediatric traumatic brain injury: a neural systems perspective. Soc Cogn Affect Neurosci. 2017;12:1414–27.
55. Ryan NP, Catroppa C, Beare R, Silk TJ, Crossley L, Beauchamp MH, et al. Theory of mind mediates the prospective relationship between abnormal social brain network morphology and chronic behavior problems after pediatric traumatic brain injury. Soc Cogn Affect Neurosci. 2016;11:683–92.
56. Bigler ED, Yeates KO, Dennis M, Gerhardt CA, Rubin KH, Stancin T, et al. Neuroimaging and social behavior in children after traumatic brain injury: findings from the Social Outcomes of Brain Injury in Kids (SOBIK) study. NeuroRehabilitation. 2013;32:707–20.
57. Levan A, Black G, Mietchen J, Baxter L, Brock Kirwan C, Gale SD. Right frontal pole cortical thickness and executive functioning in children with traumatic brain injury: the impact on social problems. Brain Imaging Behav. 2016;10:1090–5.
58. Yeates KO, Gerhardt CA, Bigler ED, Abildskov T, Dennis M, Rubin KH, et al. Peer relationships of children with traumatic brain injury. J Int Neuropsychol Soc. 2013;19:518–27.
59. Herskovits EH, Megalooikonomou V, Davatzikos C, Chen A, Bryan RN, Gerring JP. Is the spatial distribution of brain lesions associated with closed-head injury predictive of subsequent development of attention-deficit/hyperactivity disorder?Analysis with brain-image database. Radiology. 1999;213:389–94.
60. Gerring J, Brady K, Chen A, Quinn C, Herskovits E, Bandeen-Roche K, et al. Neuroimaging variables related to development of Secondary Attention Deficit Hyperactivity Disorder after closed head injury in children and adolescents. Brain Inj. 2000;14:205–18.
61. Max JE, Schachar RJ, Landis J, Bigler ED, Wilde EA, Saunders AE, et al. Psychiatric disorders in children and adolescents in the first six months after mild traumatic brain injury. J Neuropsychiatry Clin Neurosci. 2013;25:187–97.

62. Max JE, Keatley E, Wilde EA, Bigler ED, Schachar RJ, Saunders AE, et al. Depression in children and adolescents in the first 6 months after traumatic brain injury. Int J Dev Neurosci. 2012;30:239–45.
63. Max JE, Keatley E, Wilde EA, Bigler ED, Levin HS, Schachar RJ, et al. Anxiety disorders in children and adolescents in the first six months after traumatic brain injury. J Neuropsychiatry Clin Neurosci. 2011;23:29–39.
64. Grados MA, Vasa RA, Riddle MA, Slomine BS, Salorio C, Christensen J, et al. New onset obsessive-compulsive symptoms in children and adolescents with severe traumatic brain injury. Depress Anxiety. 2008;25:398–407.
65. Max JE, Lansing AE, Koele SL, Castillo CS, Bokura H, Schachar R, et al. Attention deficit hyperactivity disorder in children and adolescents following traumatic brain injury. Dev Neuropsychol. 2004;25:159–77.
66. Ryan NP, Catroppa C, Ward SC, Yeates KO, Crossley L, Hollenkamp M, et al. Association of neurostructural biomarkers with secondary attention-deficit/hyperactivity disorder (ADHD) symptom severity in children with traumatic brain injury: a prospective cohort study. Psychol Med. 2023;53:5291–300.
67. Vasa RA, Grados M, Slomine B, Herskovits EH, Thompson RE, Salorio C, et al. Neuroimaging correlates of anxiety after pediatric traumatic brain injury. Biol Psychiatry. 2004;55:208–16.
68. Herskovits EH, Gerring JP, Davatzikos C, Bryan RN. Is the spatial distribution of brain lesions associated with closed-head injury in children predictive of subsequent development of posttraumatic stress disorder? Radiology. 2002;224:345–51.
69. Sigmund GA, Tong KA, Nickerson JP, Wall CJ, Oyoyo U, Ashwal S. Multimodality comparison of neuroimaging in pediatric traumatic brain injury. Pediatr Neurol. 2007;36:217–26.
70. Lipszyc J, Levin H, Hanten G, Hunter J, Dennis M, Schachar R. Frontal white matter damage impairs response inhibition in children following traumatic brain injury. Arch Clin Neuropsychol. 2014;29:289–99.
71. Power T, Catroppa C, Coleman L, Ditchfield M, Anderson V. Do lesion site and severity predict deficits in attentional control after preschool traumatic brain injury (TBI)? Brain Inj. 2007;21:279–92.
72. Smitherman E, Hernandez A, Stavinoha PL, Huang R, Kernie SG, Diaz-Arrastia R, et al. Predicting outcome after pediatric traumatic brain injury by early magnetic resonance imaging lesion location and volume. J Neurotrauma. 2016;33:35–48.
73. Haacke EM, Mittal S, Wu Z, Neelavalli J, Cheng Y-CN. Susceptibility-weighted imaging: technical aspects and clinical applications, part 1. AJNR Am J Neuroradiol. 2009;30:19–30.
74. Sehgal V, Delproposto Z, Haacke EM, Tong KA, Wycliffe N, Kido DK, et al. Clinical applications of neuroimaging with susceptibility-weighted imaging. J Magn Reson Imaging. 2005;22:439–50.
75. Beauchamp MH, Ditchfield M, Babl FE, Kean M, Catroppa C, Yeates KO, et al. Detecting traumatic brain lesions in children: CT versus MRI versus susceptibility weighted imaging (SWI). J Neurotrauma. 2011;28:915–27.
76. Mittal S, Wu Z, Neelavalli J, Haacke EM. Susceptibility-weighted imaging: technical aspects and clinical applications, part 2. AJNR Am J Neuroradiol. 2009;30:232–52.
77. Spitz G, Maller JJ, Ng A, O'Sullivan R, Ferris NJ, Ponsford JL. Detecting lesions after traumatic brain injury using susceptibility weighted imaging: a comparison with fluid-attenuated inversion recovery and correlation with clinical outcome. J Neurotrauma. 2013;30:2038–50.
78. Tong KA, Ashwal S, Holshouser BA, Shutter LA, Herigault G, Haacke EM, et al. Hemorrhagic shearing lesions in children and adolescents with posttraumatic diffuse axonal injury: improved detection and initial results. Radiology. 2003;227:332–9.
79. Wu Z, Li S, Lei J, An D, Haacke EM. Evaluation of traumatic subarachnoid hemorrhage using susceptibility-weighted imaging. AJNR Am J Neuroradiol. 2010;31:1302–10.
80. Tong KA, Ashwal S, Holshouser BA, Nickerson JP, Wall CJ, Shutter LA, et al. Diffuse axonal injury in children: clinical correlation with hemorrhagic lesions. Ann Neurol. 2004;56:36–50.

81. Beauchamp MH, Beare R, Ditchfield M, Coleman L, Babl FE, Kean M, et al. Susceptibility weighted imaging and its relationship to outcome after pediatric traumatic brain injury. Cortex. 2013;49:591–8.
82. Baker CM, Cox AP, Hunsaker JC, Scoville J, Bollo RJ. Postoperative magnetic resonance imaging may predict poor outcome in children with severe traumatic brain injuries who undergo cranial surgery. J Neurosurg Pediatr. 2022:1–5.
83. Babikian T, Freier MC, Tong KA, Nickerson JP, Wall CJ, Holshouser BA, et al. Susceptibility weighted imaging: neuropsychologic outcome and pediatric head injury. Pediatr Neurol. 2005;33:184–94.
84. Ryan NP, Catroppa C, Cooper JM, Beare R, Ditchfield M, Coleman L, et al. Relationships between acute imaging biomarkers and theory of mind impairment in post-acute pediatric traumatic brain injury: a prospective analysis using susceptibility weighted imaging (SWI). Neuropsychologia. 2015;66:32–8.
85. Ryan NP, Catroppa C, Beare R, Coleman L, Ditchfield M, Crossley L, et al. Predictors of longitudinal outcome and recovery of pragmatic language and its relation to externalizing behaviour after pediatric traumatic brain injury. Brain Lang. 2015;142:86–95.
86. Beauchamp MH, Anderson VA, Catroppa C, Maller JJ, Godfrey C, Rosenfeld JV, et al. Implications of reduced callosal area for social skills after severe traumatic brain injury in children. J Neurotrauma. 2009;26:1645–54.
87. Giedd JN, Raznahan A, Alexander-Bloch A, Schmitt E, Gogtay N, Rapoport JL. Child psychiatry branch of the National Institute of Mental Health longitudinal structural magnetic resonance imaging study of human brain development. Neuropsychopharmacology. 2015;40:43–9.
88. Luders E, Thompson PM, Toga AW. The development of the corpus callosum in the healthy human brain. J Neurosci. 2010;30:10985–90.
89. Vander Linden C, Verhelst H, Genbrugge E, Deschepper E, Caeyenberghs K, Vingerhoets G, et al. Is diffuse axonal injury on susceptibility weighted imaging a biomarker for executive functioning in adolescents with traumatic brain injury? Eur J Paediatr Neurol. 2019;23:525–36.
90. Ryan NP, Catroppa C, Cooper JM, Beare R, Ditchfield M, Coleman L, et al. The emergence of age-dependent social cognitive deficits after generalized insult to the developing brain: a longitudinal prospective analysis using susceptibility-weighted imaging. Hum Brain Mapp. 2015;36:1677–91.
91. Ryan NP, van Bijnen L, Catroppa C, Beauchamp MH, Crossley L, Hearps S, et al. Longitudinal outcome and recovery of social problems after pediatric traumatic brain injury (TBI): contribution of brain insult and family environment. Int J Dev Neurosci. 2016;49:23–30.
92. Basser PJ. Inferring microstructural features and the physiological state of tissues from diffusion-weighted images. NMR Biomed. 1995;8:333–44.
93. Hulkower MB, Poliak DB, Rosenbaum SB, Zimmerman ME, Lipton ML. A decade of DTI in traumatic brain injury: 10 years and 100 articles later. AJNR Am J Neuroradiol. 2013. https://doi.org/10.3174/ajnr.A3395.
94. Zhang H, Schneider T, Wheeler-Kingshott CA, Alexander DC. NODDI: practical in vivo neurite orientation dispersion and density imaging of the human brain. Neuroimage. 2012;61:1000–16.
95. Yeh F-C, Liu L, Hitchens TK, Wu YL. Mapping immune cell infiltration using restricted diffusion MRI. Magn Reson Med. 2017;77:603–12.
96. Schilling KG, Janve V, Gao Y, Stepniewska I, Landman BA, Anderson AW. Histological validation of diffusion MRI fiber orientation distributions and dispersion. Neuroimage. 2018;165:200–21.
97. Ressel V, O'Gorman Tuura R, Scheer I, van Hedel HJA. Diffusion tensor imaging predicts motor outcome in children with acquired brain injury. Brain Imaging Behav. 2017;11:1373–84.

98. Wilde EA, Newsome MR, Bigler ED, Pertab J, Merkley TL, Hanten G, et al. Brain imaging correlates of verbal working memory in children following traumatic brain injury. Int J Psychophysiol. 2011;82:86–96.
99. Genc S, Anderson V, Ryan NP, Malpas CB, Catroppa C, Beauchamp MH, et al. Recovery of white matter following pediatric traumatic brain injury depends on injury severity. J Neurotrauma. 2017;34:798–806.
100. Ressel V, Berati D, Raselli C, Birrer K, Kottke R, van Hedel HJ, et al. Magnetic resonance imaging markers reflect cognitive outcome after rehabilitation in children with acquired brain injury. Eur J Radiol. 2020;126:108963.
101. Mayer AR, Ling JM, Dodd AB, Stephenson DD, Pabbathi Reddy S, Robertson-Benta CR, et al. Multicompartmental models and diffusion abnormalities in paediatric mild traumatic brain injury. Brain. 2022;145:4124–37.
102. Juranek J, Johnson CP, Prasad MR, Kramer LA, Saunders A, Filipek PA, et al. Mean diffusivity in the amygdala correlates with anxiety in pediatric TBI. Brain Imaging Behav. 2012;6:36–48.
103. Max JE, Wilde EA, Bigler ED, Thompson WK, MacLeod M, Vasquez AC, et al. Neuroimaging correlates of novel psychiatric disorders after pediatric traumatic brain injury. J Am Acad Child Adolesc Psychiatry. 2012;51:1208–17.
104. Levin HS, Wilde EA, Chu Z, Yallampalli R, Hanten GR, Li X, et al. Diffusion tensor imaging in relation to cognitive and functional outcome of traumatic brain injury in children. J Head Trauma Rehabil. 2008;23:197–208.
105. Wozniak JR, Krach L, Ward E, Mueller BA, Muetzel R, Schnoebelen S, et al. Neurocognitive and neuroimaging correlates of pediatric traumatic brain injury: a diffusion tensor imaging (DTI) study. Arch Clin Neuropsychol. 2007;22:555–68.
106. Yuan W, Treble-Barna A, Sohlberg MM, Harn B, Wade SL. Changes in structural connectivity following a cognitive intervention in children with traumatic brain injury. Neurorehabil Neural Repair. 2017;31:190–201.
107. Königs M, Pouwels PJ, Ernest van Heurn LW, Bakx R, Jeroen Vermeulen R, Carel Goslings J, et al. Relevance of neuroimaging for neurocognitive and behavioral outcome after pediatric traumatic brain injury. Brain Imaging Behav. 2018;12:29–43.
108. Kurowski B, Wade SL, Cecil KM, Walz NC, Yuan W, Rajagopal A, et al. Correlation of diffusion tensor imaging with executive function measures after early childhood traumatic brain injury. J Pediatr Rehabil Med. 2009;2:273–83.
109. Faber J, Wilde EA, Hanten G, Ewing-Cobbs L, Aitken ME, Yallampalli R, et al. Ten-year outcome of early childhood traumatic brain injury: diffusion tensor imaging of the ventral striatum in relation to executive functioning. Brain Inj. 2016;30:1635–41.
110. Volpe DSJ, Oliveira NCAC, Santos AC, Linhares MBM, Carlotti APCP. Neuropsychological outcome of children with traumatic brain injury and its association with late magnetic resonance imaging findings: a cohort study. Brain Inj. 2017;31:1689–94.
111. Wilde EA, Hyseni I, Lindsey HM, Faber J, McHenry JM, Bigler ED, et al. A Preliminary DTI Tractography Study of Developmental Neuroplasticity 5-15 Years After Early Childhood Traumatic Brain Injury. Front Neurol. 2021;12:734055.
112. Glover GH. Overview of functional magnetic resonance imaging. Neurosurg Clin N Am. 2011;22:133–9, vii.
113. Medaglia JD. Functional neuroimaging in traumatic brain injury: from nodes to networks. Front Neurol. 2017;8:407.
114. Barkovich MJ, Li Y, Desikan RS, Barkovich AJ, Xu D. Challenges in pediatric neuroimaging. Neuroimage. 2019;185:793–801.
115. Raschle N, Zuk J, Ortiz-Mantilla S, Sliva DD, Franceschi A, Grant PE, et al. Pediatric neuroimaging in early childhood and infancy: challenges and practical guidelines. Ann N Y Acad Sci. 2012;1252:43–50.
116. Karunanayaka PR, Holland SK, Yuan W, Altaye M, Jones BV, Michaud LJ, et al. Neural substrate differences in language networks and associated language-related behavioral

impairments in children with TBI: a preliminary fMRI investigation. NeuroRehabilitation. 2007;22:355–69.
117. Kramer ME, Chiu C-YP, Shear PK, Wade SL. Neural correlates of verbal associative memory and mnemonic strategy use following childhood traumatic brain injury. J Pediatr Rehabil Med. 2009;2:255–71.
118. Tlustos SJ, Chiu C-YP, Walz NC, Holland SK, Bernard L, Wade SL. Neural correlates of interference control in adolescents with traumatic brain injury: functional magnetic resonance imaging study of the counting stroop task. J Int Neuropsychol Soc. 2011;17:181–9.
119. Scheibel RS, Newsome MR, Wilde EA, McClelland MM, Hanten G, Krawczyk DC, et al. Brain activation during a social attribution task in adolescents with moderate to severe traumatic brain injury. Soc Neurosci. 2011;6:582–98.
120. Chiu C-YP, Tlustos SJ, Walz NC, Holland SK, Eliassen JC, Bernard L, et al. Neural correlates of risky decision making in adolescents with and without traumatic brain injury using the balloon analog risk task. Dev Neuropsychol. 2012;37:176–83.
121. Morgan AT, Masterton R, Pigdon L, Connelly A, Liégeois FJ. Functional magnetic resonance imaging of chronic dysarthric speech after childhood brain injury: reliance on a left-hemisphere compensatory network. Brain. 2013;136:646–57.
122. Fufaeva EV. Significance of executive functioning in children with severe traumatic brain injury at different stages of consciousness recovery. Neuropsychological approach. Zh Vopr Neirokhir Im N N Burdenko. 2014;78:50–5; discussion 55–6.
123. Tlustos SJ, Peter Chiu CY, Walz NC, Wade SL. Neural substrates of inhibitory and emotional processing in adolescents with traumatic brain injury. J Pediatr Rehabil Med. 2015;8:321–33.
124. Stephens JA, Salorio CF, Barber AD, Risen SR, Mostofsky SH, Suskauer SJ. Preliminary findings of altered functional connectivity of the default mode network linked to functional outcomes one year after pediatric traumatic brain injury. Dev Neurorehabil. 2018;21:423–30.
125. Ricker JH, Hillary FG, DeLuca J. Functionally activated brain imaging (O-15 PET and fMRI) in the study of learning and memory after traumatic brain injury. J Head Trauma Rehabil. 2001;16:191–205.
126. Lee MH, Smyser CD, Shimony JS. Resting-state fMRI: a review of methods and clinical applications. AJNR Am J Neuroradiol. 2013;34:1866–72.
127. Raichle ME, Snyder AZ. A default mode of brain function: a brief history of an evolving idea. Neuroimage. 2007;37:1083–90; discussion 1097–9.
128. Lv H, Wang Z, Tong E, Williams LM, Zaharchuk G, Zeineh M, et al. Resting-state functional MRI: everything that nonexperts have always wanted to know. AJNR Am J Neuroradiol. 2018;39:1390–9.
129. Zang Y-F, He Y, Zhu C-Z, Cao Q-J, Sui M-Q, Liang M, et al. Altered baseline brain activity in children with ADHD revealed by resting-state functional MRI. Brain Dev. 2007;29:83–91.
130. Zou Q-H, Zhu C-Z, Yang Y, Zuo X-N, Long X-Y, Cao Q-J, et al. An improved approach to detection of amplitude of low-frequency fluctuation (ALFF) for resting-state fMRI: fractional ALFF. J Neurosci Methods. 2008;172:137–41.
131. Zang Y, Jiang T, Lu Y, He Y, Tian L. Regional homogeneity approach to fMRI data analysis. Neuroimage. 2004;22:394–400.
132. Beckmann CF, DeLuca M, Devlin JT, Smith SM. Investigations into resting-state connectivity using independent component analysis. Philos Trans R Soc Lond Ser B Biol Sci. 2005;360:1001–13.
133. Wang J, Zuo X, He Y. Graph-based network analysis of resting-state functional MRI. Front Syst Neurosci. 2010;4:16.
134. Bullmore E, Sporns O. Complex brain networks: graph theoretical analysis of structural and functional systems. Nat Rev Neurosci. 2009;10:186–98.
135. Zhang X, Tokoglu F, Negishi M, Arora J, Winstanley S, Spencer DD, et al. Social network theory applied to resting-state fMRI connectivity data in the identification of epilepsy networks with iterative feature selection. J Neurosci Methods. 2011;199:129–39.

136. Shimony JS, Zhang D, Johnston JM, Fox MD, Roy A, Leuthardt EC. Resting-state spontaneous fluctuations in brain activity: a new paradigm for presurgical planning using fMRI. Acad Radiol. 2009;16:578–83.
137. Koch W, Teipel S, Mueller S, Benninghoff J, Wagner M, Bokde ALW, et al. Diagnostic power of default mode network resting state fMRI in the detection of Alzheimer's disease. Neurobiol Aging. 2012;33:466–78.
138. Vanhaudenhuyse A, Noirhomme Q, Tshibanda LJ-F, Bruno M-A, Boveroux P, Schnakers C, et al. Default network connectivity reflects the level of consciousness in non-communicative brain-damaged patients. Brain. 2010;133:161–71.
139. Bassett DS, Nelson BG, Mueller BA, Camchong J, Lim KO. Altered resting state complexity in schizophrenia. Neuroimage. 2012;59:2196–207.
140. Craddock RC, Holtzheimer PE 3rd, Hu XP, Mayberg HS. Disease state prediction from resting state functional connectivity. Magn Reson Med. 2009;62:1619–28.
141. Anderson JS, Nielsen JA, Froehlich AL, DuBray MB, Druzgal TJ, Cariello AN, et al. Functional connectivity magnetic resonance imaging classification of autism. Brain. 2011;134:3742–54.
142. Zhu C-Z, Zang Y-F, Cao Q-J, Yan C-G, He Y, Jiang T-Z, et al. Fisher discriminative analysis of resting-state brain function for attention-deficit/hyperactivity disorder. Neuroimage. 2008;40:110–20.
143. Thomason ME, Grove LE, Lozon TA Jr, Vila AM, Ye Y, Nye MJ, et al. Age-related increases in long-range connectivity in fetal functional neural connectivity networks in utero. Dev Cogn Neurosci. 2015;11:96–104.
144. Smyser CD, Inder TE, Shimony JS, Hill JE, Degnan AJ, Snyder AZ, et al. Longitudinal analysis of neural network development in preterm infants. Cereb Cortex. 2010;20:2852–62.
145. Fair DA, Dosenbach NUF, Church JA, Cohen AL, Brahmbhatt S, Miezin FM, et al. Development of distinct control networks through segregation and integration. Proc Natl Acad Sci USA. 2007;104:13507–12.
146. Fair DA, Cohen AL, Dosenbach NUF, Church JA, Miezin FM, Barch DM, et al. The maturing architecture of the brain's default network. Proc Natl Acad Sci USA. 2008;105:4028–32.
147. Thomason ME, Chang CE, Glover GH, Gabrieli JDE, Greicius MD, Gotlib IH. Default-mode function and task-induced deactivation have overlapping brain substrates in children. Neuroimage. 2008;41:1493–503.
148. Kelly AMC, Di Martino A, Uddin LQ, Shehzad Z, Gee DG, Reiss PT, et al. Development of anterior cingulate functional connectivity from late childhood to early adulthood. Cereb Cortex. 2009;19:640–57.
149. Stephens JA, Salorio CF, Gomes JP, Nebel MB, Mostofsky SH, Suskauer SJ. Response inhibition deficits and altered motor network connectivity in the chronic phase of pediatric traumatic brain injury. J Neurotrauma. 2017;34:3117–23.
150. Tuerk C, Dégeilh F, Catroppa C, Dooley JJ, Kean M, Anderson V, et al. Altered resting-state functional connectivity within the developing social brain after pediatric traumatic brain injury. Hum Brain Mapp. 2020;41:561–76.
151. Botchway E, Kooper CC, Pouwels PJW, Bruining H, Engelen M, Oosterlaan J, et al. Resting-state network organisation in children with traumatic brain injury. Cortex. 2022;154:89–104.
152. Vaughn KA, DeMaster D, Kook JH, Vannucci M, Ewing-Cobbs L. Effective connectivity in the default mode network after paediatric traumatic brain injury. Eur J Neurosci. 2022;55:318–36.
153. Holshouser BA, Tong KA, Ashwal S. Proton MR spectroscopic imaging depicts diffuse axonal injury in children with traumatic brain injury. AJNR Am J Neuroradiol. 2005;26:1276–85.
154. Moffett JR, Ross B, Arun P, Madhavarao CN, Namboodiri AMA. N-Acetylaspartate in the CNS: from neurodiagnostics to neurobiology. Prog Neurobiol. 2007;81:89–131.
155. Alessandri B, Doppenberg E, Zauner A, Woodward J, Choi S, Bullock R. Evidence for time-dependent glutamate-mediated glycolysis in head-injured patients: a microdialysis study. Acta Neurochir Suppl. 1999;75:25–8.

156. Bullock R, Zauner A, Woodward JJ, Myseros J, Choi SC, Ward JD, et al. Factors affecting excitatory amino acid release following severe human head injury. J Neurosurg. 1998;89:507–18.
157. Brand A, Richter-Landsberg C, Leibfritz D. Multinuclear NMR studies on the energy metabolism of glial and neuronal cells. Dev Neurosci. 1993;15:289–98.
158. Barker PB, Bizzi A, De Stefano N, Gullapalli R, Lin DDM. Clinical MR spectroscopy: techniques and applications. Cambridge: Cambridge University Press; 2010.
159. Bartnik-Olson BL, Alger JR, Babikian T, Harris AD, Holshouser B, Kirov II, et al. The clinical utility of proton magnetic resonance spectroscopy in traumatic brain injury: recommendations from the ENIGMA MRS working group. Brain Imaging Behav. 2021;15:504–25.
160. Maudsley AA, Andronesi OC, Barker PB, Bizzi A, Bogner W, Henning A, et al. Advanced magnetic resonance spectroscopic neuroimaging: experts' consensus recommendations. NMR Biomed. 2021;34:e4309.
161. Oz G, Alger JR, Barker PB, Bartha R, Bizzi A, Boesch C, et al. Clinical proton MR spectroscopy in central nervous system disorders. Radiology. 2014;270:658–79.
162. Wilson M, Andronesi O, Barker PB, Bartha R, Bizzi A, Bolan PJ, et al. Methodological consensus on clinical proton MRS of the brain: review and recommendations. Magn Reson Med. 2019;82:527–50.
163. Ashwal S, Holshouser BA, Shu SK, Simmons PL, Perkin RM, Tomasi LG, et al. Predictive value of proton magnetic resonance spectroscopy in pediatric closed head injury. Pediatr Neurol. 2000;23:114–25.
164. Babikian T, Freier MC, Ashwal S, Riggs ML, Burley T, Holshouser BA. MR spectroscopy: predicting long-term neuropsychological outcome following pediatric TBI. J Magn Reson Imaging. 2006;24:801–11.
165. Yeo RA, Phillips JP, Jung RE, Brown AJ, Campbell RC, Brooks WM. Magnetic resonance spectroscopy detects brain injury and predicts cognitive functioning in children with brain injuries. J Neurotrauma. 2006;23:1427–35.
166. Babikian T, Marion SD, Copeland S, Alger JR, O'Neill J, Cazalis F, et al. Metabolic levels in the corpus callosum and their structural and behavioral correlates after moderate to severe pediatric TBI. J Neurotrauma. 2010;27:473–81.
167. Walz NC, Cecil KM, Wade SL, Michaud LJ. Late proton magnetic resonance spectroscopy following traumatic brain injury during early childhood: relationship with neurobehavioral outcomes. J Neurotrauma. 2008;25:94–103.
168. Parry L, Shores A, Rae C, Kemp A, Waugh M-C, Chaseling R, et al. An investigation of neuronal integrity in severe paediatric traumatic brain injury. Child Neuropsychol. 2004;10:248–61.
169. Babikian T, Alger JR, Ellis-Blied MU, Giza CC, Dennis E, Olsen A, et al. Whole brain magnetic resonance spectroscopic determinants of functional outcomes in pediatric moderate/severe traumatic brain injury. J Neurotrauma. 2018;35:1637–45.
170. Aaen GS, Holshouser BA, Sheridan C, Colbert C, McKenney M, Kido D, et al. Magnetic resonance spectroscopy predicts outcomes for children with nonaccidental trauma. Pediatrics. 2010;125:295–303.
171. Ashwal S, Holshouser B, Tong K, Serna T, Osterdock R, Gross M, et al. Proton MR spectroscopy detected glutamate/glutamine is increased in children with traumatic brain injury. J Neurotrauma. 2004;21:1539–52.
172. Brenner T, Freier MC, Holshouser BA, Burley T, Ashwal S. Predicting neuropsychologic outcome after traumatic brain injury in children. Pediatr Neurol. 2003;28:104–14.
173. Ashwal S, Holshouser B, Tong K, Serna T, Osterdock R, Gross M, et al. Proton Spectroscopy Detected Myoinositol in Children with Traumatic Brain Injury. Pediatr Res. 2004;56:630–8.
174. Berger L, Holshouser B, Nichols JG, Pivonka-Jones J, Ashwal S, Bartnik-Olson B. White matter metabolite ratios predict cognitive outcome in pediatric traumatic brain injury. Meta 2023;13. https://doi.org/10.3390/metabo13070778.

175. Holshouser B, Pivonka-Jones J, Nichols JG, Oyoyo U, Tong K, Ghosh N, et al. Longitudinal metabolite changes after traumatic brain injury: a prospective pediatric magnetic resonance spectroscopic imaging study. J Neurotrauma. 2019;36:1352–60.
176. Dennis EL, Babikian T, Alger J, Rashid F, Villalon-Reina JE, Jin Y, et al. Magnetic resonance spectroscopy of fiber tracts in children with traumatic brain injury: a combined MRS—diffusion MRI study. Hum Brain Mapp. 2018. https://doi.org/10.1002/hbm.24209.
177. Hunter JV, Thornton RJ, Wang ZJ, Levin HS, Roberson G, Brooks WM, et al. Late proton MR spectroscopy in children after traumatic brain injury: correlation with cognitive outcomes. AJNR Am J Neuroradiol. 2005;26:482–8.
178. Kawamata T, Katayama Y, Hovda DA, Yoshino A, Becker DP. Lactate accumulation following concussive brain injury: the role of ionic fluxes induced by excitatory amino acids. Brain Res. 1995;674:196–204.
179. Krishnappa IK, Contant CF, Robertson CS. Regional changes in cerebral extracellular glucose and lactate concentrations following severe cortical impact injury and secondary ischemia in rats. J Neurotrauma. 1999;16:213–24.
180. Makoroff KL, Cecil KM, Care M, Ball WS Jr. Elevated lactate as an early marker of brain injury in inflicted traumatic brain injury. Pediatr Radiol. 2005;35:668–76.
181. Harris AD, Amiri H, Bento M, Cohen R, Ching CRK, Cudalbu C, et al. Harmonization of multi-scanner in vivo magnetic resonance spectroscopy: ENIGMA consortium task group considerations. Front Neurol. 2022;13:1045678.
182. Rostami E, Nilsson P, Enblad P. Cerebral blood flow measurement in healthy children and children suffering severe traumatic brain injury-what do we know? Front Neurol. 2020;11:274.
183. Alsop DC, Detre JA, Golay X, Günther M, Hendrikse J, Hernandez-Garcia L, et al. Recommended implementation of arterial spin-labeled perfusion MRI for clinical applications: a consensus of the ISMRM perfusion study group and the European consortium for ASL in dementia. Magn Reson Med. 2015;73:102–16.
184. Lindner T, Bolar DS, Achten E, Barkhof F, Bastos-Leite AJ, Detre JA, et al. Current state and guidance on arterial spin labeling perfusion MRI in clinical neuroimaging. Magn Reson Med. 2023;89:2024–47.
185. Paniukov D, Lebel RM, Giesbrecht G, Lebel C. Cerebral blood flow increases across early childhood. Neuroimage. 2020;204:116224.
186. Satterthwaite TD, Shinohara RT, Wolf DH, Hopson RD, Elliott MA, Vandekar SN, et al. Impact of puberty on the evolution of cerebral perfusion during adolescence. Proc Natl Acad Sci USA. 2014;111:8643–8.
187. Newsome MR, Scheibel RS, Chu Z, Hunter JV, Li X, Wilde EA, et al. The relationship of resting cerebral blood flow and brain activation during a social cognition task in adolescents with chronic moderate to severe traumatic brain injury: a preliminary investigation. Int J Dev Neurosci. 2012;30:255–66.
188. Quigg M, Kundu B. Dynamic FDG-PET demonstration of functional brain abnormalities. Ann Clin Transl Neurol. 2022;9:1487–97.
189. Bergsneider M, Hovda DA, Shalmon E, Kelly DF, Vespa PM, Martin NA, et al. Cerebral hyperglycolysis following severe traumatic brain injury in humans: a positron emission tomography study. J Neurosurg. 1997;86:241–51.
190. Bergsneider M, Hovda DA, Lee SM, Kelly DF, McArthur DL, Vespa PM, et al. Dissociation of cerebral glucose metabolism and level of consciousness during the period of metabolic depression following human traumatic brain injury. J Neurotrauma. 2000;17:389–401.
191. Hattori N, Huang S-C, Wu H-M, Liao W, Glenn TC, Vespa PM, et al. Acute changes in regional cerebral (18)F-FDG kinetics in patients with traumatic brain injury. J Nucl Med. 2004;45:775–83.
192. Worley G, Hoffman JM, Paine SS, Kalman SL, Claerhout SJ, Boyko OB, et al. 18-Fluorodeoxyglucose positron emission tomography in children and adolescents with traumatic brain injury. Dev Med Child Neurol. 1995;37:213–20.

193. Ianof JN, Anghinah R. Traumatic brain injury: an EEG point of view. Dement Neuropsychol. 2017;11:3–5.
194. Peitz GW, Wilde EA, Grandhi R. Magnetoencephalography in the detection and characterization of brain abnormalities associated with traumatic brain injury: a comprehensive review. Med Sci (Basel) 2021;9. https://doi.org/10.3390/medsci9010007.
195. Tate DF, Dennis EL, Lindsey HM, Wilde EA. Harmonization of neuropsychological and other clinical endpoints: pitfalls and possibilities. Neuropsychology. 2023;37:233–6.
196. Levin HS, Culhane KA, Mendelsohn D, Lilly MA, Bruce D, Fletcher JM, C, et al. Cognition in relation to magnetic resonance imaging in head-injured children and adolescents. Arch Neurol. 1993; 50(9):897–905.
197. Benavidez DA, Fletcher JM, Hannay HJ, Bland ST, Caudle SE, Mendelsohn DB, et al. Corpus callosum damage and interhemispheric transfer of information following closed head injury in children. Cortex. 1999; 35(3):315–36.
198. Wright KL, Hopkins RO, Robertson FE, Bigler ED, Taylor HG, Rubin KH, et al. Assessment of White Matter Integrity after Pediatric Traumatic Brain Injury. J Neurotrauma. 2020; 37(20):2188–2197.
199. Janas AM, Qin F, Hamilton S, Jiang B, Baier N, Wintermark M, et al. Diffuse Axonal Injury Grade on Early MRI is Associated with Worse Outcome in Children with Moderate-Severe Traumatic Brain Injury. Neurocrit Care. 2022;36(2):492–503.
200. Dennis EL, Jin Y, Villalon-Reina JE, Zhan L, Kernan CL, Babikian T, et al. White matter disruption in moderate/severe pediatric traumatic brain injury: advanced tract-based analyses. Neuroimage Clin. 2015;7:493–505.
201. Dennis EL, Ellis MU, Marion SD, Jin Y, Moran L, Olsen A, et al. Callosal Function in Pediatric Traumatic Brain Injury Linked to Disrupted White Matter Integrity. J Neurosci. 2015;35(28):10202–11.
202. Ryan NP, Catroppa C, Beare R, Silk TJ, Crossley L, Beauchamp MH, et al. Theory of mind mediates the prospective relationship between abnormal social brain network morphology and chronic behavior problems after pediatric traumatic brain injury. Soc Cogn Affect Neurosci. 2016;11(4):683–92..
203. Anderson V, Beauchamp MH, Yeates KO, Crossley L, Ryan N, Hearps SJC, et al. Social Competence at Two Years after Childhood Traumatic Brain Injury. J Neurotrauma. 2017;34(14):2261–2271.

Chapter 4
Sleep-Wake Disorders Among Older Adults Following Traumatic Brain Injury

Mira H. Ghneim, Meaghan Broderick, and Deborah M. Stein

Abstract Traumatic brain injuries (TBI) are increasingly common in older adults and represent a substantial source of morbidity and mortality for this population. In addition to the impact from the primary insult, TBI can lead to a variety of chronic neurocognitive conditions including dementia, depression, and sleep disturbances. When caused by TBI, these conditions differ importantly from their non-TBI-related counterparts. Much about how TBI relates to the development of these conditions is unknown, and more research is needed to further elucidate optimal treatment strategies.

Keywords Older adults · TBI · Sleep disturbances · Dementia · Depression

Abbreviations

TBI Traumatic Brain Injuries

4.1 Introduction

While aging itself brings about changes in sleep patterns, understanding the distinction between normal age-related shifts and pathological sleep-wake disorders becomes paramount. These disorders, ranging from insomnia to obstructive sleep apnea, not only disrupt sleep architecture but also exacerbate cognitive and functional impairments, especially in the context of TBI. By exploring the prevalence,

M. H. Ghneim (✉) · M. Broderick · D. M. Stein
R Adams Cowley Shock Trauma, The University of Maryland School of Medicine, Baltimore, MD, USA
e-mail: mira.ghneim@som.umaryland.edu; Meaghan.broderick@som.umaryland.edu; dstein@som.umaryland.edu

L. J. Noble-Haeusslein, D. M. Schnyer (eds.), *Traumatic Brain Injuries*, Advances in Neurobiology 42, https://doi.org/10.1007/978-3-031-69832-3_4

risk factors, and underlying mechanisms of sleep disturbances post-TBI, this chapter underscores the critical need for targeted interventions and further research to optimize outcomes and enhance the overall well-being of older adults facing the dual challenges of aging and TBI.

4.2 Sleep-Wake Disorders

4.2.1 The Physiological Changes of the Sleep-Wake Cycle vs. Sleep-Wake Disorder

Sleep-wake disturbances are not an inherent part of the aging process, and it is essential to distinguish between them properly. As individuals age, they experience an advanced circadian tendency, manifesting as an earlier bedtime and wake-up time. Additionally, sleep architecture changes. This includes an increased proportion of time spent in the lighter stages of sleep, known as stages N1 and N2; a decreased proportion of time spent in the deeper stage of sleep, known as stage N3; and a decrease in rapid eye movement (REM) sleep. As an individual ages, these changes lead to a reduction in deep and restorative sleep, an increase in light and transitory sleep, and an overall decrease in the total amount of time asleep. As a result, older adults tend to spend less time asleep when compared to younger individuals [1].

On the other hand, sleep-wake disorders (SWD) in older adults are multifactorial (long or short sleep duration, sleep fragmentation, sleep-disordered breathing) and consequently contribute to daytime napping, excessive daytime sleeping, acute sleep deprivation (single night of insufficient sleep), and chronic sleep deprivation (repetitive bouts of poor night sleep over prolonged periods) leading to an interruption of the regular sleep-wake cycles [1, 2]. SWDs manifest as insomnia, obstructive sleep apnea (OSA), REM sleep behavior disorder, parasomnias (nightmares and nightmare disorders, night terrors, bedwetting, sleepwalking), sleep paralysis, restless leg syndrome, circadian rhythm disorders, and narcolepsy.

4.2.2 SWDs and TBI in Older Adults

TBI results in a cascade of events, which includes neuronal circuit disruption, inflammation, and neuroplasticity signaling. Consequently, TBI patients develop alterations in their sleep-wake cycles that lead to the development of SWDs. SWDs post-TBI have significant implications due to their detrimental effect on recovery and impairment of quality-of-life postinjury.

4.2.2.1 Prevalence

SWDs after traumatic brain injury include OSA, central sleep apnea, insomnia, pleiosomnia, excessive daytime sleepiness, narcolepsy, restless leg syndrome, periodic leg movements sleep, REM behavior disorders, and circadian rhythm disorder [3, 4]. Very few studies have assessed the prevalence of SWDs in older adults post-TBI. A recent study by Albrecht et al., using a large commercial insurance database, assessed the pre-TBI and 12-month post-TBI prevalence of SWDs in a large cohort of older adults with TBI compared to older adults without TBI. The study showed that the TBI cohort demonstrated a significantly higher prevalence of any diagnosed SWDs pre-TBI (14% vs. 9%), with insomnia and OSA being the most commonly diagnosed pre-TBI SWDs [insomnia (6.3% vs. 3.8%), OSA (6.4% vs. 4.8%)]. This trend persisted at 12 months post-TBI with an overall significantly higher prevalence of SWD (22.7% vs. 14.1%), OSA (9.4% vs. 6.7%), and insomnia (12% vs. 6.6%) in older adults with TBI compared to the non-TBI group. Additionally, TBI was found to increase the risk of new insomnia diagnoses and the composite of any sleep disorder by 17% and 9%, respectively, at 12 months post-TBI [3].

4.2.2.2 Risk Factors [5]

The majority of the available research has focused on post-TBI SWDs in children, adolescents, and younger adults [6–8]. To mitigate this knowledge gap, a recent study by Wei and colleagues examined the prevalence and risk factors for the development of SWDs in older adults with TBI compared to a gender and age-matched non-TBI cohort. Similar to the findings by Albrecht et al., this study identified insomnia and OSA as the two most common SWD in older adults with and without TBI, with a significantly increased prevalence of both in the TBI cohort. The primary risk factorsRisk factors for post-TBI OSA were increasing age and male gender.

Sleep-disordered breathing is known to be more frequent in older adults and males [9, 10]. OSA is characterized by episodes of partial or total upper airway obstruction during sleep, leading to transient blood gas abnormalities and sleep fragmentation [11]. Its primary symptoms include snoring, daytime sleepiness, and decreased cognitive functioning [12]. Those who experience OSA often report excessive daytime sleepiness and fatigue and deficits in memory and cognition and are at high risk for accelerated cerebrovascular and cardiovascular atherosclerosis [13]. While the exact mechanism of post-TBI OSA development is not fully understood, one theory that is extrapolated from the stroke literature suggests that TBI-mediated brain lesions involving the respiratory control neurocircuitry in the brainstem or the cortex lead to pharyngeal muscle weakness resulting in the higher OSA prevalence in older adults with TBI.

Insomnia is characterized by an increase in neuroexcitation, increasing alertness, a rise in metabolic activity, hypothalamic-pituitary-adrenal axis involvement, increased heart rate, and overactivation of the sympathetic nervous system. This leads to high stimulation, frequent arousals, fragmented sleep, intermittent hypoxia, and excessive daytime sleepiness. Those who experience insomnia often report difficulty initiating and maintaining sleep, early morning awakening, and inability to return to sleep. The primary risk factors for post-TBI insomnia identified by Wei and colleagues were increased severity of pain and depression symptoms. Bidirectional relations of sleep problems with depression and pain have been extensively reported in the literature [14, 15]. It is worth noting that prior prospective research in young adults with TBI has reported that the severity levels of depression and pain are independent and substantial contributors to unfavorable sleep manifestations. The findings by Wei and colleagues suggest that a similar relationship may exist in older adults with TBI [5].

4.2.2.3 Pathophysiology of SWDs in TBI

Although SWDs have been long recognized as a post-TBI consequence, the underlying neurologic mechanisms have yet to be established. However, the pathophysiology and neuropathology of SWDs after TBI are under investigation in both animal models and humans.

The association between TBI and SWDs is complex. It is thought to be attributed to direct injury to the sleep circuitry, nonselective disruption of the nonsleep circuitry and brain function, or both. Nonselective disruption of brain function can manifest as other conditions such as chronic pain, posttraumatic stress disorder, headaches, autonomic dysfunction, and endocrine disorders of the pituitary gland, all of which can have their own deleterious effects on sleep [4].

Direct Injury to the Sleep Modulating Circuitry

Prowake Circuit Physiology and Molecular Modulators

Wake-promoting circuits in the rostral brainstem and hypothalamus, along with the far-reaching axonal projections of the locus coeruleus, serotonergic neurons in the dorsal raphe nucleus, histaminergic neurons in the tuberomammillary nucleus, dopaminergic neurons of the ventral tuberomammillary nucleus, and glutaminergic neurons of the medial pontine reticular formation that disperse throughout the forebrain, thalamus, and limbic system, are required for arousal. These networks utilize the neurotransmitters/neuropeptide systems, including histamine, serotonin, norepinephrine, acetylcholine, dopamine, glutamate, and orexin/hypocretin. These neuronal networks have specific roles in cognition, behavior, sensory processing, and autonomic control during waking and contribute to overall arousal. Differences in the pattern and intensity of neuronal activity across sleep-waking states among these systems likely account for the varied manifestations of SWDs post-TBI [16, 17].

Orexin is a neuropeptide expressed by neurons in the perifornical lateral hypothalamus. These neurons have extensive projections to other hypothalamic nuclei, the limbic system, the thalamus, the cortex, and the spinal cord, exciting several monoaminergic and cholinergic wake-promoting systems. Circadian oscillations in orexin within the CSF correlate with arousal cycles, and exogenous administration of orexin promotes wakefulness (68). The lack of orexinergic neurons and subsequent orexin deficiency is associated with sleepiness, cataplexy, hypnopompic and hypnagogic hallucinations, and sleep paralysis in patients with narcolepsy. A dysfunction in the orexin neurons and levels expressed, as seen in narcolepsy, is postulated to be one of the mechanisms that facilitate post-TBI SWDs and is currently being studied in TBI animal models [16, 17].

The neurophysiologic action of serotonin and its receptors is complex as it has dual opposing effects on arousal and REM sleep. The outcome of serotonergic receptor agonism or antagonism is a function of the type of nearby receptors and neurons and the timing within the circadian rhythm at which the release of serotonin occurs. The functions of the serotonin receptors and pathways are the most widely studied and understood components of the sleep-wake cycle. Therefore, they are the primary target for pharmacological induction of physiological sleep [4]. However, the influence of TBI on serotonergic pathways that result in SWDs remains to be understood.

Prosleep Circuit Physiology and Molecular Modulators

The sleep-promoting circuits are housed in the median preoptic nucleus (MPN), which activates in response to sleep deprivation, and the ventrolateral preoptic nucleus (VLPO), which activates only after recoveryRecovery sleep has begun. This is mediated by the gamma-aminobutyric acid and glaninergic neurons. The axonal projections within the MPN and VLPO strongly inhibit the wake-promoting circuitry. Additionally, the wake state demands increased metabolic neuronal activity, leading to increased adenosine production. The more time one is awake, the greater the concentration of adenosine. Adenosine is an antagonist of the histaminergic receptors, which inhibits the wake-promoting circuitry and activates the sleep-promoting circuitry in the MPN and VLPO nucleus [18, 19].

Melatonin is a serotonin-derived hormone that is synthesized in the pineal gland. Since melatonin secretion follows a circadian rhythm, the highest melatonin levels are secreted in the evening and night. Melatonin affects the sleep-wake cycle and other processes, such as memory, addiction, fear, pain, and mood. Pineal melatonin activates MT1 and MT2 receptors in the suprachiasmic nucleus of the hypothalamus, the thalamus, the pars tuberalis within the pituitary gland, and peripheral tissues. Recent advancements using selective MT1 and MT2 ligands and MT1/MT2 knockout mice have suggested that the activation of MT1 receptors regulates REM sleep, whereas MT2 receptors increase non-REM sleep. Combined, both receptors function to maintain proper circadian rhythm in response to changes in seasons and duration of exposure to daylight [18]. While clinical data from TBI survivors indicate that melatonin secretion is altered compared to controls at both acute and chronic time points postinjury, the effects of melatonin on the MPN and VLPO nucleus remain poorly understood [19, 20].

Although no neuropathology study in deceased TBI patients has shown the primary mechanism by which TBI leads to SWDs, the effects of TBI on these prowake and prosleep circuitry is thought to be mediated through the structural brain changes secondary to TBI that disrupts neural circuitry directly via diffuse axonal injury. It is postulated that TBI either disrupts neuronal axonal projections or the cell bodies that make up the MPN and the VLPO sleep-inducing circuits, as lesions in this region are known to cause insomnia or disrupts the wake-promoting neurons that reside in the brainstem and hypothalamus. Additionally, TBI can indirectly affect brain structure and function via trauma-induced inflammation, edema, compression, and ischemia mediated through IL-6, IL-1, and TNF-α [19, 20].

Nonselective Disruption of the Nonsleep Circuitry and Brain Function

Endocrine Dysfunction and Immune-Endocrine Interactions

The vascular and anatomical characteristics of the hypothalamic-pituitary axis (HPA) make it particularly vulnerable to TBI. Hormonal alterations resulting from disruptions in the HPA are frequently observed following TBI and are associated with long-term sequelae. It should be noted that neuroendocrine dysregulation has been observed even in those injuries defined as mild and/or lacking overt histological damage. Hyperresponsiveness to stress and dysregulation of the HP-adrenal axis has been reported following experimental TBI and is associated with poor sleep efficiency and increased wake time after sleep. Disruptions in the HP-somatotopic axis following TBI result in growth hormone deficiency. Growth hormone is primarily secreted during slow-wave sleep. Accordingly, a reduction in the amount of growth hormone released in the setting of TBI results in increased sleep fragmentation and poor subjective sleep [20].

Involvement of the hypothalamus attests to the reciprocal nature of the sleep-wake and endocrine systems. Hormonal release is extensively regulated by sleep and is accordingly affected in individuals with SWDs. Furthermore, the secretion of subsets of hormones is associated with particular sleep stages. Thus, it is not unreasonable to speculate that SWDs post-TBI contributes to hormonal dysregulation [20].

This bidirectional relationship between sleep and neuroendocrine function has widespread physiological and functional effects, such as depression and anxiety, which are frequently reported following TBI and influenced by hormones and sleep. In addition, sleep critically interacts with several neuroendocrine axes that impact immune responses via hormonal release [20].

The immune system is activated after TBI, and elevated inflammation can suppress the production of hormones and worsen endocrine dysfunction. Similarly, chronic endocrine dysfunction and hypopituitarism can increase neuropathology following a TBI, which subsequently activates the immune response and increases inflammation. Both inflammation and endocrine dysfunction may contribute to TBI-induced SWDs; thus, the bidirectional relationship between the immune and endocrine systems acts as a positive feedback loop that initiates and exacerbates SWDs. Inflammation leads to endocrine dysfunction and altered sleep, and chronic endocrine dysregulation leads to inflammation and SWDs [20].

Comorbid Chronic Pain and Psychiatric Disorders

Comorbidities such as depression, anxiety, PTSD, chronic pain, and headaches likely contribute to, and perhaps in some cases cause, SWDs in TBI patients. The currently available evidence suggests that post-TBI SWDs is precipitated by or can worsen psychiatric disorders. The mechanism is probably multifactorial and is partially related to a hyperarousal state. Similarly, it has been shown that chronic pain predicts SWDs, and SWDs predict chronic pain. This bidirectional link between central pain disorders and SWDs is thought to be mediated through biopsychological mechanisms that include the dopamine and opioid systems and emotional dysregulation [4, 20–22].

4.2.2.4 Screening for TBI-SWD

Older adults who experience a TBI should be screened for SWDs. Screening tools are divided into subjective and objective strategies. Subjective tools include a sleep diary, clinical history taking, or questionnaires. These tools' advantages include low cost, ease of use, and ease of standardization. The primary disadvantage is reliance on patient recall. Three standardized questionnaires have been validated in TBI patients. The Epworth Sleepiness Scale (ESS) screens for excessive nighttime sleepiness, OSA, central sleep apnea, and circadian rhythm disorders. The Pittsburgh Sleep Quality Index (PSQI) screens for insomnia, excessive daytime sleepiness, and circadian rhythm disorders. The Insomnia Severity Index (ISI) is a seven-item questionnaire that asks respondents to describe the nature and symptoms of their sleep problems using a Likert-type scale [4, 23].

Objective methods include polysomnogram (PSG), actigraphy (an accelerator worn on the nondominant wrist), and the multiple sleep latency test (MSLT). PSG utilizes an electroencephalogram, electrooculogram, electromyogram, electrocardiogram, pulse oximetry, airflow, and respiratory effort to evaluate underlying causes of sleep disturbances. PSG is considered to be the gold standard for diagnosing OSA, central sleep apnea, and sleep-related hypoventilation/hypoxia [24]. Actigraphy records and integrates the occurrence and degree of limb movement activity over time. Mathematical algorithms are then applied to the data collected over days to weeks to estimate wakefulness and sleep and evaluate central hypersomnolence disorders [25]. MLST is based on the notion that sleep latency reflects underlying physiological sleepiness. It measures the tendency to fall asleep under controlled conditions. Following a standardized procedure, this is achieved through four to five naps given 2 h apart during the day. The mean sleep latency from all naps is used to measure sleepiness. The test is valid and reliable and is part of the diagnostic criteria for narcolepsy and idiopathic hypersomnia [24]. The advantages of these objective tests are the accuracy and quantity of data collected. The disadvantages include the cost of equipment, the expertise to interpret the data, and the lack of standardization [4].

4.2.2.5 Therapeutic Approaches to TBI-SWD

There are no approved therapies for the management of post-TBI SWDs. This can be attributed to the fact that in the absence of TBI, SWDs such as insomnia are challenging to treat. Moreover, the existence of comorbid conditions such as posttraumatic stress disorder, chronic pain, and cognitive impairments, which depend on the severity of the TBI, can complicate the treatment regimen.

Over the last 20 years, 24 randomized controlled trials (RCTs) have been conducted to assess the effect of various pharmacological and nonpharmacological treatment options on all post-TBI SWDs. Recently, Stewart et al. published a systematic review detailing the study designs, populations, SWDs, TBI severity, treatment options, and effectiveness across all SWDs post-TBI studied in these 24 RCTs. The authors concluded that post-TBI sleep problems are vastly understudied. The reasons for this are likely multifactorial and may include a lack of routinely incorporating a detailed sleep history in assessing TBI patients. Additionally, despite some interventions showing promising results, patient sample heterogeneity and lack of long-term efficacy data limits their uncontested application in clinical practice [26].

Given that OSA and insomnia have been identified as the two most common post-TBI SWDs in older adults, this section focuses on what is known so far regarding the available treatment options for the management of post-TBI insomnia and OSA based on RCTs discussed by Stewart and colleagues.

Management of OSA

There are no pharmacological treatment options for OSA. The most common nonpharmacological option is continuous positive airway pressure (CPAP). Kataria et al. conducted a text message reminder service vs. no reminder to assess CPAP adherence in 19 newly diagnosed OSA veterans with chronic TBI. OSA was assessed using the ESS scale. Mean overall percent compliance during the first 7 days was significantly greater in the reminder group than in the standard of care group (83.9% vs. 55.4%, $p = 0.04$). Mean overall percent compliance at 30 days was also greater in the reminder group but was not statistically significant (58.9% vs. 36.9%, $p = 0.22$). Compared to the baseline, the reminder group ESS scores improved to the normal range with a medium effect size ($d = -0.39$) [27]. No additional RCTs have been published to date assessing the beneficial effects of CPAP on post-TBI OSA. Moreover, there need to be more studies documenting an evidence-based approach to the management of OSA post-TBI.

Management of Insomnia

The current available RCTs assessing the effect of treatment options on post-TBI insomnia have focused on melatonin as the primary pharmacological therapy and cognitive behavioral therapy (CBT) as the primary nonpharmacological treatment modality.

Grima et al. published a randomized, double-blind, placebo-controlled, two-period, two-treatment (melatonin and placebo) crossover study of patients with a history of mild to severe TBI who experienced post-TBI SWDs. Participants were given melatonin and placebo capsules for 4 weeks and separated by a 48-h washout period. Treatment was taken nightly, 2 h before bedtime. Melatonin supplementation was found to reduce the PSQI score relative to placebo significantly. While melatonin did not affect sleep latency or daytime sleepiness on the ESS score, it did increase sleep efficiency on actigraphy. No serious adverse events were reported [28].

Two RCTs implemented CBT as a treatment for insomnia post-TBI [29, 30]. The CBT therapy in both studies focused on improving sleep hygiene through developing a better understanding of the characteristics of insomnia and sleep habits, learning and utilization of relaxation training, sleep interventions and meditation, and implementation of environment modifications over a 6–8-week time period. Each study enrolled 24 patients with mild to severe TBI. Participants' ages ranged between 30 and 60 years of age. Improvement in insomnia symptoms was assessed using the PSQI and ESS tools. The CBT groups in both studies reported better sleep quality, reduced daily fatigue levels, reduced sleep disturbances, and overall improvement in insomnia symptoms. The PSQI at ESS scores improved significantly after the treatment and at the two-month follow-up in the CBT group when compared to the control group. Overall, both studies reported robust and sustained treatment effects in reducing post-TBI insomnia.

These results suggest that CBT and melatonin are promising treatments with the potential to improve sleep quality in post-TBI insomnia. However, it is worth noting that overall, the current available studies assessing post-TBI SWD treatment options are limited by small sample sizes and failure to implement and control for intergroup variations due to age, head injury severity, time since injury, sleep symptom prevalence, and severity and the presence of comorbidities such as depression, anxiety, pain, and posttraumatic stress disorder limiting its generalizability. Therefore, to advance our understanding and identify optimal treatment strategies for post-TBI SWDs, designing and conducting future RCTs with robust methodology and sample sizing to allow for generalizability of findings in the post-TBI group, and comparison with the non-TBI population is warranted. This is especially true for older adults, who are often underrepresented in TBI research [26].

4.2.2.6 The Bidirectional Link Between SWDs and TBI Symptoms and the Role of the Glymphatic System

The primary function of sleep is to restore cellular and neuronal processes. Even in the absence of injury, a clear relationship exists between lack of sleep and worsening psychomotor function, impaired judgment, and somatic disturbances such as immune dysregulation and impaired cardiovascular function. Importantly, there is a well-documented bidirectional relationship between poor sleep and the development of depression and anxiety. Recently, a similar relationship has been observed in the setting of mild TBI, with SWD resulting in impaired recovery after mild TBI. This has been linked to impaired brain homeostatic processes, cerebral temperature maintenance, and cerebral glycogen storage maintenance. In addition, post-mild TBI SWDs can cause the release of proinflammatory cytokines IL-1β, IL-6, and TNF-α. Neuroinflammation can lead to impairment of neuronal function and delayed recovery [31].

Recent evidence shows that glymphatic exchange is significantly increased during sleep and suppressed during wakefulness. The glymphatic system is a network of perivascular pathways along which CSF and brain interstitial fluid exchange. Functionally connected to the meningeal lymphatic vascular drainage via the cisternal CSF compartment, glymphatic exchange along perivascular pathways plays a role in eliminating waste generated by the brain. Importantly, in rodents, glymphatic clearance increases during sleep and becomes impaired after moderate TBI. Both in animal models and humans, the aberrant accumulation of cerebral waste, such as amyloid-β and tau, has been linked with neurodegeneration and cerebral dysfunction. Glymphatic dysfunction thus offers a potential biological mechanism for the observed association between sleep disturbances and the persistence of post-TBI symptoms [31].

Glymphatic dysfunction may link mild TBI and neurobehavioral symptoms by two distinct mechanisms. First, loss of perivascular AQP4 integral membrane proteins post mild TBI can directly slow cellular waste clearance from the interstitial space. Second, mild TBI may indirectly reduce sleep-active glymphatic function by disrupting sleep phases in which glymphatic exchange is most active. Thus, by reducing both the opportunity for sleep-active glymphatic exchange and the efficacy of glymphatic function, mild TBI may promote the accumulation of cellular waste, neuropeptides, and inflammatory mediators in the interstitial compartment, promoting neuronal dysfunction and propagation/persistence of symptoms [31].

4.2.2.7 Implications of SWD in Older Adults Following TBI

Healthy sleep is associated with positive brain function, including synaptic growth, neural plasticity, and glymphatic clearance of neurotoxins. Poor sleep is associated with cognitive dysfunction, functional impairment, and mood disturbances. Therefore, sleep disturbances and clinical sleep disorders can impede recovery from TBI both directly (i.e., by inhibiting neural recovery) as well as indirectly (i.e., increasing risk for cognitive, functional, and mood disturbances). Furthermore, poor sleep among older adults is a risk factor for falls, increasing the risk of repetitive injury [3].

Acute and chronic sleep deprivation leads to impairment of daily functioning, lapses of attention, slowed working memory, slower reaction times, slower cognitive processing, and reduced neurocognitive and motor performance [32]. This decline in performance has been reported to increase the risk of falls. A recent systematic review assessing the relationship between sleep disturbances and falls identified subjective sleep fragmentation, OSA, and insomnia as the primary risk factors associated with an increased risk of falls in older adults [32]. Specifically, older adults who reported waking up two or more times at night were 2.5 times more likely to experience a fall than those who did not report nighttime awakenings [33]. The apnea-hypopnea index is a scale used to assess the degree of OSA. Older adults with a score of 15 or greater were found to be at double risk of having two or more falls in a year [34]. Finally, moderate insomnia (symptoms 1–5 nights per week) and severe insomnia (symptoms ≥ six nights per week) were associated with a 47% and 86% probability of experiencing a fall within 6 months, respectively [35].

Sleep complaints increase with age and are common in older adults. Insomnia and OSA are the most common SWDs reported in older adults and are risk factors for TBI in older adults. Moreover, insomnia and OSA represent the most common SWDs in older adults post-TBI. Insomnia is a known risk factor for impaired daytime functioning, substance use, depression, chronic pain, obesity, high blood pressure, cardiovascular disease, and dementia in older adults. It is associated with poorer quality of life and poor physical and psychological outcomes [36, 37]. Similarly, OSA is associated with an increased cardiovascular and cerebrovascular risk, increased all-cause mortality, poorer quality of life, and accelerated cognitive decline (reduced reaction time, processing speed, and executive function) due to intermittent episodes of hypoxia [13].

The primary treatment modality for OSA is CPAP, but there is limited evidence regarding its efficacy in treating post-TBI OSA. In terms of post-TBI insomnia, it is important to consider opportunities for treatment research. In other disease states, treatment of insomnia has been shown to improve health-related outcomes and quality of life. Although the FDA has approved multiple medications for the treatment of insomnia, these medications are generally ill-advised among older adults due to unfavorable risk-benefit ratios due to an increased risk of falls. Furthermore, there has been limited evaluation of sleep medications among individuals with TBI, and a dearth of evidence is available to evaluate these medications among older adults with TBI. As a result, nonpharmacological treatments such as CBT for insomnia, considered first-line treatments among older adults, should be evaluated for efficacy among older adults with TBI. Similarly, newer insomnia medications (i.e., orexin receptor antagonists) warrant study, as they might present more favorable risk/benefit profiles among TBI patients.

Overall, it is crucial to recognize that SWDs are a common sequela of TBI associated with poorer recovery. This is important among older adults who, at baseline and in the absence of SWDs, fare worse following TBI relative to their younger counterparts and have a higher prevalence of SWDs. Therefore, it is essential to screen older adults for SWDs post-TBI and incorporate prevention and treatment strategies to mitigate the adverse outcomes associated with post-TBI SWDs.

In conclusion, the intricate interplay between physiological changes in the sleep-wake cycle and the emergence of SWDs following TBI among older adults underscores the pressing need for targeted interventions and comprehensive management strategies. SWDs, including insomnia and obstructive sleep apnea, significantly impact recovery trajectories and quality of life post-TBI, emphasizing the importance of tailored screening protocols and therapeutic approaches. Moreover, the bidirectional link between SWDs and TBI symptoms, coupled with emerging insights into the role of the glymphatic system, underscores the multifaceted nature of sleep disturbances in older adults with TBI. Future research efforts should prioritize elucidating mechanistic pathways, refining treatment modalities, and fostering interdisciplinary collaboration to optimize outcomes and enhance well-being in this vulnerable population.

Acknowledgments None.

Disclosure Statement The authors do not have any relevant financial conflict of interest or funding sources to disclose.

References

1. Rodriguez JC, Dzierzewski JM, Alessi CA. Sleep problems in the elderly. Med Clin North Am. 2015;99:431–9.
2. Choshen-Hillel S, Ishqer A, Mahameed F, Reiter J, Gozal D, Gileles-Hillel A, et al. Acute and chronic sleep deprivation in residents: cognition and stress biomarkers. Med Educ. 2021;55:174–84.
3. Albrecht JS, Wickwire EM. Sleep disturbances among older adults following traumatic brain injury. Int Rev Psychiatry. 2020;2(32):31–8.
4. Werner J, Baumann C. TBI and sleep–wake disorders: pathophysiology, clinical management, and moving towards the future. Semin Neurol. 2017;37(04):419–32.
5. Wei L, Wen YT, Thompson HJ, Liu CY, Su YK, Chen PY, et al. Sleep disturbances following traumatic brain injury in older adults: a comparison study. J Head Trauma Rehabil. 2020;35:288–95.
6. Baumann CR, Werth E, Stocker R, Ludwig S, Bassetti CL. Sleep-wake disturbances 6 months after traumatic brain injury: a prospective study. Brain J Neurol. 2007;130:1873–83.
7. Pillar G, Averbooch E, Katz N, Peled N, Kaufman Y, Shahar E. Prevalence and risk of sleep disturbances in adolescents after minor head injury. Pediatr Neurol. 2003;29:131–5.
8. Tham SW, Palermo TM, Vavilala MS, Wang J, Jaffe KM, Koepsell TD, et al. The longitudinal course, risk factors, and impact of sleep disturbances in children with traumatic brain injury. J Neurotrauma. 2012;29:154–61.
9. Brennan C, Worrall-Davies A, McMillan D, Gilbody S, House A. The hospital anxiety and depression scale: a diagnostic meta-analysis of case-finding ability. J Psychosom Res. 2010;69(4):371–8.
10. Bursac Z, Gauss CH, Williams DK, Hosmer DW. Purposeful selection of variables in logistic regression. Source Code Biol Med. 2008;16(3):17.

11. Meira E, Cruz M, Kryger MH, Morin CM, Palombini L, Salles C, Gozal D. Comorbid Insomnia and Sleep Apnea: mechanisms and implications of an underrecognized and misinterpreted sleep disorder. Sleep Med. 2021;84:283–8.
12. Monti A, Doulazmi M, Nguyen-Michel VH, Pautas E, Mariani J, Kinugawa K. Clinical characteristics of sleep apnea in middle-old and oldest-old inpatients: symptoms and comorbidities. Sleep Med. 2021;82:179–85.
13. Osorio RS, Martínez-García MÁ, Rapoport DM. Sleep apnoea in the elderly: a great challenge for the future. Eur Respir J. 2022;59:2101649.
14. Chen PY, Tsai PS, Chen NH, Chaung LP, Lee CC, Chen CC, et al. Trajectories of sleep and its predictors in the first year following traumatic brain injury. J Head Trauma Rehabil. 2015;30:E50–5.
15. Ponsford JL, Parcell DL, Sinclair KL, Roper M, Rajaratnam SMW. Changes in sleep patterns following traumatic brain injury: a controlled study. Neurorehabil Neural Repair. 2013;27(7):613–21.
16. Sandsmark DK, Elliott JE, Lim MM. Sleep-wake disturbances after traumatic brain injury: synthesis of human and animal studies. Sleep [Internet]. 2017 Mar 17 [cited 2023 Nov 29]; Available from: https://academic.oup.com/sleep/article-lookup/doi/10.1093/sleep/zsx044
17. Saper CB, Scammell TE, Lu J. Hypothalamic regulation of sleep and circadian rhythms. Nature. 2005;27(437):1257–63.
18. Gobbi G, Comai S. Differential function of melatonin MT1 and MT2 receptors in REM and NREM sleep. Front Endocrinol. 2019;1(10):87.
19. Liu J, Clough SJ, Hutchinson AJ, Adamah-Biassi EB, Popovska-Gorevski M, Dubocovich ML. MT_1 and MT_2 melatonin receptors: a therapeutic perspective. Annu Rev Pharmacol Toxicol. 2016;(56):361–83.
20. Rowe RK, Griesbach GS. Immune-endocrine interactions in the pathophysiology of sleep-wake disturbances following traumatic brain injury: a narrative review. Brain Res Bull. 2022;185:117–28.
21. Sinha SS. Trauma-induced insomnia: a novel model for trauma and sleep research. Sleep Med Rev. 2016;25:74–83.
22. Koffel E, Kroenke K, Bair MJ, Leverty D, Polusny MA, Krebs EE. The bidirectional relationship between sleep complaints and pain: analysis of data from a randomized trial. Health Psychol. 2016;35:41–9.
23. Bastien CH, Vallières A, Morin CM. Validation of the Insomnia Severity Index as an outcome measure for insomnia research. Sleep Med. 2001;2:297–307.
24. Rundo JV, Downey R. Polysomnography. In: Handbook of clinical neurology [Internet]. Elsevier; 2019 [cited 2023 Nov 29]. p. 381–92. Available from: https://linkinghub.elsevier.com/retrieve/pii/B9780444640321000254
25. Smith MT, McCrae CS, Cheung J, Martin JL, Harrod CG, Heald JL, et al. Use of actigraphy for the evaluation of sleep disorders and circadian rhythm sleep-wake disorders: an American Academy of Sleep Medicine clinical practice guideline. J Clin Sleep Med. 2018 Jul;15(14):1231–7.
26. Stewart K, Shakarishvili N, Michalak A, Maschauer EL, Jenkins N, Riha RL. Treating sleep disorders following traumatic brain injury in adults: time for renewed effort? Sleep Med Rev. 2022;63:101631.
27. Kataria L, Sundahl C, Skalina L, Shah M, Pfeiffer M, Balish M, et al. 0526 ANNIE: the Veterans Health Administration's personalized text message application promotes compliance with positive airway pressure. Sleep. 2017;(40):A196–6.
28. Grima NA, Rajaratnam SMW, Mansfield D, Sletten TL, Spitz G, Ponsford JL. Efficacy of melatonin for sleep disturbance following traumatic brain injury: a randomised controlled trial. BMC Med. 2018;16:8.
29. Nguyen S, McKay A, Wong D, Rajaratnam SM, Spitz G, Williams G, et al. Cognitive behavior therapy to treat sleep disturbance and fatigue after traumatic brain injury: a pilot randomized controlled trial. Arch Phys Med Rehabil. 2017;98:1508–17.e2.

30. Theadom A, Barker-Collo S, Jones K, Dudley M, Vincent N, Feigin V. A pilot randomized controlled trial of on-line interventions to improve sleep quality in adults after mild or moderate traumatic brain injury. Clin Rehabil. 2018;32:619–29.
31. Piantino JA, Iliff JJ, Lim MM. The bidirectional link between sleep disturbances and traumatic brain injury symptoms: a role for glymphatic dysfunction? Biol Psychiatry. 2022;91:478–87.
32. Knechel NA, Chang P. The relationships between sleep disturbance and falls: a systematic review. J Sleep Res. 2022;31:e13580.
33. Hill EL, Cumming RG, Lewis R, Carrington S, Couteur DGL. Sleep disturbances and falls in older people. J Gerontol A Biol Sci Med Sci. 2007;62:62–6.
34. Kaushik S, Wang JJ, Mitchell P. Sleep apnea and falls in older People. J Am Geriatr Soc. 2007;55:1149–50.
35. Avidan AY, Fries BE, James ML, Szafara KL, Wright GT, Chervin RD. Insomnia and hypnotic use, recorded in the minimum data set, as predictors of falls and hip fractures in Michigan nursing homes. J Am Geriatr Soc. 2005;53:955–62.
36. Grandner MA, Perlis ML. Pharmacotherapy for insomnia disorder in older adults. JAMA Netw Open. 2019;27(2):e1918214.
37. Gardner RC, Dams-O'Connor K, Morrissey MR, Manley GT. Geriatric traumatic brain injury: epidemiology, outcomes, knowledge gaps, and future directions. J Neurotrauma. 2018;35:889–906.

Chapter 5
Dementia and Depression Among Older Adults Following Traumatic Brain Injury

Mira H. Ghneim, Meaghan Broderick, and Deborah M. Stein

Abstract Traumatic brain injuries are increasingly common in older adults and represent a substantial source of morbidity and mortality for this population. In addition to the impact from the primary insult, TBI can lead to a variety of chronic neurocognitive conditions including dementia, depression, and sleep disturbances. When caused by TBI, these conditions differ importantly from their non-TBI-related counterparts. Much about how TBI relates to the development of these conditions is unknown, and more research is needed to further elucidate optimal treatment strategies.

Keywords Dementia · Depression · Risk factors · Diagnostics · Modalities · Pathophysiology · Management

5.1 Introduction

Older adults, 65 years and older, represent the fastest-growing population in the United States and are expected to double to more than 80 million by 2050 [1]. Compared to previous generations, today's older adults have more active lifestyles and thus are at an increased risk of injury. Traumatic brain injury (TBI) is widespread, with a lifetime prevalence of 40%, and the highest incidence is in older adults [2]. Between 2016 and 2019, older adults accounted for 44% of TBI-related hospitalizations and 39% of TBI-related deaths [3, 4]. The highest hospitalization rates (per 100,000 population) were among older adults aged ≥75 (313.4 in 2016 and 320.8 in 2017). The highest death rates (per 100,000 population) were among

M. H. Ghneim (✉) · M. Broderick · D. M. Stein
R Adams Cowley Shock Trauma, The University of Maryland School of Medicine, Baltimore, MD, USA
e-mail: mira.ghneim@som.umaryland.edu; Meaghan.broderick@som.umaryland.edu; dstein@som.umaryland.edu

L. J. Noble-Haeusslein, D. M. Schnyer (eds.), *Traumatic Brain Injuries*, Advances in Neurobiology 42, https://doi.org/10.1007/978-3-031-69832-3_5

both the ≥75 (75.7 in 2018 and 76.7 in 2019) and the 65–74 (104.8 in 2016 and 102.7 in 2017) age groups [5].

The consequences of TBI in older adults are severe, with high mortality rates and significant cognitive and functional impairment for survivors [6–8]. Many who were previously independent in their daily activities become dependent on assistance after sustaining a TBI [9]. Moreover, older adults are at a high risk of experiencing long-term sequela of TBI, which include both an accelerated cognitive decline (dementia) and neuropsychiatric symptoms (sleep-wake disturbances and depression) following a TBI, which can further hinder their recovery [5, 10].

Historically, TBI research in older adults has predominantly focused on the cognitive and functional outcomes during the immediate peri-injury period and the short-term posthospitalization phase, with minimal emphasis on the TBI-related neurologic and neuropsychiatric sequelae of dementia, depression, and sleep-wake disturbances. All three entities are known to impede recovery through a negative impact on brain processes, including attention, executive function, and working memory, independent of the TBI itself. However, they are frequently overlooked, underdiagnosed, and inadequately studied in the literature, posing a significant challenge to their effective management.

Therefore, the objective of this is to shed some light on what we know so far regarding the prevalence, pathophysiology, diagnosis, treatment options, and implications of dementia and depression in older adults following a TBI.

5.2 Dementia

5.2.1 Definition, Risk Factors, and Implications in the General Older Adult Population

5.2.1.1 Definition and Classification

The *DSM-5 defines dementia*—or major neurocognitive disorder—as significant cognitive impairment in at least one of the following cognitive domains, representing a decline from a previous level of functioning: learning and memory, language, executive function, complex attention, perceptual-motor function, and social cognition. This decline interferes with independent functioning, leading to severe disability and increased morbidity and mortality [11]. The most common cause of dementia in later life is Alzheimer's disease (AD), a slow and insidious disorder that affects memory first and foremost, followed by a decline in other domains as the disease progresses. After AD, other important neurodegenerative dementias include frontotemporal lobar dementias, Lewy body dementias, and vascular cognitive dementia [12].

Alzheimer's dementia is characterized by progressive loss of memory and other cognitive functions. Pathologically, it is the most well-studied dementia and is defined by abundant amyloid-β (Aβ) neuritic plaques and paired helical filaments-tau neurofibrillary tangles in the cerebral cortex. AD can be challenging to recognize in its earliest stages, given that the principal complaint is an increase in forgetfulness with the preservation of daily functioning. This stage is often referred to as the prodromal or mild cognitive impairment stage of AD. The diagnosis of dementia due to AD becomes clinically apparent as episodes of forgetfulness increase and progressive loss of other cognitive domains begins to affect daily functioning. As the symptoms of AD worsen, patients may become incapable of attending to basic needs such as feeding, dressing, and self-care. Progression from the first warning signs to the final stages of the disease can span a decade or more, creating a vast social and financial burden on society and extracting an immeasurable emotional toll on the patient and their family members [12].

Frontal lobe dementia is characterized by progressive behavioral and language deficits, with neurodegeneration involving the frontal and temporal lobes. Symptom profiles include changes in personality and behavior and deficits in expressive and receptive language. Lewy body dementia is characterized by varying degrees of parkinsonism, fluctuating cognition and alertness, visual hallucinations, autonomic dysfunction, and REM sleep changes. Vascular cognitive dementia is characterized by a broad spectrum of cognitive and behavioral changes associated with cerebrovascular pathology. Individuals often present with attentional and executive impairment ranging from mild cognitive impairment to severe dementia. Vascular dementia primarily affects the deep white matter pathways, eventually reaching a clinically significant threshold, causing cognitive decline within the domains of attention, processing speed, and higher-order executive functioning [12].

5.2.1.2 Stages and Grading Systems of Dementia

Dementia is a progressive disease with a variable clinical course and is an entity that defines an individual's life expectancy independent of other comorbidities. While stages of dementia have been classically described as mild (early), moderate (middle), and severe (late), more specific grading systems exist that provide a comprehensive description of symptom progression (Table 5.1) [13]. Using these grading systems as a guide to develop a better global understanding of cognitive and functional decline allows physicians to better communicate with patients and their caregivers to determine optimal treatment options during acute and critical illness (Table 5.2).

5.2.1.3 Diagnostic Modalities

While the definitive diagnosis of dementia can only be made by an examination of brain tissue at autopsy, the two current modalities utilized for clinical diagnosis are biomarkers and neuroimaging [12].

Table 5.1 Stages of dementia

	Mild	Moderate	Severe
Duration of symptoms	2–4 years	2–10 years	1–3 years
Activities of daily living (ADLs)	Problems with routine tasks	Need help with basic ADL (feeding, dressing, bathing)	Progress to total dependence on caregiver
Behavior	Changes in personality	Anxiety, suspicion, pacing, insomnia, depression, agitation, and wandering	Crying, screaming, and groaning
Cognition	Confusion with memory loss Misplacing objects Forgetting names Disorientation	Difficulty recognizing family and friends Chronic loss of recent memory	Loss of speech Misidentifies or is unable to recognize familiar people
Assistance needed	Minimal and focused on daily to do lists to assist with memory loss	Hands on daily assistance with ADLs	24-h daily supervision

Adapted from the National Institute on Aging. National Institutes of Health; 2003. NIH publication 02-3782. Available at: http: www.alzheimers.org/unraveling/index.htm. Dementia Care Central; 2020

Table 5.2 Dementia grading systems

	Descriptions
Global Deterioration Scale/Reisberg Scale (GDS) [13]	Most commonly used scale Seven stages based on cognitive decline Stages 1–3—Predementia Stage 4–7—Dementia
Functional Assessment Staging Test (FAST) [14]	Seven stages (1–7) based on level of functioning and ability to perform ADLs. Stage 4—Mild dementia Stage 5—Moderate dementia Stage 6—Moderately severe dementia Stage 7—Severe dementia (associated with the highest morality rate)
Clinical Dementia Rating (CDR) [15]	Assess cognitive and functional abilities Used mainly as a research tool

Adapted from the Dementia Care Central;2020. Available at https://www.dementiacarecentral.com/aboutdementia/facts/stages/#

Biomarkers are either based on imaging, biofluid sampling, or electrophysiological tracings. Imaging-based biomarker measurements are derived from magnetic resonance imaging (MRI), which allows the identification of specific structural changes associated with the various types of dementia, as well as positron emission tomography (PET), single-photon emission computed tomography (SPECT), and infrared optical imaging. Biofluid-based markers are measured in biological samples such as blood, CSF, saliva, or urine. Electrophysiological biomarkers include electroencephalography (EEG), EEG-based event-related potentials (ERPs), and magnetoencephalography [12].

Neuroimaging methods have rapidly advanced in their utility for assessing brain changes associated with dementia. Different MRI techniques in high-field-strength MRI scanners provide insight into the structural integrity of brain tissue, enabling the identification of patterns of cerebral atrophy, white matter tract abnormalities, and degenerative and vascular-related tissue damage. Functional neuroimaging approaches (arterial spin labeling and blood-oxygen-level-dependent MRI and fluorodeoxyglucose [FDG]-PET), on the other hand, provide information about regional patterns of normal and abnormal blood flow and cellular metabolism, and PET with new radioligands can estimate neuropathological burden through radioligands' selective binding to pathological proteins in the brain [12].

5.2.1.4 Prevalence and Projected Life Expectancy

At present, 6.7 million older adults have dementia, and 73% are 75 years of age or older. By 2050, the number of people aged 65 and older with dementia is projected to increase to 13 million. AD accounts for 60–80% of the cases of dementia, and the remainder account for 10% each. The life expectancy from the onset of dementia is a function of age at the time of diagnosis and dementia type. For older adults who are diagnosed with AD, the average life expectancy is 10 years. However, individuals who are diagnosed at the age of 65 have been reported to survive 20–25 years after initial diagnosis, older adults in their 80s have been reported to survive up to 10 years after initial diagnosis, and older adults in their 90s have been reported to survive for 4–5 years after initial diagnosis. The average life expectancy is 7–13 years, 5–8 years, and 5 years from the time of diagnosis for frontal lobe dementia, Lewy body dementia, and vascular dementia, respectively [16].

5.2.1.5 Risk Factors

Factors associated with the development of dementia at all stages of life are classified into modifiable and nonmodifiable risk factors. The primary nonmodifiable risk factors are age, inherited genetic risk, and sex. Increasing age and being female are associated with an increased risk of developing dementia. Modifiable risk factors include (1) early life education quality; (2) midlife onset of hearing loss, hypertension or obesity, large volume alcohol consumption (>21 units weekly), and experiencing a TBI; (3) late-life onset of diabetes, smoking, or depression, social isolation, physical inactivity, and exposure to air pollution [17].

5.2.1.6 Implications in the General Older Adult Population

The consequences of dementia in the older adult patient population are complex and include the financial burden on the healthcare system, the economic, emotional, and psychological burden on the caregivers, and the clinical consequences in the older adult with dementia.

From a healthcare system standpoint, in 2023, Alzheimer's and other dementias cost the nation $345 billion in healthcare expenses. By 2050, these costs are expected to rise to nearly $1 trillion. It is also projected that by 2050, dementia will result in 1.6 million deaths [16, 18].

From a caregiver standpoint, over 11 million Americans provide unpaid care for older adults with dementia. In 2022, unpaid caregivers provided an estimated 18 billion hours of care valued at $339.5 billion. Additionally, studies have shown that levels of psychological distress and stress are significantly higher, and levels of self-efficacy, subjective well-being, and physical health are markedly lower in dementia caregivers than in other caregivers; these differences are even more prominent when compared with noncaregivers [19].

Dementia is associated with an increased risk of morbidity, mortality, and poor quality of life in older adults. Dementia has been shown to increase the risk of death in older patients hospitalized for acute illnesses. For those who survive, especially older patients with advanced dementia, mortality after discharge is even higher. Morrison et al. reported a 53% six-month mortality for patients with end-stage dementia and pneumonia versus 13% for cognitively intact patients. In the same cohort, there was a 55% six-month mortality in patients with hip fractures and end-stage dementia versus 12% mortality for cognitively intact patients. Of persons in the United States who die with dementia, approximately 16% die in the hospital. Between 2000 and 2009, 19% of nursing home patients in the United States who died with advanced dementia experienced burdensome hospitalization near the end of life. This consisted of hospitalizations within the last 3 days of life, increased intensive care unit (ICU) utilization, and extended ICU lengths of stay in the last 30–90 days of life [20].

5.2.2 Dementia in Older Adults Following TBI

5.2.2.1 Prevalence

TBI has been established as a modifiable and preventable risk factor for the development of dementia. While the majority of research has focused primarily on the impact of early-life TBI on late-life dementia development, a few recent studies have focused on the association between experiencing a TBI at an older age and subsequent development of dementia [17]. A Swedish cohort of over 3 million people aged 50 years or older found TBI increased 1-year dementia risk (OR 3.5, 95% CI 3.2–3.8), and risk remained elevated, albeit attenuated over 30 years (OR1.3,

95% CI1.1–1.4). A nationwide Danish cohort study of nearly 3 million people aged 50 or older, followed for a mean of 10 years, found increased dementia (HR 1.2, 95% CI 1.2–1.3) and Alzheimer's disease risk (HR 1.2, 95% CI 1.1–1.2). Dementia risk was highest in 6 months after TBI (HR 4.1, 3.8–4.3) [17, 21, 22].

5.2.2.2 Pathophysiology

The mechanisms by which TBI leads to the development of dementia remain not very well understood. One major theory is that TBI leads to the triggering of a progressive neurodegenerative cascade that is thought to be modulated through a combination of axonal and cytoskeletal alterations in the injured brain as well as neuroinflammation, impaired homeostasis, and disruption of the blood-brain barriers that cause subsequent neurodegeneration and brain atrophy [23]. Alternatively, TBI may accelerate an already ongoing neurodegenerative process via increased accumulation of Aβ and tau neurofibrillary tangles postinjury [24]. Brain regions with the most prominent atrophy after TBI—hippocampus, amygdala, precuneus, and parietal and frontal cortices—overlap closely with regions of predominant Aβ deposition, decreased glucose use, and progressive atrophy in AD, and a single, severe TBI is associated with widespread hyperphosphorylated tau pathology in humans and mouse models [25–27]. A final theory suggests that a static brain injury leads to white matter degeneration that reduces the cognitive reserve of the aging brain [28]. Autopsy and neuroimaging studies have shown that a single episode of moderate to severe TBI can cause marked cerebral atrophy 6 months postinjury that may progress for many years [29].

5.2.2.3 Risk Factors

The increased risk of neurodegeneration in older adults following TBI has been linked to age, TBI severity, ε4 allele of apolipoprotein E (*APOE*), sex, and repetitive injury resulting in multiple TBIs [21, 23, 30].

Females are at an increased risk of development of dementia overall, although males appear to experience an increased risk of developing dementia post-TBI. A history of TBI and inheritance of an APOE ε4 allele were associated with a tenfold increased risk of dementia. In contrast, APOE ε4 in the absence of TBI resulted in only a twofold increased risk of dementia [31]. Increasing age and injury severity have been associated with an increased risk of developing dementia in older adults when compared to their younger counterparts [28]. A Finnish study of 40,639 individuals presenting with mild, moderate, and severe TBI revealed that mild TBI increased the risk of dementia less than severe TBI, and multiple TBIs increased the risk further (OR 1.6, 95% CI 1.6–1.7 for single TBI; OR 2.1, 95% CI 2.0–2.2 for more severe TBI; and OR 2.8, 95%CI 2.5–3.2 for multiple TBI) [32]. A study of 178,779 older veterans with TBI with propensity-matched veterans without TBI found that dementia risk was associated with TBI severity (HR 2.4, 95% CI 2.1–2.7

for mild TBI without loss of consciousness; HR 2·5, 95% CI 2·3–2·8 for mild TBI with loss of consciousness; and HR 3.8, 95% CI 3.6–3.9 for moderate to severe TBI) [29]. A nested case-control study of early-onset clinically diagnosed AD within a cohort of older adults also found that the magnitude of association between TBI and the risk of dementia development was a function of lag time between exposure and outcome, with a 5-year lag time for brain injury associated with a higher risk of AD (OR 1.23; 95% CI 1.18–1.29). Additionally, a dose-response relationship was identified such that increasing number and severity of TBI was associated with an increased risk of developing AD [23].

5.2.2.4 Consequences of Dementia Post-TBI

No studies have been conducted to assess the long-term consequences of dementia following TBI in older adults. This is partly attributed to the challenge of distinguishing between the cognitive, physical, and emotional outcomes associated with moderate to severe TBI, which may mirror those observed in dementia. Additionally, the current post-TBI dementia research is fraught with challenges (discussed below). Consequently, pinpointing the specific cause of decline becomes more intricate. Despite this, the prevailing hypothesis suggests that TBI may accelerate the onset of dementia, leading to an earlier manifestation of clinical symptoms in the life course of older adults post-TBI. The expression of these symptoms is influenced by factors such as age of onset, dementia type, stage of dementia, and accompanying baseline comorbidities. Cognitive decline manifests as deficits in memory, attention, and executive function, while functional decline is evident in the inability to perform basic activities of daily living (ADLs) and instrumental ADLs. Emotional and behavioral changes include mood swings, irritability, aggression, and social withdrawal. Regarding the impact on the healthcare system, as well as the financial and caregiver burden associated with post-TBI dementia, it is conceivable that the burden is compounded by the sequelae of long-term TBI consequences and the effects of dementia.

5.2.2.5 Management

The etiology of dementia is complex, encompassing both modifiable and nonmodifiable risk factors. Notably, TBI falls within the modifiable risk factor category. Although the direct impact of reducing TBI incidence on diminishing dementia risk in older adults has not been extensively studied, and TBI alone does not singularly cause dementia, there is a conceivable connection between decreased TBI occurrences and a lower risk of dementia. Given that falls are the primary cause of TBI in older adults, injury prevention emerges as a pivotal method to mitigate this risk. Consequently, efforts aimed at preventing injuries, particularly those stemming from falls, may play a crucial role in reducing the overall risk of dementia in this patient demographic.

5.2.2.6 Limitations of the Current Literature

Research on whether TBI is linked to later cognitive decline has proven challenging. Neurodegenerative dementias have an insidious onset and may evolve over decades. Because of this, findings from nearly all longitudinal studies examining whether older individuals proceed to develop dementia following TBI are limited, as these studies cover relatively short durations (e.g., almost all within 10 years) of follow-up compared to the overall neurodegenerative process. That is, an underlying risk or neurodegenerative process may have already been present at the time of injury, making it difficult to determine the role of TBI in the development of dementia. Studies typically have relied on retrospective examination of TBI information gathered from older individuals presenting for dementia evaluations. One of the biggest concerns is that many of these studies have only collected a limited history of TBI details, mostly whether there is a lifetime history of TBI. As a result, the injury's severity and age are often not or poorly documented. Second, there is no universal standard for classifying TBI, making generalization of the findings and overall interpretation of the literature even more difficult. Second, although several epidemiological studies exist, all have relied on examining a population's medical records, which raises concerns about the accuracy of the diagnoses since most TBIs are diagnosed and documented outside of specialized dementia clinics [27].

5.3 Depression

5.3.1 Definition, Risk Factors, and Implications in the General Older Adult Population

It is important to recognize that depression is not part of the normal aging process. Therefore, to develop a better understanding of the implications of depression in older adults following a TBI, it is essential to understand the etiology, risk factors, and implications of depression in the general population of older adults.

5.3.1.1 Definition

According to the *Diagnostic and Statistical Manual of Mental Health Disorders*, fifth edition (DSM-5), a major depression episode (MDE) requires five or more symptoms to be present within a 2-week period. One of the symptoms should, at least, be either a depressed mood or anhedonia (loss of interest or pleasure). Secondary symptoms of an MDE are appetite or weight changes, sleep difficulties (insomnia or hypersomnia), psychomotor agitation or retardation, fatigue or loss of energy, diminished ability to think or concentrate, feelings of worthlessness or excessive guilt, and suicidality [11].

5.3.1.2 Prevalence

The Centers for Disease Control and Prevention estimates that the prevalence of major depression in older adults is distributed as follows: 1–5% for older adults living in the general community, 11.5% for older adults who are hospitalized, and 13.5% for older adults who require home care [33]. Despite the lower incidence of depression in older adults compared to their younger counterparts, the associated outcomes are markedly worse. This holds not only in comparisons to younger adults with depression but also when contrasting older adults with depression to nondepressed older adults. Older adults with depression experience markedly higher rates of suicide, accelerated cognitive decline, and increased risk of dementia, death, and disability compared to younger adults with depression and nondepressed older adults [34].

5.3.1.3 Risk Factors and Etiology

Unlike their younger counterparts, diagnosing depression in older adults is complicated by a host of factors, such as the ability to distinguish between normal grief and clinical depression, poor physical health, a higher incidence of medical illnesses that may overlap with somatic symptoms of depression, and older adults' tendency to mask or underreport symptoms of depression [35].

The etiology of depression in older adults is multifactorial and has been suggested to be secondary to genetic factors, biological risk factors, physical illness, age-related neurobiological changes in the brain (arteriosclerosis and inflammatory, endocrine, and immune changes that compromise the integrity of frontostriatal pathways, the amygdala, and the hippocampus), psychological risk factors, social risk factors, late-life onset of dementia, and the type of stressful events that occur with greater frequency later in life [35, 36]. Additionally, there are gender-based differences, with females experiencing an increased risk of depression. However, unlike the younger population, the risk factors leading to gender differences in MDE in older adults are poorly understood.

5.3.1.4 Symptomatology and Consequences

Older adults who experience MDE often report symptoms of sadness, irritability, loss of interest, fatigue, sleep disturbances, psychomotor retardation, poor concentration, and memory dysfunction [37]. Clinically, depression in older adults is characterized by poor or impaired cognitive performance, episodic memory, visuospatial skills, processing speed, and reduced executive function, affecting verbal fluency, response inhibition, working memory, and problem-solving [38].

Depression in older adults is often chronic and recurrent. Therefore, even with appropriate therapies, approximately 35–40% of depressed older adults experience a recurrence of MDE within 2 years, and 50% experience a recurrence within

4 years of initial diagnosis [38]. As a result, depression increases disability, contributes to poorer outcomes, and is associated with an increased suicide risk and mortality in older adults [38].

5.3.2 *Depression in Older Adults Following a TBI*

5.3.2.1 Prevalence

Depression is a well-documented consequence of TBI [34]. The prevalence of depression post-TBI is higher than the general population for all age groups, especially in older adults, and has been reported to range from 21% to 42% [2, 39]. Additionally, studies have shown that depression continues to be a significant long-term consequence of mild, moderate, and severe TBI for at least 5 years after injury [34].

5.3.2.2 Risk Factors

The presentation of post-TBI depression is a multifactorial response to several underlying factors associated with the TBI and patient characteristics. While depression may exacerbate poor outcomes experienced by older adults following TBI, much of the available literature has focused primarily on risk factors for the development of post-TBI depression in younger adults, and little is known about the risk factors of depression following TBI among older adults. To mitigate this knowledge gap, two Medicare beneficiary-based studies focused on elucidating the risk factors associated with the development of post-TBI depression in older adults. These studies identified the following factors to be associated with an increased risk of post-TBI depression in older adults:

1. Age: Older adults (≥75 years of age) experiencing an increased risk of developing post-TBI depression compared to both the <75 years of age general older adult and post-TBI older adult population.
2. Sex: Women are more likely to develop depression compared to men in the general population, but men experience a high risk of developing depression post-TBI.
3. Inpatient hospitalization postinjury.
4. Treatment by a critical care physician.
5. Increased hospital length of stay.
6. Discharge to a skilled nursing facility.
7. Preinjury depression

Preinjury depression has been identified as the strongest predictor for the development of post-TBI depression. Hospitalization, admission to a critical care unit, long hospital length of stay, and discharge to a skilled nursing facility serve as proxies for injury severity, which suggest that head injury severity may have a more considerable impact on the development and diagnosis of postinjury depression than previously understood.

5.3.2.3 Pathophysiology

Biological Theory

There are numerous biological mechanisms through which depression can develop after TBI [40]. Currently, most of the evidence, based on research in major depressive disorders in the general population and early work on depression post-TBI, supports the inflammatory theory as the primary culprit for the development of post-TBI depression. In older adults, inflammatory mediators, including cytokines and inflammatory cell surface markers, are elevated within the central nervous system in the setting of injury during TBI and have been linked to the development of depression [40]. Additionally, inflammation occurring early after TBI may contribute to long-term, chronic inflammation even several years postinjury, leading to a chronic risk for depression.

Acute cerebral spinal fluid (CSF) inflammation is associated with post-TBI depression in the first 6 months postinjury. Specifically, higher critical CSF levels of the cell surface markers such as sVCAM-1, sICAM-1, and sFAS are associated with a significant increase in risk for depression in the first year post-TBI. Higher acute CSF levels of IL-12 are associated with post-TBI depression at 12 months. These inflammatory markers are known to be associated with inflammation and white matter changes in the dorsolateral prefrontal cortex among older adults with depression [40].

IL-1, IL-6, and tumor necrosis factor-α (TNF-α) are the primary cytokines linked to major depressive disorder. Elevated levels of all three cytokines have been linked to increased fatigue, lack of interest, and dysregulation of the CSF. IL-6 and TNF-α are thought to potentiate mood changes through their effect on the brain-derived neurotrophic factor (BDNF). BDNF is a protein that plays a crucial role in the growth, development, and maintenance of neurons in the brain. It has been linked to mood regulation, and its levels are often found to be altered in conditions such as depression and anxiety. Higher levels of BDNF are generally associated with better mood and a lower risk of developing mood disorders [41].

Increases in these proinflammatory cytokines result in the dampening of the BDNF effect in the preclinical models, which may then contribute to depression. BDNF levels are decreased in untreated depression but increased with antidepressant treatment. Serum BDNF levels during the first week post-TBI may be indicative of the risk for the development of post-TBI depression by 1 year. Overall, BDNF represents a potential biomarker for risk stratification and a potential therapeutic target for both the prevention and treatment of post-TBI depression in older adults [40].

IL-7 is another cytokine that has shown preliminary associations for the development of post-TBI depression. IL-7 has been implicated in the lymphoproliferative processes needed to generate an adaptive immune response to injury. This connection to generating an effective immune response may explain why low levels of CSF IL-7 in the first week postinjury were inversely related to the development of post-TBI depression at 1-year postinjury. Patients with severe TBI and deficient IL-7 levels have been reported to have a 3.4-fold greater risk for depression. These results suggest that the disruption of T-cell homeostasis is potentially a pathway for the inflammation-mediated development of post-TBI depression [40].

It is postulated that the interplay of the different inflammatory cascade components, detailed above, leads to what is known as the "sickness behavior" model of post-TBI depression. The initial phase of the model is known as the acute adaptive phase. It is marked by a combination of symptoms such as malaise, increased pain sensitivity, feelings of personal failure and guilt, depressed mood, lack of interest or decreased pleasure, psychomotor slowing, poor concentration, fatigue, and weight loss. While sickness behavior is initially adaptive and helps the body respond to acute injury, it becomes maladaptive if it persists beyond 3–6 weeks. The transition to a maladaptive phase is driven by the inflammatory cytokines and protein dysregulation in the setting of a traumatic brain injury, leading to the development of the chronic progressive phase. It is during this phase that post-TBI depression develops, which presents with the typical MDE symptoms, including suicidal ideations. This maladaptive phase is marked by persistent neuroinflammation and neurodegeneration, potentiating the inflammatory cascade through a chronic positive feedback loop. This leads to the persistence of post-TBI depression.

Psychosocial Adjustment Theory

Psychosocial risk factors, such as preinjury medical comorbidities, social determinants of health, and changes in functional abilities and community-based participation after injury, are thought to contribute to an adjustment-based, rather than biologically based, mechanism for the development of depression after TBI. Risk factors that lead to maladaptive adjustment/coping include increasing age, race, functional disability postinjury, sleep disturbances and fatigue, unemployment, substance use history, preinjury psychiatric conditions, and impoverishment at the time of injury.

According to the adjustment model, the degree of change in overall abilities to perform ADLs and achieve goal-directed problem-solving behaviors postinjury is directly related to the development of postinjury depression. This is influenced by two main factors. The first is the degree of structural injury to the brain. The second is the individual's perceived disability due to the injury. For those who suffer more severe cognitive impairment due to a more severe TBI, the degree of brain injury post-TBI limits the ability to engage in goal-directed, problem-solving behavior, leading to a poor quality of life and subsequent onset of depression. However, individuals who experience less severe injuries but engage in wishful thinking or avoidance behaviors after injury have been shown to experience impairments associated with executive function, leading to the development of depressive symptoms when compared to individuals who do engage in goal-directed problem-solving postinjury. The latter group often have higher self-efficacy and less perceived stress and are less likely to experience postinjury depression [40].

Adjustment-based depression is characterized more by low self-worth, feelings of guilt, agitation, and suicidal endorsement when compared to depression resulting from maladaptive sickness behavior. This point has implications for treatment decisions, as particular subtypes of post-TBI depression, such as inflammatory-induced versus adjustment-based, may be more or less responsive to pharmacological versus behavioral interventions.

5.3.2.4 Screening Modalities

The Patient Health Questionnaire-9 (PHQ-9)

The PHQ-9 is a self-administered questionnaire that scores each of the 9 DSM-5 criteria as "0" (not at all) to "3" (nearly every day). Major depression is diagnosed if five or more of the nine depressive symptoms have been present at least "more than half the days" in the past 2 weeks, and one of the symptoms is depressed mood or anhedonia. The PHQ-9 score ranges from 0 to 27. It is one of the most used screening measures after TBI, as it has been shown to have good psychometric properties in adults with TBI. However, it is important to note that there are also some disagreements in the literature regarding whether screening instruments for individuals with TBI should assess somatic symptoms, as the PHQ-9 does. The primary concern is that somatic symptoms such as fatigue and sleep disturbances may be attributable to the TBI itself rather than to depression. Instead, assessment of the emotional symptoms of depression may have more diagnostic utility after TBI than somatic symptoms [42, 43].

The Traumatic Brain Injury Quality of Life (TBI-QOL)

TBI-QOL is an assessment instrument (item banks) developed with item response theory methods. It is psychometrically linked with two other modern measurement systems—the Patient Reported Outcome Measurement Information System (PROMIS) and the Neurological Quality of Life (Neuro-QOL) measurement system. TBI-QOL consists of 20 subdomains of 4 more significant health-related quality of life domains (physical health, emotional health, cognitive health, and social functioning). Each domain and subdomain have a specific number of items and a range of scores. A global quality of life score is a composite of all four main domains. For constructs like depression that are assessed with a TBI-QOL item bank and with a PROMIS or Neuro-QOL item bank, the TBI-QOL score can be transformed to the PROMIS metric, which permits comparison against the PROMIS normative sample, which mirrors the demographics of the 2000 US Census. Because of the uncertainty about assessing somatic symptoms of depression after TBI, the TBI-QOL Depression item bank only includes items related to the cognitive and emotional aspects of depression [42, 44].

Recent studies assessing the correlation between PHQ-9 and the TBI-QOL depression subdomain scores with the emotional health domain have shown equivalent performance between the two tools in the ability to screen for depressive symptoms post-TBI [42, 45].

5.3.2.5 Treatment Approaches

The current literature does not support any gold standard for post-TBI depression treatment. No studies on effective treatment have been able to differentiate potential subgroups that may have differential responses to treatment. Additionally, none of the available literature is specific to older adults. However, provided below is a summary of what is known so far regarding the effectiveness of pharmacological and cognitive-based behavioral therapy for the management of depression post-TBI in the general population.

Pharmacologic Interventions: Selective Serotonin Reuptake Inhibitors (SSRIs)

Treatment with SSRIs is currently the most evidence-based approach for the management of post-TBI depression. However, it is crucial to recognize that the presently available literature is based on retrospective and widely heterogeneous cohort studies. Currently, sertraline may be the most effective pharmacological treatment option. This is because fluoxetine, citalopram, paroxetine, and escitalopram have side effects that may compound TBI symptoms. Additionally, there is some growing evidence that these SSRIs may be less effective among individuals with TBI compared to those without TBI [46]. The decreased effectiveness is attributed to the high levels of inflammation postinjury. An animal model of postepilepsy depression, another condition associated with high levels of inflammation, revealed that treatment with fluoxetine alone yielded no antidepressant effects. Still, when paired with an IL-1Ra (an IL-1 antagonist), nearly all depressive indicators were eradicated [47]. Therefore, combination therapy with an SSRI and antiinflammatory agent for the management of post-TBI depression is a promising venue that warrants further exploration.

Nonpharmacological Interventions: Behavioral and Psychological Interventions

Nonpharmacological interventions often focus on the management of both the cognitive and emotional symptoms of depression. In the setting of post-TBI depression, there is a growing body of evidence that both cognitive behavioral and problem-solving treatments are effective for the management of the executive dysfunction (cognitive sequelae) of post-TBI depression [42]. Behavioral interventions focused on problem-solving have been shown to have the potential to improve mood, goal attainment, and executive function. Setting and attaining goals are critical components of the rehabilitation process and may be compromised by executive impairment after TBI. If goal attainment after TBI is compromised, it further contributes to the discrepancy between the appraisal of one's pre- and postinjury abilities, leading to depression. Therefore, interventions that promote goal attainment and address impairments in executive function may prevent or improve post-TBI depression. This is particularly true if the etiology of the depression is adjustment-based [40, 48, 49].

Rumination is a style of thinking that involves repetitive thinking about personal problems and has been found to impact the onset of MDE in adults [50]. Metacognition strategy training is a problem-solving-based approach that focuses on addressing self-regulatory executive dysfunction and a cognitive attention syndrome of rumination to treat depression. Metacognition therapy is an effective modality to treat depression in individuals with executive dysfunction that manifests as rumination. Given that rumination has been reported as a frequent symptom in individuals post-TBI, metacognition therapy could be effective in the management of post-TBI depression [50, 51].

Based on the available literature, current recommendations of the American Congress of Rehabilitation Brain Injury Interdisciplinary Special Interest Group (ACRM BI-ISG) are to (1) incorporate training in formal problem-solving strategies applied to everyday activities as the cornerstone for cognitive rehabilitation in depression post-TBI [40] and (2) provide treatment in a top-down manner, focusing on functional and activity-based goals rather than improving symptoms or bodily functions, to ensure generalization to nontrained goals and activities. While these strategy-training and self-management approaches are recommended guidelines for treating executive impairments after TBI, the extent to which they improve post-TBI depression and its impact on other outcomes remains unclear. Therefore, further work is needed to assess the efficacy of the interventions listed above and to determine which component of the interventions is effective for specific subgroups of individuals, such as older adults with post-TBI depression, to inform a future-tailored approach.

5.3.2.6 Implications of Post-TBI Depression in Older Adults

Older adults are among those with the highest risk and rates of TBI and, therefore, are at an increased risk of developing depression post-TBI. The incidence and prevalence of depression are notably higher in older adults residing in nursing homes, skilled nursing facilities, and other long-term care compared to those in the community post-TBI [10, 52].

Post-TBI depression is associated with poorer health-related quality of life, increased cognitive and physical disability, concurrent anxiety, frustration, and aggression, sleeping difficulties and fatigue, poor social functioning, worse concentration, worse executive functioning, reduced prefrontal grade matter volume, and higher levels of caregiver burden and distress [34]. In older adults, post-TBI depression is also associated with higher rates of suicide, cognitive decline, dementia, death, and disability. Dementia risk is exceptionally high in those with depression and cooccurring cognitive impairment or structural brain changes.

While the incidence and consequences of post-TBI depression in older adults are significant, this condition remains underdiagnosed in this patient population due to a lack of routine screening. Consequently, post-TBI depression is undertreated in older adults. To address this issue, we must enhance our understanding of the topic to guide screening and increase referral practices in the rehabilitation and postacute care settings. Moreover, given the limited attention in the literature to the psychological consequences of TBI in older adults, future research should prioritize gaining insights into the cognitive and somatic symptoms of depression following TBI in older adults, elucidating distinctions between younger adults and the general non-TBI population. Assessing the effectiveness of both pharmacological and nonpharmacological interventions is also crucial, particularly since depression in older adults is often undertreated. Notably, the conventional approaches involving SSRIs for the treatment of post-TBI depression warrant caution, as older adults with baseline comorbidities and polypharmacy face a heightened risk of adverse drug events related to SSRIs.

In summary, TBI frequently precipitates an increased susceptibility to dementia and depression among older adults, engendering substantial ramifications for their quality of life and functional outcomes. Nonetheless, there is currently a lack of understanding surrounding the mechanistic underpinnings, targeted interventions, and preventive strategies tailored to older adults, leading to a critical gap in current knowledge. Consequently, future investigations should prioritize elucidating the underlying mechanisms linking TBI to dementia and depression, exploring innovative interventions specific for older adults, and enhancing support systems for this vulnerable population to improve long-term outcomes and well-being.

Acknowledgments None.

Disclosure Statement The authors do not have any relevant financial conflict of interest or funding sources to disclose.

References

1. Administration on Aging. 2021 Profile of older Americans [Internet]. 2022 Nov p. 1–20. Available from: https://acl.gov/sites/default/files/Profile%20of%20OA/2021%20Profile%20of%20OA/2021ProfileOlderAmericans_508.pdf
2. Gardner RC, Dams-O'Connor K, Morrissey MR, Manley GT. Geriatric traumatic brain injury: epidemiology, outcomes, knowledge gaps, and future directions. J Neurotrauma. 2018;35:889–906.

3. Peterson AB, Zhou H, Thomas KE, Daugherty J. Surveillance report: traumatic brain injury-related hospitalizations and deaths by age group, sex, and mechanism of injury, United States 2016/2017 [Internet]. Centers for Disease Control and Prevention; 2016. Available from: https://www.cdc.gov/traumaticbraininjury/pdf/TBI-surveillance-report-2016-2017-508.pdf
4. Peterson AB, Thomas KE, Zhou H. Surveillance Report [Internet]. Centers for Disease Control and Prevention; 2018. Available from: https://www.cdc.gov/traumaticbraininjury/pdf/tbi-surveillance-report-2018-2019-508.pdf
5. Van Der Vlegel M, Mikolić A, Lee Hee Q, Kaplan ZLR, Retel Helmrich IRA, Van Veen E, et al. Health care utilization and outcomes in older adults after traumatic brain injury: a CENTER-TBI study. Injury. 2022 Aug;53:2774–82.
6. Livingston DH, Lavery RF, Mosenthal AC, Knudson MM, Lee S, Morabito D, et al. Recovery at one year following isolated traumatic brain injury: a Western Trauma Association prospective multicenter trial. J Trauma Inj Infect Crit Care. 2005;59:1298–304.
7. Mosenthal AC, Lavery RF, Addis M, Kaul S, Ross S, Marburger R, et al. Isolated Traumatic brain injury: age is an independent predictor of mortality and early outcome J Trauma Inj Infect Crit Care 2002; 52:907–911.
8. Mosenthal AC, Livingston DH, Lavery RF, Knudson MM, Lee S, Morabito D, et al. The effect of age on functional outcome in mild traumatic brain injury: 6-month report of a prospective multicenter trial. J trauma Inj infect Crit Care. 2004;56:1042–8.
9. Lilley EJ, Williams KJ, Schneider EB, Hammouda K, Salim A, Haider AH, et al. Intensity of treatment, end-of-life care, and mortality for older patients with severe traumatic brain injury. J Trauma Acute Care Surg. 2016;80:998–1004.
10. Albrecht JS, Kiptanui Z, Tsang Y, Khokhar B, Liu X, Simoni-Wastila L, et al. Depression among older adults after traumatic brain injury: a national analysis. Am J Geriatr Psychiatry. 2015;23:607–14.
11. American Psychiatric Association, editor. Diagnostic and statistical manual of mental disorders: DSM-5-TR. Fifth edition, text revision. Washington, DC: American Psychiatric Association Publishing; 2022. 1 p.
12. Koenig AM, Nobuhara CK, Williams VJ, Arnold SE. Biomarkers in Alzheimer's, frontotemporal, Lewy body, and vascular dementias. Focus. 2018;16:164–72.
13. Reisberg B, Ferris SH, de Leon MJ, Crook T. The Global Deterioration Scale for assessment of primary degenerative dementia. Am J Psychiatry. 1982;139:1136–9.
14. Sclan SG, Reisberg B. Functional assessment staging (FAST) in Alzheimer's disease: reliability, validity, and ordinality. Int Psychogeriatr. 1992;4(Suppl 1):55–69.
15. O'Bryant SE, Lacritz LH, Hall J, Waring SC, Chan W, Khodr ZG, et al. Validation of the new interpretive guidelines for the clinical dementia rating scale sum of boxes score in the national Alzheimer's coordinating center database. Arch Neurol. 2010;67:746–9.
16. Alzheimer's Association. Alzheimer's Disease and Dementia. [cited 2023 Nov 29]. Alzheimer's disease facts and figures. Available from: https://www.alz.org/alzheimers-dementia/facts-figures
17. Livingston G, Huntley J, Sommerlad A, Ames D, Ballard C, Banerjee S, et al. Dementia prevention, intervention, and care: 2020 report of the Lancet Commission. Lancet. 2020;396:413–46.
18. Ghneim M, Diaz JJ. Dementia and the critically ill older adult. Crit Care Clin. 2021;37:191–203.
19. Pinquart M, Sörensen S. Differences between caregivers and noncaregivers in psychological health and physical health: a meta-analysis. Psychol Aging. 2003;18:250–67.
20. Morrison RS. Survival in end-stage dementia following acute illness. JAMA. 2000;284:47.
21. Nordström A, Nordström P. Traumatic brain injury and the risk of dementia diagnosis: a nationwide cohort study. Menon D, editor. PLOS Med. 2018;15(1):e1002496.
22. Fann JR, Ribe AR, Pedersen HS, Fenger-Grøn M, Christensen J, Benros ME, et al. Long-term risk of dementia among people with traumatic brain injury in Denmark: a population-based observational cohort study. Lancet Psychiatry. 2018;5:424–31.
23. Tolppanen A, Taipale H, Hartikainen S. Head or brain injuries and Alzheimer's disease: a nested case-control register study. Alzheimers Dement. 2017;13:1371–9.
24. Johnson VE, Stewart W, Smith DH. Widespread τ and amyloid-β pathology many years after a single traumatic brain injury in humans. Brain Pathol Zurich Switz. 2012;22:142–9.

25. Cao J, Gaamouch FE, Meabon JS, Meeker KD, Zhu L, Zhong MB, et al. ApoE4-associated phospholipid dysregulation contributes to development of Tau hyper-phosphorylation after traumatic brain injury. Sci Rep. 2017;7:11372.
26. Zanier ER, Bertani I, Sammali E, Pischiutta F, Chiaravalloti MA, Vegliante G, et al. Induction of a transmissible tau pathology by traumatic brain injury. Brain J Neurol. 2018;141:2685–99.
27. LoBue C, Cullum CM, Didehbani N, Yeatman K, Jones B, Kraut MA, et al. Neurodegenerative dementias after traumatic brain injury. J Neuropsychiatry Clin Neurosci. 2018;30:7–13.
28. Gardner RC, Burke JF, Nettiksimmons J, Kaup A, Barnes DE, Yaffe K. Dementia risk after traumatic brain injury vs nonbrain trauma: the role of age and severity. JAMA Neurol. 2014;71:1490.
29. Barnes DE, Byers AL, Gardner RC, Seal KH, Boscardin WJ, Yaffe K. Association of mild traumatic brain injury with and without Loss of consciousness with dementia in US military veterans. JAMA Neurol. 2018;(75):1055.
30. Sundström A, Nilsson LG, Cruts M, Adolfsson R, Van Broeckhoven C, Nyberg L. Increased risk of dementia following mild head injury for carriers but not for non-carriers of the APOE ε4 allele. Int Psychogeriatr. 2007;19:159.
31. Plassman BL, Havlik RJ, Steffens DC, Helms MJ, Newman TN, Drosdick D, et al. Documented head injury in early adulthood and risk of Alzheimer's disease and other dementias. Neurology. 2000;(55):1158–66.
32. Raj R, Kaprio J, Korja M, Mikkonen ED, Jousilahti P, Siironen J. Risk of hospitalization with neurodegenerative disease after moderate-to-severe traumatic brain injury in the working-age population: a retrospective cohort study using the Finnish national health registries. Brohi K, editor. PLOS Med. 2017;14: e1002316.
33. Centers for Disease Control and Prevention. Depression is not a normal part of growing older | Alzheimer's disease and healthy aging | CDC [Internet]. 2022 [cited 2023 Nov 29]. Available from: https://www.cdc.gov/aging/depression/index.html
34. Passler JS, Sander AM, Temkin NR, Barber J, Gardner RC, Manley GT. Depression in older adults 12 months after traumatic brain injury: a TRACK-TBI study. Arch Phys Med Rehabil. 2022;103:83–9.
35. Alexopoulos GS. Depression in the elderly. Lancet Lond Engl. 2005;365(9475):1961–70.
36. Fiske A, O'Riley AA. Toward an understanding of late life suicidal behavior: the role of lifespan developmental theory. Aging Ment Health. 2016;20:123–30.
37. Seel RT, Kreutzer JS. Depression assessment after traumatic brain injury: an empirically based classification method11No commercial party having a direct financial interest in the results of the research supporting this article has or will confer a benefit upon the author(s) or upon any organization with which the author(s) is/are associated. Arch Phys Med Rehabil. 2003;84:1621–8.
38. Szymkowicz SM, Gerlach AR, Homiack D, Taylor WD. Biological factors influencing depression in later life: role of aging processes and treatment implications. Transl Psychiatry. 2023;13:160.
39. Albrecht JS, Barbour L, Abariga SA, Rao V, Perfetto EM. Risk of depression after traumatic brain injury in a large national sample. J Neurotrauma. 2019 Jan;15(36):300–7.
40. Juengst SB, Kumar RG, Wagner AK. A narrative literature review of depression following traumatic brain injury: prevalence, impact, and management challenges. Psychol Res Behav Manag. 2017;10:175–86.
41. Yang T, Nie Z, Shu H, Kuang Y, Chen X, Cheng J, et al. The role of BDNF on neural plasticity in depression. Front Cell Neurosci. 2020;14:82.
42. Cohen ML, Holdnack JA, Kisala PA, Tulsky DS. A comparison of PHQ-9 and TBI-QOL depression measures among individuals with traumatic brain injury. Rehabil Psychol. 2018;63:365–71.
43. Kroenke K, Spitzer RL, Williams JBW. The PHQ-9: validity of a brief depression severity measure. J Gen Intern Med. 2001;16:606–13.
44. Tulsky DS, Kisala PA, Victorson D, Carlozzi N, Bushnik T, Sherer M, et al. TBI-QOL: development and calibration of item banks to measure patient reported outcomes following traumatic brain injury. J Head Trauma Rehabil. 2016;31:40–51.

45. Boulton AJ, Tyner CE, Choi SW, Sander AM, Heinemann AW, Bushnik T, et al. Linking the GAD-7 and PHQ-9 to the TBI-QOL anxiety and depression item banks. J Head Trauma Rehabil. 2019;34:353–63.
46. Neurobehavioral Guidelines Working Group, Warden DL, Gordon B, McAllister TW, Silver JM, Barth JT, et al. Guidelines for the pharmacologic treatment of neurobehavioral sequelae of traumatic brain injury. J Neurotrauma. 2006;23:1468–501.
47. Pineda EA, Hensler JG, Sankar R, Shin D, Burke TF, Mazarati AM. Interleukin-1β causes fluoxetine resistance in an animal model of epilepsy-associated depression. Neurother J Am Soc Exp Neurother. 2012;9:477–85.
48. Hudak AM, Hynan LS, Harper CR, Diaz-Arrastia R. Association of depressive symptoms with functional outcome after traumatic brain injury. J Head Trauma Rehabil. 2012;27:87–98.
49. Arundine A, Bradbury CL, Dupuis K, Dawson DR, Ruttan LA, Green REA. Cognitive behavior therapy after acquired brain injury: maintenance of therapeutic benefits at 6 months post-treatment. J Head Trauma Rehabil. 2012;27:104–12.
50. Pedersen H, Grønnæss I, Bendixen M, Hagen R, Kennair LEO. Metacognitions and brooding predict depressive symptoms in a community adolescent sample. BMC Psychiatry. 2022;22:157.
51. Ownsworth T, Gooding K, Beadle E. Self-focused processing after severe traumatic brain injury: relationship to neurocognitive functioning and mood symptoms. Br J Clin Psychol. 2019;58:35–50.
52. Menzel JC. Depression in the elderly after traumatic brain injury: a systematic review. Brain Inj. 2008;22:375–80.

Chapter 6
The Effects of Early Life History of TBI on the Progression of Normal Brain Aging with Implications for Increased Dementia Risk

Makenna B. McGill and David M. Schnyer

Abstract There is increasing interest in the risk conferred on neurological health by a traumatic brain injury (TBI) and how that influences the lifespan trajectory of brain aging. This chapter explores the importance of this issue, population, and methodological considerations, including injury documentation and outcome assessment. We then explore some of the findings in the neuroimaging and neuropsychological research literature examining the interaction between an earlier life history of TBI and the normal aging process. Finally, we consider the limitations of our current knowledge and where the field needs to go in the future.

Keywords Traumatic brain injury · Aging · Neuroimaging · Dementia risk

Abbreviations

ADNI	Alzheimer's Disease Neuroimaging Initiative
AOC	Alterations of consciousness
CCN	Cognitive control network
CDR	Clinical Dementia Rating
CSF	Cerebral spinal fluid
CT	Computed tomography
DMN	Default mode network
DOD	Department of Defense
DWI	Diffusion weighted image
FA	Fractional anisotropy
FDG PET	Fluorodeoxyglucose PET
fMRI	Functional magnetic resonance imaging

M. B. McGill (✉) · D. M. Schnyer
Department of Psychology, The University of Texas at Austin, Austin, TX, USA
e-mail: makennamcgill@utexas.edu; schnyer@utexas.edu

L. J. Noble-Haeusslein, D. M. Schnyer (eds.), *Traumatic Brain Injuries*, Advances in Neurobiology 42, https://doi.org/10.1007/978-3-031-69832-3_6

GM	Gray matter
GOSE	Glasgow Outcome Scale-Extended
LOC	Loss of consciousness
MD	Mean diffusivity
MRI	Magnetic resonance imaging
NIH	National Institutes of Health
PET	Positron emission tomography
PTA	Posttraumatic amnesia
PTSD	Posttraumatic stress disorder
RS fMRI	Resting-state fMRI
SN	Salience network
TBIs	Traumatic brain injuries
VA	Veterans Administration
WM	White matter
WMHs	White matter hyperintensities

6.1 Human Aging

The elderly population in the United States is rapidly growing and will place increased demands on the healthcare system [1]. The number of US adults aged 65 and older is expected to reach 89 million by 2050, which is more than double the number of elderly people in 2010 [2]. As the aging population increases, the prevalence of chronic diseases is projected to increase [1, 3]. Importantly, the number of people living with dementia will also rise [4, 5]. Considerable research has been focused on understanding how health conditions across the lifespan may influence normal aging and present in older adults to manage disease burden in this growing population.

6.2 Traumatic Brain Injury and Human Aging

One such health condition is traumatic brain injury (TBI). The lifetime prevalence of TBI in the general adult population was found to be 12% by a meta-analysis of 15 studies from various developed countries [6]. Data analyzed by the Centers for Disease Control and Prevention (CDC) revealed that in the United States, there were a total of 2.8 million TBI-related emergency department visits, hospitalizations, and deaths in 2013 [7]. The effects of TBI are not only acute but can have long-lasting consequences on health and functioning. Global data from 2016 suggests that 759 per 100,000 people are living with a disability from experiencing a TBI [8]. TBI was historically viewed as a singular event with a finite recovery; however, as our understanding is expanding, many have begun to conceptualize TBI as a chronic disease [9].

Elderly adults are at increased risk for experiencing an TBI. In fact, the highest incidence of TBI-related visits to hospital emergency rooms and admission to the hospital is in elderly adults [7]. Once individuals reach the age of 65, the majority of TBI incidents are fall related, and this proportion increases to nearly 80% by age 85+ [10]. Given this increased incidence and the multitude of detrimental health effects of a late life TBI, the topic has garnered increased interest in the health research community [11]. While not discounting the importance of studying the effects of late life TBI on the aging adult, a much less understood and researched domain is the interaction between earlier life TBI and an individual's aging health trajectory. It is this which is the focus of the current chapter.

Investigating the effects of earlier life TBI on late-life brain health is particularly important given that TBI has been identified as a risk factor for dementia [12]. A meta-analysis of 21 studies found that individuals were 1.96 times more likely to be diagnosed with dementia if they had previously sustained a mild TBI [13]. Sustaining a moderate-severe TBI and/or multiple TBIs has been associated with even greater risk of developing dementia in both civilian and veteran cohorts [14–16]. However, it is important to note that many studies on this topic are retrospective, and a causal link between TBI and dementia has not been clearly delineated. Furthermore, some studies present inconclusive or even negative findings about TBI and dementia risk [17], likely due to methodological difficulties in capturing injuries that vary in severity, frequency, and mechanism [18]. Further research is needed to better understand the mechanism(s) underlying this complex relationship between TBI and dementia.

Given the heterogeneity of TBI and the methodological challenges this presents, it should be distinguished that the focus of this chapter will be on individuals who experienced a closed head TBI, resulting from an external force strong enough to move the brain within the skull. While penetrating injuries that pierce the skull and dura also fall under the TBI umbrella, they are less prevalent [19] and typically result in damage to specific regions of the brain, and the pathophysiology can differ considerably from closed head injury [20].

From considerable research, we know that the primary injury to the brain in a closed head TBI initiates physiological and biochemical processes that then can lead to multiple secondary injury mechanisms, many of which may persist for years [21]. No singular mechanism is responsible for secondary injury; rather, contributing factors include glutamate excitotoxicity, mitochondrial dysfunction, oxidative stress, neuroinflammation, and axonal degeneration, among others. This secondary injury can have subsequent chronic effects on the central nervous system as well as on other body systems (e.g., cardiovascular, endocrine) [22].

Many of the mechanisms by which secondary injury after TBI negatively affects the brain have also been implicated in normal brain aging [23]. The brain reserve model offers a framework for understanding how these processes may converge in older adults with a history of TBI. The model posits that there is a threshold for the amount of brain damage that can be sustained before an individual's functioning is affected [24]. TBI may reduce the amount of brain reserve (e.g., synaptic count, white matter integrity) at the time of injury, such that individuals with a history of TBI enter older age with diminished reserve. It is also possible that early life TBI initiates neurodegenerative cascades that not only reduce reserve at the time of injury but result in an interaction with naturally occurring aging processes, thereby

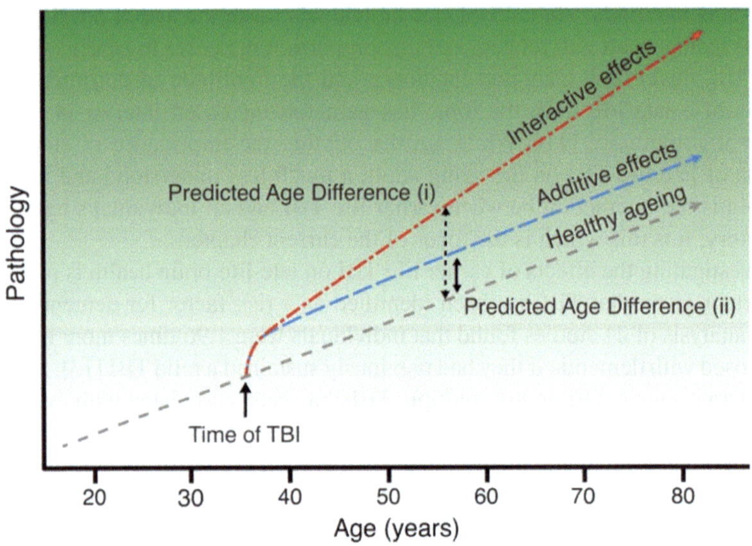

Fig. 6.1 Three possible aging trajectories following a TBI. (Reprinted from Cole et al. [25])

accelerating brain and cognitive aging. Figure 6.1, reprinted from Cole and colleagues [25], depicts three possible aging trajectories following a TBI: (1) unaltered, healthy aging, (2) additive effects of TBI history and normal aging, and (3) interactive effects of TBI history and normal aging, resulting in an acceleration of brain aging. When considering how a history of TBI injury interacts with the health of aging populations, it is essential to consider the systemic, cumulative nature of TBI-related damage throughout the lifespan and the most effective methods to understand that.

6.3 Causes of TBI

Estimates of lifetime prevalence of some form of head trauma have varied widely from nearly 30% [26] to as high as 42% [27]. While most of these injuries will be relatively mild, more severe injury that results in hospitalization is a significant health problem. In 2020, according to the Center for Disease Control and Prevention, there were 214K TBI-related hospitalizations, and this does not include TBIs that were treated without hospital admission in emergency or primary care settings [28]. The injuries that do not result in hospital admission are typically milder TBI or concussions and given the estimates cited above are likely of much greater prevalence. For instance, in a web panel survey of 6427 US adults, 27% reported a lifetime concussion [29], and among those, only about half sought medical evaluation/ care. This statistic indicates that if anything, concussion incidence is probably underestimated. Some common causes of TBI include motor vehicle accidents, falls, recreational accidents, violence, work-related accidents, combat-related injuries, and both contact and noncontact sports-related injuries. While a wide variety of accidents can result in a traumatic brain injury, the heterogenous nature of the injury type that results from these different accident causes has not been well studied.

One source of injuries is those associated with combat, which almost exclusively affect the military and veteran populations. While service members can experience similar sources of TBI as civilians, such as motor vehicle accidents and falls, they can also experience a TBI from blast exposure—a uniquely military-related cause of injury. Reporting from the Defense and Veterans Brain Injury Center (DVBIC) indicated that over 380K service members sustained a TBI between 2001 and 2018 and more than 30% of those were from a blast event [30]. TBIs resulting from blast exposure have some unique characteristics that should be considered when looking at long-term outcomes associated with aging (Bryden, Tilghman, & Hinds, 2019).

Sport-related TBIs appear to be one of the most common causes of mild TBI or concussion. Average annual incidence estimate of concussion is 31.5 per 100K, with a steady rise from 19.7 in 1999 to 45.6 in 2011 [31]. One question is whether this increase represents an actual increase in incidence as opposed to increased awareness of sports-related concussions and TBI such that more are referred for evaluation by healthcare professionals. In 2019, about 15% of all high school students reported a sports-related concussion incident experienced in the previous 12 months [28]. Sports-related head trauma will often not result in seeking of medical care, so the long-term impact of these injuries has been significantly understudied.

One area of sports-related TBI that has gained increased scrutiny is understanding the long-term consequences of repetitive head trauma exposure such as seen in professional contact sport athletes. First noticed in professional boxers, there has been increased scientific attention toward chronic traumatic encephalopathy (CTE)—a condition that appears to be a potential long-term outcome of repetitive head impacts (RHI). Given that evidence points to CTE as being highly related to the effect of RHI on the brain over time, CTE is an important consideration in studying the interaction between TBI and aging [32]. However, since evidence indicates that CTE is not present in a higher-than-normal population in individuals with a single TBI exposure [33], it seems to be a unique and very complex subcategory of the interaction between TBI and aging. While the RHI-CTE relationship lies at one end of a continuum, the issues of whether a previous TBI increases vulnerability to future TBI [17] and whether multiple TBIs lead to worse outcomes [34] are also very important issues to consider when examining influences on the aging process. Overall, a life history of multiple TBIs complicates our interest in how TBI interacts with the aging process and will need to be incorporated into any study.

To summarize, the prevalence of TBI across the lifespan is high. The long-term outcome of these injuries has been a topic of recent investigation and is a significant public health issue. What is less well understood is how a TBI during earlier life interacts with the aging process. Exploring what is currently known about this interaction and where research needs to be focused going forward is the aim of the current chapter.

6.4 Measurement Difficulties

To better understand the long-term interaction between a history of TBI and the aging process, there are a broad range of methodological considerations. Given the unique case of understanding how a lifetime history of TBI interacts with the aging process, there are two important windows for which reliable data is needed (see Fig. 6.2). The first data collection period is the time of injury and acute recovery phase. Obtaining reliable measures from this time is complicated by the fact that injuries will often have occurred decades before a person is enrolled in a study of its effects on aging. Further, unless this time window is part of a well-designed longitudinal study, the data available is likely to be unstandardized, making it difficult to integrate across individuals. While information about the injury severity is critical, the length and nature of the acute recovery period is also important [35]. Therefore, even the scope of this initial window can differ significantly from individual to individual. The second time window for data collection would be when the previously injured individual is elderly. This is likely to be when they are enrolled in a study looking at the interaction between early life TBI and aging, so there is a greater ability to collect current status data in a standardized manner.

To improve the standardization of data collected in TBI studies and clinical trials, the Federal Interagency Traumatic Brain Injury Research (FITBIR) Informatics System was created as a collaboration between NIH and the DOD. FITBIR is a

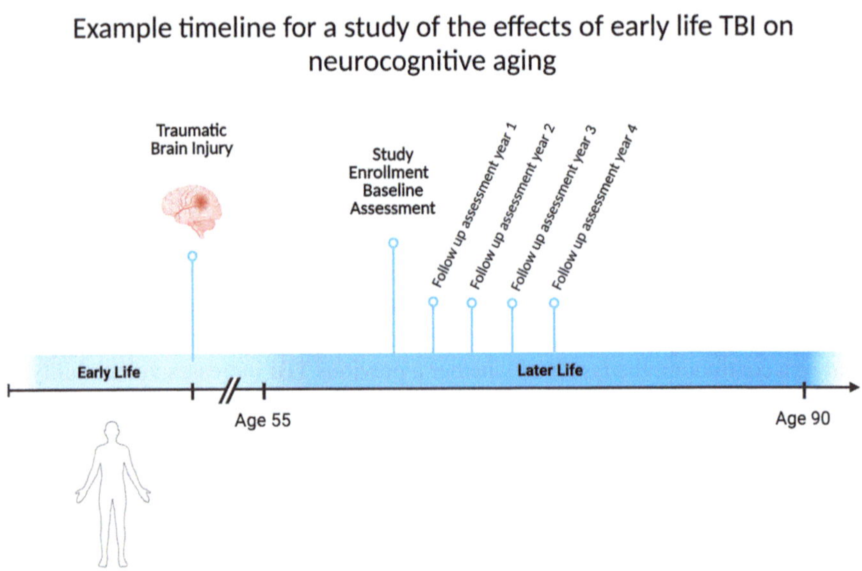

Fig. 6.2 Model study design to assess the effects of early life traumatic brain injury on the aging process. Typical injury time window would be earlier than 55 years old. Targeting enrollment during the early aging time period after 60 will then allow for periodic longitudinal assessments as the individual passes through the ages most at risk for dementia onset. (Created with BioRender.com)

multimodal informatics platform for TBI-relevant data, including medical imaging, clinical assessment, behavioral history, and assessments of functional, cognitive, and emotional status. The major motivation behind FITBIR was the ability to share data across the entire TBI research field and to facilitate multisite collaborations and clinical trials. Given the difficulty in aligning the earlier life TBI and aging data collection time windows, the ideal design to examine the interaction between a lifetime history of TBI and the aging process would be a well-designed longitudinal study leveraging the FITBIR data standards. Some of those may be underway now, including the TRACK and CENTRE TBI networks, but will be years from being able to answer important questions about the influence of TBI history on aging.

6.4.1 Time of Injury and Acute Phase Assessment

One of the main considerations is the source and accuracy of data about the injury itself, since this is very often remote from studying its influence on the aging process. Paramount is how to determine the severity of injury, particularly if there is not access to objective medical records and/or detailed information surrounding the recovery and outcome of injury. The issue of severity and access to objective medical records are intimately linked. In many cases, information about a remote TBI will be collected through self-report of the individual. Not surprising, self-reported history of TBI suffers from accuracy problems due to changes in memory over time as well as changes in our knowledge and understanding of TBI [36]. Whether someone is admitted to a hospital or whether imaging such as a CT scan is obtained after injury can be dictated by residence and socioeconomic factors, rather than severity of injury per se [37, 38]. Ideally, a standardized measure of severity is collected at the time of injury, such as the Glasgow Coma Scale [39] or the Injury Severity Score [40]. However, when not available, self-report and/or available medical records could be queried for information about length of changes in consciousness associated with the injury and/or whether the injury resulted in, and length of, hospitalization. The availability of hospital records should provide a more accurate source of information than self-report. While many studies reply on a simple report of previous injury, there exists a systematic instrument designed to capture self-reported TBI information: The Ohio State University TBI Identification method (OSU TBI-ID). This is standardized, structured interview designed to elicit lifetime history of TBI [41]. The survey consists of five questions designed to elicit recall of information about head injury, including questions about loss or alterations of consciousness and the age(s) at which the injury occurred. Finally, one issue not captured by the acute injury data is information about the individuals' course of recovery. A recent report from the TRACK-TBI study revealed that while recovery from TBI can take up to a year to stabilize, a subset of both mild and moderate-severe injury patients remains impaired 5 years after injury [35]. This study focused on three standardized outcome measures: the Glasgow Outcome Scale-Extended (GOSE), which estimates how injuries have affected daily function [42], the

Rivermead Post Concussion Symptoms Questionnaire (RPQ), which assesses the presence and severity of 16 symptoms that are new or worsened since the injury [43], and the Quality of Life After Brain Injury-Overall Scale (QOLIBRI-OS), which assesses cognitive, emotional, and physical functioning to estimate health-related quality of life [44]. These three instruments would add important variables for understanding the outcome of the recovery period. A comprehensive picture of both injury severity and length and dimensions of recovery would be important to understand how that TBI experience interacts with aging.

6.4.2 Aging Adult Assessment and Experimental Measures: The ADNI Model

Assessing elderly adults to examine the interaction of early life TBI with the aging process would likely follow best practices for any study of brain and cognitive aging. Some of the common decision points for a study of this type begin with questions of what constitutes "elderly" and what population of elderly to recruit from. One of the oldest longitudinal studies of aging is the Baltimore Longitudinal Study on Aging (BLSA), a prospective longitudinal study of psychological and physical aging that was initiated in 1958 [45]. While it is the longest longitudinal study of aging, the BLSA does not collect data about earlier life TBI but does ask about falls with injury to the head "within the last 12 months"—so in that regard, it is tracking new injuries associated with the aging process. Another large-scale longitudinal study, the Alzheimer's Disease Neuroimaging Initiative (ADNI), enrolls participants between the ages of 55 and 90 who are recruited at 57 sites in the United States and Canada [46]. While the study is focused on using neuroimaging to predict diseases of aging, such as Alzheimer's disease and other dementias, the study collects a broad range of data including blood, MRI, PET, neuropsychological testing, and medical history and current status. Although the ADNI study collected data from the broad population, a more specialized offshoot funded by the Department of Defense (DOD-ADNI) is focused on enrolling aging veterans with and without an earlier life history of TBI [47]. The latter is the first study of its type attempting to look at the interaction between earlier, service-related TBI and how these veterans age. The TBI inclusion criteria included: moderate-severe nonpenetrating TBI, defined as LOC (any length), PTA > 24 h, or AOC > 24 h. The control group excluded mild/moderate-severe TBI, any history of head trauma associated with injury-onset cognitive complaints, and/or LOC > 5 min. While this criterion is rigorous, it was primarily determined by self-report and therefore subject to potential reporting accuracy and biases. Finally, the sample is male only and restricted to the population of veterans that could be recruited through the VA healthcare system. The ADNI longitudinal studies are among the most promising to examine the question of how early life history of TBI interacts with the aging process and whether and how it increases risk for dementia because the outcome measures being collected span multiple brain imaging techniques and cognitive and psychosocial assessments.

The ADNI study is designed as a longitudinal study and is focused on the issue of brain aging and vulnerability to dementia. From that perspective, it provides a model of the type of data that should be collected to support a study of the interaction between earlier life TBI and aging with a particular focus on increased risk for dementia. Baseline assessments include a comprehensive battery of neuropsychological tests to assess memory and learning, executive functioning, neuropsychiatric index, and mood status [48]. In addition, standardized protocols for structural and functional MR imaging of the brain, FDG PET imaging, and biospecimens including blood (for genetic and proteomic analysis) and cerebral spinal fluid (Aβ-42 levels) are collected. The neuroimaging, blood, CSF, and abbreviated neuropsychological testing is collected at 12, 24, and 36 months. As the study has continued longitudinally, the neuroimaging protocols have been updated, and the fourth phase of the study (ADNI4) began in the summer of 2023 [49].

6.5 Neuroimaging Measures

Likely due to the methodological challenges associated with studying the effects of TBI history on late-life brain health, there is a dearth of work published on this topic, particularly using advanced neuroimaging methods. While the findings on the acute and postacute effects of TBI on brain structure and function are more robust, less is known about the interaction between an earlier life history of TBI and brain aging. It is on this relationship that we will focus our discussion of neuroimaging findings.

6.5.1 T1- and T2-Weighted MRI

Noninvasive neuroimaging techniques provide measures of brain structure and function that can reveal insights into the long-term effects of TBI and its interaction with the aging process. The most common methods are those carried out with magnetic resonance imaging (MRI). One technique is the T1-weighted (T1w) MRI that allows for the study of changes in brain structure. The T1w image is optimal for separating gray and white matter as well as cerebral spinal fluid in a manner that allows for identification and quantitative measurement of brain structures, such as volume and cortical thickness. Within the aging literature, measures of brain volumes (e.g., whole-brain, gray matter [GM], white matter [WM]) and cortical thickness (the width of GM, measured as the distance between the gray-white matter boundary and the outer cortical surface) are commonly used to detect age-related brain changes. Previous findings suggest that with increasing age, there is a reduction in global and regional brain volumes [50] and cortical thinning, particularly in frontal regions [51]. T2-weighted (T2w) MRI is optimal for examining macrolevel white matter characteristics and can also be employed to detect age-related damage. WM hyperintensities (WMHs) are seen on both T2w and fluid-attenuated inversion recovery (FLAIR) sequences and indicate injury to cerebral WM and have been associated with cognitive aging [52, 53]. Both normal aging processes and TBI involve the deterioration of GM and WM [21, 54, 55]; however, further clarity is needed on how these brain changes may be distinct and/or interact.

The secondary injury cascades that are initiated following a TBI can result in cerebral atrophy that persists well into the chronic stage of injury [56]. A review paper suggests that while findings regarding atrophy following mild TBI remain inconclusive, both generalized and focal brain atrophies are consistently found in studies of moderate-to-severe TBI. Key features of cerebral atrophy following a TBI include widening of sulci, enlargement of ventricles, shrinking of the hippocampus, and cortical thinning [57]. Longitudinal studies show greater rates of global, GM, and WM atrophy in middle-aged adults with a history of TBI compared to those without [58–61], with some work suggesting that WM atrophy is most pronounced [59, 60].

Previous work has also found that relative to those without TBI, adults with TBI display widespread cortical thinning during the postacute (typically defined as 2–6 months postinjury) and chronic (typically defined as more than 6 months postinjury) phases of injury [62–65]. Mazaharally and colleagues examined cortical thinning at four time points, with the final time point occurring 3–7 years postinjury. Compared to adults without TBI, those who sustained a TBI 5 months prior displayed cortical thinning primarily in frontal regions. Throughout the chronic phase of injury, this group showed widespread cortical thinning in parietal regions and an increase in cortical thickness in frontal regions. The authors suggest that while early cortical thinning may be attributable to focal brain injury damage, chronic cortical thinning may be indicative of diffuse injury (i.e., transneuronal degeneration) [64]. Collectively, these studies identified cortical thinning in parietal and temporal regions, which is a different pattern than that seen with age-related cortical thinning [51]. It is possible that patterns of age-related cortical thinning could be exacerbated in older adults with a history of TBI; however, it remains difficult to examine this in samples largely consisting of middle-aged adults.

Furthermore, while the aforementioned studies examined cerebral atrophy during the chronic phase of injury, the study periods lasted for a maximum of 7 years postinjury, with most studies occurring during the first 1–2 years of injury. Given that the chronic TBI can persist for decades, the longitudinal designs employed by these studies are limited in their ability to inform our understanding of the longer-term effects of TBI on age-related changes in brain structure. It is also important to note that while some studies have identified an interaction between TBI and age of injury on cerebral atrophy, such that older age at injury is associated with greater atrophy [65, 66], this is a different research question than one focused on how an earlier life TBI interacts with aging. As such, the chronicity of the TBI (i.e., the time spent aging with a history of TBI) is more informative to this question than the age at injury.

Very little work has examined cerebral atrophy in older adults with a remote history of TBI. As discussed earlier in the chapter, the ADNI study design is particularly well suited to examine brain aging in older adults with a history of TBI. Leveraging the DOD-ADNI dataset, Mohamed and colleagues found that older veterans with a remote history of TBI displayed lower total brain volumes and regional gray matter volumes (e.g., precuneus, paracentral gyrus, postcentral gyrus, subcallosal gyrus, hippocampus, and thalamus) compared to veterans without TBI, suggesting that damage related to early life TBI persists into older age [67]. However, these findings are contradicted by recent work conducted in the same dataset by Weiner et al., who found no significant differences in hippocampal, amygdala, entorhinal, temporal, or parietal cortex volumes between veterans with and without a history of TBI [68]. One possible explanation for this discrepancy is the limited size of the TBI group in both studies (n = 23 and n = 43, respectively); as noted by Weiner and colleagues, this may have limited their ability to detect statistically significant differences. Conversely, the smaller sample size of the study conducted by Mohamed et al. may have resulted in false positives. Weiner et al. also conducted longitudinal analyses in a subset of the sample. Among 65-year-old cognitively unimpaired veterans, hippocampal volume significantly declined over a 1-year period only in the TBI group; however, the difference in rates of decline between the groups did not reach significance [68]. Further examination of these effects in additional, larger samples of older adults with a remote history of TBI is warranted. Another study examined cortical thinning in the ADNI cohort and found that among older adults with preclinical Alzheimer's disease, those with a history of mild TBI displayed greater cortical thinning than those without a history of TBI [69]. While they found no significant differences between cognitively unimpaired individuals with and without a history of TBI, this subset of the sample was small (n = 16), and their analyses were focused on regions vulnerable to Alzheimer's disease and therefore may not have captured normal and/or other pathological aging trajectories.

Turning to studies that focused on macro-level WM pathologies, previous work has identified greater WMH volume in middle-aged adults with a remote history of moderate-to-severe TBI (mean time since injury = 8.6 years) compared to those without a history of TBI [70]. Similar results were found in a study of adults who sustained a TBI 2 months to 13 years prior (mean time since injury = 2.4 years), primarily of mild severity (83%). Notably, in a subset of older adults (ages 45–81), those with a history of TBI displayed greater subcortical but not deep (excluding periventricular) WMH volumes compared to those without a history of TBI, whereas both types of lesions were significantly greater in the younger subset's TBI group relative to the control group. This pattern of findings suggests that subcortical WMHs may be more indicative of TBI-related damage (e.g., macroscopic traumatic axonal injury, gliosis) than deeper WMHs, particularly in older age as overall WMH burden tends to increase [71]. In contrast to these findings, however, DOD-ADNI data revealed no differences in total WMH volume between older veterans with and without a history of TBI [68]. Future studies should employ a longitudinal design to examine if greater regional and/or global WMH burden persists and potentially worsens in older age following TBI-related macroscopic damage to WM.

Longitudinal neuroimaging studies that allow researchers to follow aging trajectories of adults that have sustained a TBI into and throughout older age remain crucial. However, advancements in machine learning have allowed for examination of brain aging in the relative absence of such datasets. "Brain age" studies have been increasingly employed over the past decade [72]. Typically, the input to the regression machine learning model is structural neuroanatomical features in healthy individuals, such as cortical volumes or thickness. The model is then calculated to predict normal individuals' actual age. Next, the model is applied to a clinical sample to predict their brain age, and a "brain age gap" can then be calculated as the difference between brain-predicted age and chronological age in each individual, with a higher brain age gap indicating an acceleration of the normal aging trajectory. Due to the high number of neuroanatomical features (e.g., regional brain volumes, regional cortical thickness) included in such models, brain age analyses capture the distributed rather than focal nature of secondary injury cascades and may be uniquely informative in the examination of brain aging following TBI.

Cole and colleagues conducted one of the first brain age analyses in a sample of adults with a history of TBI (mean time since injury = 2.4 years) and identified a brain age gap of 4.7 years using structural GM metrics and 6.0 years using structural WM metrics. This suggests that the brains of those with a history of TBI appear to be 5–6 years older than their chronological age. Importantly, the brain age gap was significantly, positively associated with time since injury, indicating that cerebral changes increase throughout the chronic phase of injury [25]. While subsequent brain age studies employing T1w imaging have identified similarly elevated brain age gaps in TBI samples, time since injury has not been found to be associated with this metric [73–76].

Employing the same machine learning model as Cole et al., Spitz and colleagues [77] examined brain age in a sample that was older (mean age = 57 years) and had a greater time since injury (mean = 22 years). The TBI group, who had sustained moderate-to-severe TBIs, displayed a significantly greater brain age gap (1.6 years) compared to a noninjured control group. The authors raise that given the characteristics of their sample, the theory of TBI-accelerated brain aging would suggest that they should see a greater brain age gap than that identified by Cole et al. (5–6 years), whose sample was younger and had less remote injuries. Spitz and colleagues also note that their sample displayed a large range in brain age gap (−14–22 years), which is important to consider in addition to the mean brain age gap and suggests that some individuals may be more vulnerable to changes in brain aging as a result of TBI than others. Lastly, brain age gap was not significantly associated with time since injury in this study. It is possible that accelerated brain aging is confined to some initial period of the chronic phase (e.g., <5 or <10 years) and then returns to a normal rate. However, Spitz et al. conducted a sensitivity analysis by examining a subset of their sample who had sustained a TBI 2 months to 5 years prior and found no association with time since injury. They argue that longitudinal studies are needed to affirm or disaffirm the theory of accelerated aging following TBI.

Taken together, work conducted using T1w, T2w, and FLAIR imaging suggests that chronic TBI is associated with cerebral atrophy and WMH burden. However, it remains unclear if these neuroanatomical changes reflect accelerated changes in the trajectory of brain aging that persists into late adulthood. While brain age studies may be more informative than shorter-term (e.g., 1–2 year) longitudinal studies of chronic TBI in middle-aged adults, large-N longitudinal studies conducted in older adults who are decades removed from injury are best suited to answering this important research question.

6.5.2 Diffusion-Weighted MRI

As suggested by previously noted studies of cerebral atrophy [59, 60], WM may be particularly vulnerable to TBI-related damage. Primary injury to the brain causes shearing and strain of cerebral WM, which can result in diffuse axonal injury that can initiate a secondary injury cascade that breaks down the axonal cytoskeleton and impairs axonal transportation. Progressive axonal damage is reflected in decreased WM volumes and impaired communication between brain structures [21, 78]. Therefore, studying damage to the microstructure of the WM may uniquely inform our understanding of the effects of TBI on brain aging.

Diffusion-weighted imaging (DWI) is an MRI technique that measures the diffusion of water along axons to assess the organization of WM tracts. Scalar metrics such as fractional anisotropy (FA), the degree of directionality of water diffusion, and mean diffusivity (MD), the overall water diffusion, are commonly used to characterize WM microstructure and are thought to represent properties such as myelination and axonal density [79]. Lower FA and higher MD have been associated both with normal aging [80] and with chronic TBI [62, 81–83]. Longitudinal work conducted in middle-aged veterans who sustained a TBI identified decreases in FA and increases in MD across multiple WM tracts that did not resolve at a 2-year follow-up visit [84]. This suggests that TBI-related damage to WM may persist years after injury and could have the potential to worsen in older age.

Perhaps to an even greater extent than in work conducted using T1w imaging, very few DWI studies are focused on older adults with a remote history of TBI. A recent study leveraged data from multiple cohorts and included older veterans from the DOD-ADNI and Vietnam Era Twin Study of Aging (VESTA) studies. They identified age-dependent effects of TBI on WM organization such that older age was associated with lower FA to a greater extent in the TBI group relative to a non-injured control group [85]. These findings are consistent with prior work conducted using the DOD-ADNI dataset, which revealed lower FA and higher MD across numerous WM tracts in older veterans with a history of TBI compared to those without [67]. Taken together, these findings suggest that older individuals with chronic TBI display disrupted WM organization.

One informative study analyzed DWI data in older adults (mean age = 61.2 years) who sustained either a recent (mean time since injury = 1.8 years) or remote (mean time since injury = 36.9 years) mild TBI. They found higher MD, particularly in frontal WM tracts, in the remote but not recent TBI group relative to noninjured controls [86]. These findings indicate that despite risks associated with sustaining a TBI in older age [87], aging with a history of TBI may be associated with greater damage to WM due to the initiation of neurodegenerative cascades that continue their effects across decades. The theory of accelerated WM aging following a TBI is supported by work conducted in a sample of veterans with a wide age range (23–77 years). A nonlinear model revealed numerous WM tracts in which trajectories of age-related differences in WM organization became more pronounced in mid-to-late adulthood in those with a history of TBI (mean time since injury = 17.9 years) compared to those without. Notably, a linear model, which assumes year-to-year changes are static across the full age range, did not reveal significant interaction effects [88]. This pattern of findings suggests that the effects of TBI on WM aging may accelerate in late life.

As with T1w imaging, brain age studies can be conducted using DWI data. In one of the only studies to employ this approach, Gan et al. estimated the brain age gap of adults with mild TBI using a model based on FA in the WM tracts of noninjured control participants. In this longitudinal design, TBI participants were assessed during the acute and chronic (6–12 months postinjury) phases of injury. The TBI group showed a brain age gap of 2.59 years during the acute phase, which was significantly greater than that of the control group. Notably, the TBI group's brain age gap was not statistically different in the chronic phase compared to the acute phase, suggesting that there may be a finite period of brain acceleration that stabilizes within the first year after injury [89]. The implications of this study contrast previously discussed DWI findings that support the notion of WM aging accelerating into late adulthood. However, as noted by Gan and colleagues, their study captures a relatively early phase of chronic TBI. It is possible that further WM aging processes may occur in the years or decades following TBI, particularly as individuals reach older age. Additionally, this study focused on mild TBI, which may not accelerate age-related WM damage to the same extent as moderate-to-severe TBI. Future brain age studies should investigate these effects in a sample with a history of TBIs that are more chronic and of variable severity.

6.5.3 Functional MRI

One final noninvasive brain imaging technique that could be used to examine the interaction between earlier life TBI and aging is functional MRI (fMRI). Functional MRI is most commonly applied using two different approaches. The first reflects some of the earliest applications of fMRI, namely, task based. These studies have participants engage in a cognitive task while recording the fMRI signal. Analysis is focused on which brain regions are engaged during the task and to what extent. A second approach commonly taken is to look at brain interactions while a participant lies still in the MRI machine and "does nothing." This has been referred to as "resting-state" fMRI (RS fMRI), and analysis is focused on looking at the activity patterns between regions which are interpreted to reflect intrinsic connectivity patterns across brain networks during awake rest. Considerable research has been done using fMRI to examine functional brain changes associated with TBI [90].

Like the corpus of work examining structural brain changes associated with TBI, little of the task and resting-state fMRI research examines the long-term changes in brain aging associated with an early life TBI. Instead, the most common use of fMRI is to examine functional connectivity-related changes during the acute phase following injury. These studies have implicated changes in multiple brain networks such as the default mode network (DMN), the cognitive control network (CCN), salience network (SN), and thalamic and motor networks (for a review of this literature, see Hayes et al. [90]). Few studies examine the effects of TBI in the chronic period. One study looked at the connectivity between multiple networks in moderate-to-severe patients at approximately 2 years postinjury where they found reduced DMN connectivity from the posterior cingulate and increased anticorrelations with portions of the CCN. The authors asserted that this pattern of results was consistent with a "normalization" of network connectivity over time [91]. Most of these studies were conducted in younger individuals.

A study that compared older patients to younger ones applied RS fMRI to assess brain connectivity and examine the effects of age of injury during the acute and recovery phases [92]. The recovery was assessed at approximately 18 months, and the research found widespread increases in connectivity of the medial prefrontal and posterior cingulate regions in younger patients, but limited increases of resting-state functional connectivity (rsFC) were noted at ~18 months in older patients. While this study suggested that elderly patients are less amendable to brain recovery after injury, it did not address whether that would be the case across a longer-term recovery period from an earlier life TBI. A more recent study [93] examined the influence of history, number of TBIs, and time since injury in male Special Operations Forces combat soldiers (mean age = 33) on brain functional connectivity using a graph theoretical approach [94]. The study found that a greater number of mild TBI incidents were associated with weaker brain network "resilience"—or the extent to which a network can resist failures in its main components. Time since injury was coarsely measured as "past month," "past year," and "year+," but they found no effect of this variable in their analysis, pointing to a relatively stable pattern of differences across this small time window.

While there appears to be no resting-state nor task-based fMRI studies that examine elderly individuals with an early life history of TBI, it is clear from the work looking at white matter changes and those examining brain connectivity using RS fMRI that examining brain connectivity by combining modalities—structural and functional—will likely be a powerful approach to understanding how a life history of TBI influences brain connectivity during aging [80].

6.6 Behavioral Measures

6.6.1 Functional

History of an earlier life TBI is associated not only with effects on brain structure and function but also with changes in cognition and behavior. While a significant portion of TBI patients improve during the first 6 months postinjury and achieve favorable behavioral outcomes (e.g., as measured by the GOSE) within the first year, recent work focused on chronic TBI suggests that outcomes may change—both worsen or improve—in the years following injury [95]. For example, a study of the long-term effects of TBI found that approximately 25% of individuals who sustained a TBI displayed functional decline 7–13 years postinjury [96]. Importantly, these effects may be exacerbated with older age. A recent study identified a significant interaction between age and TBI chronicity, such that the negative effects of greater time since injury on functional outcomes were worsened in those ages 75 and older [97].

6.6.2 Psychiatric

Of note is the psychiatric sequelae of TBI; one study found that one-third of individuals with a history of TBI reported unfavorable quality of life associated with increased risk of mental health disorders (e.g., anxiety, depression) [98]. Our understanding of the psychiatric consequences of TBI and of the mechanistic processes underlying this relationship, which is likely bidirectional, remains limited [99]. However, there is a growing recognition of the importance of examining the long-term effects of TBI on mental health status. In fact, studies such as the DOD-ADNI were designed specifically to study the effects of both TBI and posttraumatic stress disorder (PTSD) on dementia risk in older veterans [47]. Work conducted with this dataset has identified those with PTSD and with comorbid TBI and PTSD as particularly vulnerable to brain and cognitive aging [67]. Given that mental health disorders such as depression have been identified as an independent risk factor for dementia [12], it is crucial for future studies to examine the interaction between TBI history and aging and consider psychiatric contributions to this relationship.

6.6.3 Neuropsychological

Much of the literature on behavioral outcomes following TBI has focused on cognition [100, 101]. However, a limited number of studies have specifically examined the effects of earlier life TBI on cognition in older age. A twin study of World War II veterans (mean age = 67.0 years) consisting of up to 4 visits over a 12-year period found that a remote history of TBI (mean time since injury = 34.0 years) was associated with poorer cognitive performance on a dementia assessment tool. Notably, history of a TBI that involved loss of consciousness was associated with faster cognitive decline over the study period [102]. Weiner et al. conducted longitudinal analyses in a subset of the DOD-ADNI sample and found that compared to noninjured controls, older veterans with a remote history of TBI showed greater worsening of Clinical Dementia Rating (CDR) scores over the study period (mean = 1.8 years) but not of other measures of global cognitive performance, memory, and executive function [68]. Another recent longitudinal study found no evidence of worsening cognitive aging trajectories in older adults (mean age = 62.7 years) with a history of TBI (mean time since injury = 29.6 years) throughout a 4-year study period. They did identify significant differences between the TBI and noninjured control groups at the baseline visit: compared to controls, those with a history of mild TBI showed poorer performance on a measure of attention, and those with a history of moderate-to-severe TBI showed poorer performance on measures of attention, executive function, and processing speed [103]. Taken together, these studies suggest that earlier life TBI is associated with poorer cognitive outcomes in older age. However, much like findings from neuroimaging studies, it remains unclear if and when acceleration of cognitive aging trajectories might occur. Findings from the ADNI study suggest that history of TBI may accelerate the onset of cognitive impairment by two or more years; older adults with a remote history of TBI (average age at injury = 35.4 years) displayed an onset of cognitive impairment at an average age of 68.2 years, which was significantly earlier than a noninjured control group (average age at onset = 70.9 years) [104].

Ideally, neuroimaging methods can be used in conjunction with neuropsychological assessment to deepen our understanding of the neurodegenerative cascades that may be driving changes in the cognitive trajectories of older adults with a history of TBI. Of the previously described neuroimaging studies that focused on the effects of earlier life TBI on brain structure in late adulthood, many did not examine cognitive performance [69, 85, 86, 88]. Spitz and colleagues conducted a brain age analysis using T1w imaging data and found that greater brain age gap—which was identified in the TBI group relative to controls—was associated with poorer performance on measures of verbal memory but not of visuospatial ability/memory or attention/executive functioning [77]. Using DOD-ADNI data, Mohamed et al. identified significant associations between diffusion MRI metrics and cognitive performance. However, it was the PTSD and comorbid TBI and PTSD groups that displayed poorer cognitive performance than the TBI and noninjured control groups, suggesting that these effects are driven by PTSD rather than TBI history [67]. While neuroimaging and neuropsychological findings independently suggest that history of earlier life TBI can negatively influence the trajectory of aging into late adulthood, it will be particularly important for future work to examine the association between brain and cognitive aging patterns in this population.

6.7 Conclusion

A significant number of individuals will experience a TBI in their lifetime [6]. The risk is even higher for those working in specific professions such as the military or professional contact sports. While TBI can result in lifetime impairments, many individuals appear to fully recover and return to their lives normally [95]. Despite this, some recent research indicates that a life history of TBI increases risk for developing dementia later in life [13]. This increased risk—namely, the interaction between TBI history and the brain aging process—is poorly understood. This chapter outlines the importance of this issue, and the critical need for more research specifically focused on understanding the conditions and mechanisms behind this increased risk. Preclinical animal models will be important and are explored elsewhere in this book, but here we have focused on work in humans as they age. We outline some of the important methodological considerations when examining the interaction between life history of TBI and aging and then present some of the research on this question, primarily that which utilizes noninvasive brain imaging and neuropsychological measures of cognitive performance. This rather sparse research on the topic clearly points to an interaction between brain aging and life history of TBI that affects cortical gray and white matter and has important implications for cognition. Recent work on the psychiatric sequalae of TBI and on the chronic effects of TBI on multiple body systems (e.g., cardiovascular) point to important avenues of future research on how individual risk factors for dementia may be associated and become compounded following an earlier life TBI. The presented findings underscore the importance of conducting additional, well-designed longitudinal studies to continue to probe the mechanisms behind these interactions as they hold implications not only for the topic of this chapter but also for dementia risk more broadly.

Acknowledgments TRACK TBI Network.

References

1. Dall TM, Gallo PD, Chakrabarti R, West T, Semilla AP, Storm MV. An aging population and growing disease burden will require a large and specialized health care workforce by 2025. Health Aff (Millwood). 2013;32:2013–20. https://doi.org/10.1377/hlthaff.2013.0714.
2. Vincent GK, Velkoff VA. The next four decades: the older population in the United States: 2010 to 2050. In: Censusgov. 2010. https://www.census.gov/library/publications/2010/demo/p25-1138.html. Accessed 3 Oct 2023.
3. Hung WW, Ross JS, Boockvar KS, Siu AL. Recent trends in chronic disease, impairment and disability among older adults in the United States. BMC Geriatr. 2011;11:47. https://doi.org/10.1186/1471-2318-11-47.
4. Prince M, Ali G-C, Guerchet M, Prina AM, Albanese E, Wu Y-T. Recent global trends in the prevalence and incidence of dementia, and survival with dementia. Alzheimers Res Ther. 2016;8:23. https://doi.org/10.1186/s13195-016-0188-8.

5. Wu Y, Beiser AS, Breteler MMB, Fratiglioni L, Helmer C, Hendrie HC, Honda H, Ikram MA, Langa KM, Lobo A, Matthews FE, Ohara T, Pérès K, Qiu C, Seshadri S, Sjölund B, Skoog I, Brayne C. The changing prevalence and incidence of dementia over time—current evidence. Nat Rev Neurol. 2017;13:327–39. https://doi.org/10.1038/nrneurol.2017.63.
6. Frost RB, Farrer TJ, Primosch M, Hedges DW. Prevalence of traumatic brain injury in the general adult population: a meta-analysis. Neuroepidemiology. 2012;40:154–9. https://doi.org/10.1159/000343275.
7. Taylor CA, Bell JM, Breiding MJ, Xu L. Traumatic brain injury-related emergency department visits, hospitalizations, and deaths—United States, 2007 and 2013. Morb Mortal Wkly Rep Surveill Summ Wash DC 2002. 2017;66:1–16. https://doi.org/10.15585/mmwr.ss6609a1.
8. GBD. Traumatic Brain Injury and Spinal Cord Injury Collaborators (2019) Global, regional, and national burden of traumatic brain injury and spinal cord injury, 1990-2016: a systematic analysis for the Global Burden of Disease Study 2016. Lancet Neurol. 2016;18:56–87. https://doi.org/10.1016/S1474-4422(18)30415-0.
9. Dams-O'Connor K, Juengst SB, Bogner J, Chiaravalloti ND, Corrigan JD, Giacino JT, Harrison-Felix CL, Hoffman JM, Ketchum JM, Lequerica AH, Marwitz JH, Miller AC, Nakase-Richardson R, Rabinowitz AR, Sander AM, Zafonte R, Hammond FM. Traumatic brain injury as a chronic disease: insights from the United States Traumatic Brain Injury Model Systems Research Program. Lancet Neurol. 2023;22:517–28. https://doi.org/10.1016/S1474-4422(23)00065-0.
10. Harvey LA, Close JCT. Traumatic brain injury in older adults: characteristics, causes and consequences. Injury. 2012;43:1821–6. https://doi.org/10.1016/j.injury.2012.07.188.
11. Gardner RC, Dams-O'Connor K, Morrissey MR, Manley GT. Geriatric traumatic brain injury: epidemiology, outcomes, knowledge gaps, and future directions. J Neurotrauma. 2018;35:889–906. https://doi.org/10.1089/neu.2017.5371.
12. Livingston G, Huntley J, Sommerlad A, Ames D, Ballard C, Banerjee S, Brayne C, Burns A, Cohen-Mansfield J, Cooper C, Costafreda SG, Dias A, Fox N, Gitlin LN, Howard R, Kales HC, Kivimäki M, Larson EB, Ogunniyi A, Orgeta V, Ritchie K, Rockwood K, Sampson EL, Samus Q, Schneider LS, Selbæk G, Teri L, Mukadam N. Dementia prevention, intervention, and care: 2020 report of the Lancet Commission. Lancet. 2020;396:413–46. https://doi.org/10.1016/S0140-6736(20)30367-6.
13. Snowden TM, Hinde AK, Reid HMO, Christie BR. Does mild traumatic brain injury increase the risk for dementia? A systematic review and meta-analysis. J Alzheimers Dis JAD. 2020;78:757–75. https://doi.org/10.3233/JAD-200662.
14. Barnes DE, Byers AL, Gardner RC, Seal KH, Boscardin WJ, Yaffe K. Association of mild traumatic brain injury with and without loss of consciousness with dementia in US military veterans. JAMA Neurol. 2018;75:1055–61. https://doi.org/10.1001/jamaneurol.2018.0815.
15. Fann JR, Ribe AR, Pedersen HS, Fenger-Grøn M, Christensen J, Benros ME, Vestergaard M. Long-term risk of dementia among people with traumatic brain injury in Denmark: a population-based observational cohort study. Lancet Psychiatry. 2018;5:424–31. https://doi.org/10.1016/S2215-0366(18)30065-8.
16. Nordström A, Nordström P. Traumatic brain injury and the risk of dementia diagnosis: a nationwide cohort study. PLoS Med. 2018;15:e1002496. https://doi.org/10.1371/journal.pmed.1002496.
17. Dams-O'Connor K, Gibbons LE, Bowen JD, McCurry SM, Larson EB, Crane PK. Risk for late-life re-injury, dementia and death among individuals with traumatic brain injury: a population-based study. J Neurol Neurosurg Psychiatry. 2013;84:177–82. https://doi.org/10.1136/jnnp-2012-303938.
18. Barr WB. POINT/COUNTER-POINT—links between traumatic brain injury and dementia remain poorly defined. Arch Clin Neuropsychol. 2020;35:128–32. https://doi.org/10.1093/arclin/acz075.

19. Ginsburg J, Huff JS. Closed head trauma. In: StatPearls. Treasure Island: StatPearls Publishing; 2023.
20. Vakil MT, Singh AK. A review of penetrating brain trauma: epidemiology, pathophysiology, imaging assessment, complications, and treatment. Emerg Radiol. 2017;24:301–9. https://doi.org/10.1007/s10140-016-1477-z.
21. Greve MW, Zink BJ. Pathophysiology of traumatic brain injury. Mt Sinai J Med N Y. 2009;76:97–104. https://doi.org/10.1002/msj.20104.
22. Sabet N, Soltani Z, Khaksari M. Multipotential and systemic effects of traumatic brain injury. J Neuroimmunol. 2021;357:577619. https://doi.org/10.1016/j.jneuroim.2021.577619.
23. Zia A, Pourbagher-Shahri AM, Farkhondeh T, Samarghandian S. Molecular and cellular pathways contributing to brain aging. Behav Brain Funct. 2021;17:6. https://doi.org/10.1186/s12993-021-00179-9.
24. Bigler ED, Stern Y. Chapter 43—Traumatic brain injury and reserve. In: Grafman J, Salazar AM, editors. Handbook of clinical neurology. Elsevier; 2015. p. 691–710.
25. Cole JH, Leech R, Sharp DJ. Prediction of brain age suggests accelerated atrophy after traumatic brain injury. Ann Neurol. 2015;77:571–81. https://doi.org/10.1002/ana.24367.
26. Daugherty J, DePadilla L, Sarmiento K, Breiding MJ. Self-reported lifetime concussion among adults: comparison of 3 different survey questions. J Head Trauma Rehabil. 2020;35:E136–43. https://doi.org/10.1097/HTR.0000000000000534.
27. Whiteneck GG, Cuthbert JP, Corrigan JD, Bogner JA. Prevalence of self-reported lifetime history of traumatic brain injury and associated disability: a statewide population-based survey. J Head Trauma Rehabil. 2016;31:E55–62. https://doi.org/10.1097/HTR.0000000000000140.
28. Centers for Disease Control and Prevention (CDC). Surveillance report of traumatic brain injury-related deaths by age group, sex, and mechanism of injury —United States, 2018 and 2019. Centers for Disease Control and Prevention, U.S. Department of Health and Human Services; 2020.
29. Womack LS, Breiding MJ, Daugherty J. Concussion evaluation patterns among US adults. J Head Trauma Rehabil. 2022;37:303–10. https://doi.org/10.1097/HTR.0000000000000756.
30. Defense and Veterans Brain Injury Center. DoD TBI worldwide numbers. Silver Spring Maryland: Defense and Veterans Brain Injury Center; 2020.
31. Selassie AW, Wilson DA, Pickelsimer EE, Voronca DC, Williams NR, Edwards JC. Incidence of sport-related traumatic brain injury and risk factors of severity: a population-based epidemiologic study. Ann Epidemiol. 2013;23:750–6. https://doi.org/10.1016/j.annepidem.2013.07.022.
32. McKee AC, Stein TD, Huber BR, Crary JF, Bieniek K, Dickson D, Alvarez VE, Cherry JD, Farrell K, Butler M, Uretsky M, Abdolmohammadi B, Alosco ML, Tripodis Y, Mez J, Daneshvar DH. Chronic traumatic encephalopathy (CTE): criteria for neuropathological diagnosis and relationship to repetitive head impacts. Acta Neuropathol (Berl). 2023;145:371–94. https://doi.org/10.1007/s00401-023-02540-w.
33. Bieniek KF, Ross OA, Cormier KA, Walton RL, Soto-Ortolaza A, Johnston AE, DeSaro P, Boylan KB, Graff-Radford NR, Wszolek ZK, Rademakers R, Boeve BF, McKee AC, Dickson DW. Chronic traumatic encephalopathy pathology in a neurodegenerative disorders brain bank. Acta Neuropathol (Berl). 2015;130:877–89. https://doi.org/10.1007/s00401-015-1502-4.
34. Dams-O'Connor K, Spielman L, Singh A, Gordon WA, Lingsma HF, Maas AIR, Manley GT, Mukherjee P, Okonkwo DO, Puccio AM, Schnyer DM, Valadka AB, Yue JK, Yuh EL, Casey and the T-TI including: SS, Cooper SR, Cheong M, Hricik AJ, Knight EE, Menon DK, Morabito DJ, Pacheco JL, Sinha TK, Vassar MJ. The impact of previous traumatic brain injury on health and functioning: a TRACK-TBI study. J Neurotrauma. 2013;30:2014–20. https://doi.org/10.1089/neu.2013.3049.
35. Nelson LD, Temkin NR, Barber J, Brett BL, Okonkwo DO, McCrea MA, Giacino JT, Bodien YG, Robertson C, Corrigan JD, Diaz-Arrastia R, Markowitz AJ, Manley GT, Investigators TRACK-TBI. Functional recovery, symptoms, and quality of life 1 to 5 years after traumatic

brain injury: findings from the TRACK-TBI study. JAMA Netw Open. 2023;6:e233660. https://doi.org/10.1001/jamanetworkopen.2023.3660.
36. McKinlay A, Horwood LJ, Fergusson DM. Accuracy of self-report as a method of screening for lifetime occurrence of traumatic brain injury events that resulted in hospitalization. J Int Neuropsychol Soc. 2016;22:717–23. https://doi.org/10.1017/S1355617716000497.
37. Johnson LW, Diaz I. Exploring the social determinants of health and health disparities in traumatic brain injury: a scoping review. Brain Sci. 2023;13:707. https://doi.org/10.3390/brainsci13050707.
38. McQuistion K, Zens T, Jung HS, Beems M, Leverson G, Liepert A, Scarborough J, Agarwal S. Insurance status and race affect treatment and outcome of traumatic brain injury. J Surg Res. 2016;205:261–71. https://doi.org/10.1016/j.jss.2016.06.087.
39. Teasdale G, Jennett B. Assessment of coma and impaired consciousness: a practical scale. Lancet. 1974;304:81–4. https://doi.org/10.1016/S0140-6736(74)91639-0.
40. Baker SP, O'neill B, Haddon WJ, Long WB. The injury severity score: a method for describing patients with multiple injuries and evaluating emergency care. J Trauma Acute Care Surg. 1974;14:187.
41. Corrigan JD, Bogner J. Initial reliability and validity of the Ohio State University TBI Identification Method. J Head Trauma Rehabil. 2007;22:318–29. https://doi.org/10.1097/01.HTR.0000300227.67748.77.
42. Wilson JT, Pettigrew LE, Teasdale GM. Structured interviews for the Glasgow outcome scale and the extended Glasgow outcome scale: guidelines for their use. J Neurotrauma. 1998;15:573–85. https://doi.org/10.1089/neu.1998.15.573.
43. King NS, Crawford S, Wenden FJ, Moss NEG, Wade DT. The Rivermead Post Concussion Symptoms Questionnaire: a measure of symptoms commonly experienced after head injury and its reliability. J Neurol. 1995;242:587–92. https://doi.org/10.1007/BF00868811.
44. von Steinbuechel N, Wilson L, Gibbons H, Muehlan H, Schmidt H, Schmidt S, Sasse N, Koskinen S, Sarajuuri J, Höfer S, Bullinger M, Maas A, Neugebauer E, Powell J, von Wild K, Zitnay G, Bakx W, Christensen A-L, Formisano R, Hawthorne G, Truelle J-L. QOLIBRI overall scale: a brief index of health-related quality of life after traumatic brain injury. J Neurol Neurosurg Psychiatry. 2012;83:1041–7. https://doi.org/10.1136/jnnp-2012-302361.
45. Shock NW, Gerontology Research Center (U.S.).Normal human aging: the Baltimore longitudinal study of aging. U.S. Dept. of Health and Human Services, Public Health Service, National Institutes of Health, National Institute on Aging, Gerontology Research Center ; For sale by the Supt. of Docs., U.S. G.P.O. [Baltimore, Md.], Washington, D.C., [Baltimore, Md.], Washington, D.C.; 1984.
46. Mueller SG, Weiner MW, Thal LJ, Petersen RC, Jack C, Jagust W, Trojanowski JQ, Toga AW, Beckett L. The Alzheimer's disease neuroimaging initiative. Neuroimaging Clin N Am. 2005;15(869–877):xi–xii. https://doi.org/10.1016/j.nic.2005.09.008.
47. Weiner MW, Veitch DP, Hayes J, Neylan T, Grafman J, Aisen PS, Petersen RC, Jack C, Jagust W, Trojanowski JQ, Shaw LM, Saykin AJ, Green RC, Harvey D, Toga AW, Friedl KE, Pacifico A, Sheline Y, Yaffe K, Mohlenoff B, Department of Defense Alzheimer's Disease Neuroimaging Initiative. Effects of traumatic brain injury and posttraumatic stress disorder on Alzheimer's disease in veterans, using the Alzheimer's Disease Neuroimaging Initiative. Alzheimers Dement. 2014;10:S226–35. https://doi.org/10.1016/j.jalz.2014.04.005.
48. Petersen RC, Aisen PS, Beckett LA, Donohue MC, Gamst AC, Harvey DJ, Jack CR, Jagust WJ, Shaw LM, Toga AW, Trojanowski JQ, Weiner MW. Alzheimer's Disease Neuroimaging Initiative (ADNI): clinical characterization. Neurology. 2010;74:201–9.
49. Weiner MW, Veitch DP, Miller MJ, Aisen PS, Albala B, Beckett LA, Green RC, Harvey D, Jack CR Jr, Jagust W, Landau SM, Morris JC, Nosheny R, Okonkwo OC, Perrin RJ, Petersen RC, Rivera-Mindt M, Saykin AJ, Shaw LM, Toga AW, Tosun D, Trojanowski JQ, Initiative ADN. Increasing participant diversity in AD research: plans for digital screening, blood testing, and a community-engaged approach in the Alzheimer's Disease Neuroimaging Initiative 4. Alzheimers Dement. 2023;19:307–17. https://doi.org/10.1002/alz.12797.

50. Scahill RI, Frost C, Jenkins R, Whitwell JL, Rossor MN, Fox NC. A longitudinal study of brain volume changes in normal aging using serial registered magnetic resonance imaging. Arch Neurol. 2003;60:989–94. https://doi.org/10.1001/archneur.60.7.989.
51. Salat DH, Buckner RL, Snyder AZ, Greve DN, Desikan RSR, Busa E, Morris JC, Dale AM, Fischl B. Thinning of the cerebral cortex in aging. Cereb Cortex N Y N 1991. 2004;14:721–30. https://doi.org/10.1093/cercor/bhh032.
52. Maillard P, Carmichael O, Fletcher E, Reed B, Mungas D, DeCarli C. Coevolution of white matter hyperintensities and cognition in the elderly. Neurology. 2012;79:442–8. https://doi.org/10.1212/WNL.0b013e3182617136.
53. Nordahl CW, Ranganath C, Yonelinas AP, Decarli C, Fletcher E, Jagust WJ. White matter changes compromise prefrontal cortex function in healthy elderly individuals. J Cogn Neurosci. 2006;18:418–29. https://doi.org/10.1162/089892906775990552.
54. Driscoll I, Davatzikos C, An Y, Wu X, Shen D, Kraut M, Resnick SM. Longitudinal pattern of regional brain volume change differentiates normal aging from MCI. Neurology. 2009;72:1906–13. https://doi.org/10.1212/WNL.0b013e3181a82634.
55. Liu H, Yang Y, Xia Y, Zhu W, Leak RK, Wei Z, Wang J, Hu X. Aging of cerebral white matter. Ageing Res Rev. 2017;34:64–76. https://doi.org/10.1016/j.arr.2016.11.006.
56. Ross DE. Review of longitudinal studies of MRI brain volumetry in patients with traumatic brain injury. Brain Inj. 2011;25:1271–8. https://doi.org/10.3109/02699052.2011.624568.
57. Harris TC, de Rooij R, Kuhl E. The shrinking brain: cerebral atrophy following traumatic brain injury. Ann Biomed Eng. 2019;47:1941–59. https://doi.org/10.1007/s10439-018-02148-2.
58. Brezova V, Gøran Moen K, Skandsen T, Vik A, Brewer JB, Salvesen Ø, Håberg AK. Prospective longitudinal MRI study of brain volumes and diffusion changes during the first year after moderate to severe traumatic brain injury. NeuroImage Clin. 2014;5:128–40. https://doi.org/10.1016/j.nicl.2014.03.012.
59. Cole JH, Jolly A, de Simoni S, Bourke N, Patel MC, Scott G, Sharp DJ. Spatial patterns of progressive brain volume loss after moderate-severe traumatic brain injury. Brain. 2018;141:822–36. https://doi.org/10.1093/brain/awx354.
60. Farbota KDM, Sodhi A, Bendlin BB, McLaren DG, Xu G, Rowley HA, Johnson SC. Longitudinal volumetric changes following traumatic brain injury: a tensor-based morphometry study. J Int Neuropsychol Soc. 2012;18:1006–18. https://doi.org/10.1017/S1355617712000835.
61. Warner MA, Youn TS, Davis T, Chandra A, Marquez de la Plata C, Moore C, Harper C, Madden CJ, Spence J, McColl R, Devous M, King RD, Diaz-Arrastia R. Regionally selective atrophy after traumatic axonal injury. Arch Neurol. 2010;67:1336–44. https://doi.org/10.1001/archneurol.2010.149.
62. Gimbel SI, Hungerford LD, Twamley EW, Ettenhofer ML. White matter organization and cortical thickness differ among active duty service members with chronic mild, moderate, and severe traumatic brain injury. J Neurotrauma. 2023;41:818. https://doi.org/10.1089/neu.2023.0336.
63. Govindarajan KA, Narayana PA, Hasan KM, Wilde EA, Levin HS, Hunter JV, Miller ER, Patel VKS, Robertson CS, McCarthy JJ. Cortical thickness in mild traumatic brain injury. J Neurotrauma. 2016;33:1809–17. https://doi.org/10.1089/neu.2015.4253.
64. Mazaharally M, Stojanovski S, Trossman R, Szulc-Lerch K, Chakravarty MM, Colella B, Glazer J, Green RE, Wheeler AL. Patterns of change in cortical morphometry following traumatic brain injury in adults. Hum Brain Mapp. 2022;43:1882–94. https://doi.org/10.1002/hbm.25761.
65. Santhanam P, Wilson SH, Oakes TR, Weaver LK. Accelerated age-related cortical thinning in mild traumatic brain injury. Brain Behav. 2019;9:e01161. https://doi.org/10.1002/brb3.1161.
66. Schönberger M, Ponsford J, Reutens D, Beare R, O'Sullivan R. The relationship between age, injury severity, and MRI findings after traumatic brain injury. J Neurotrauma. 2009;26:2157–67. https://doi.org/10.1089/neu.2009.0939.

67. Mohamed AZ, Cumming P, Nasrallah FA. White matter alterations are associated with cognitive dysfunction decades after moderate-to-severe traumatic brain injury and/or posttraumatic stress disorder. Biol Psychiatry Cogn Neurosci Neuroimaging. 2021;6:1100. https://doi.org/10.1016/j.bpsc.2021.04.014.
68. Weiner MW, Harvey D, Landau SM, Veitch DP, Neylan TC, Grafman JH, Aisen PS, Petersen RC, Jack CR Jr, Tosun D, Shaw LM, Trojanowski JQ, Saykin AJ, Hayes J, De Carli C, for the Alzheimer's Disease Neuroimaging Initiative and the Department of Defense Alzheimer's Disease Neuroimaging Initiative. Traumatic brain injury and post-traumatic stress disorder are not associated with Alzheimer's disease pathology measured with biomarkers. Alzheimers Dement. 2023;19:884–95. https://doi.org/10.1002/alz.12712.
69. Wang M-L, Wei X-E, Yu M-M, Li P-Y, Li W-B. Self-reported traumatic brain injury and in vivo measure of AD-vulnerable cortical thickness and AD-related biomarkers in the ADNI cohort. Neurosci Lett. 2017;655:115–20. https://doi.org/10.1016/j.neulet.2017.06.055.
70. Berginström N, Nordström P, Nyberg L, Nordström A. White matter hyperintensities increases with traumatic brain injury severity: associations to neuropsychological performance and fatigue. Brain Inj. 2020;34:415–20. https://doi.org/10.1080/02699052.2020.1725124.
71. Trifan G, Gattu R, Haacke EM, Kou Z, Benson RR. MR imaging findings in mild traumatic brain injury with persistent neurological impairment. Magn Reson Imaging. 2017;37:243–51. https://doi.org/10.1016/j.mri.2016.12.009.
72. Baecker L, Garcia-Dias R, Vieira S, Scarpazza C, Mechelli A. Machine learning for brain age prediction: introduction to methods and clinical applications. EBioMedicine. 2021;72:103600. https://doi.org/10.1016/j.ebiom.2021.103600.
73. Amgalan A, Maher AS, Ghosh S, Chui HC, Bogdan P, Irimia A. Brain age estimation reveals older adults' accelerated senescence after traumatic brain injury. GeroScience. 2022;44:2509–25. https://doi.org/10.1007/s11357-022-00597-1.
74. Dennis EL, Taylor BA, Newsome MR, Troyanskaya M, Abildskov TJ, Betts AM, Bigler ED, Cole J, Davenport N, Duncan T, Gill J, Guedes V, Hinds SR, Hovenden ES, Kenney K, Pugh MJ, Scheibel RS, Shahim P-P, Shih R, Walker WC, Werner JK, York GE, Cifu DX, Tate DF, Wilde EA. Advanced brain age in deployment-related traumatic brain injury: a LIMBIC-CENC neuroimaging study. Brain Inj. 2022;36:662–72. https://doi.org/10.1080/02699052.2022.2033844.
75. Savjani RR, Taylor BA, Acion L, Wilde EA, Jorge RE. Accelerated changes in cortical thickness measurements with age in military service members with traumatic brain injury. J Neurotrauma. 2017;34:3107–16. https://doi.org/10.1089/neu.2017.5022.
76. Shida AF, Massett RJ, Imms P, Vegesna RV, Amgalan A, Irimia A, for the Alzheimer's Disease Neuroimaging Initiative. Significant acceleration of regional brain aging and atrophy after mild traumatic brain injury. J Gerontol Ser A. 2023;78:1328–38. https://doi.org/10.1093/gerona/glad079.
77. Spitz G, Hicks AJ, Roberts C, Rowe CC, Ponsford J. Brain age in chronic traumatic brain injury. NeuroImage Clin. 2022;35:103039. https://doi.org/10.1016/j.nicl.2022.103039.
78. Narayana PA. White matter changes in patients with mild traumatic brain injury: MRI perspective. Concussion Lond Engl. 2017;2:CNC35. https://doi.org/10.2217/cnc-2016-0028.
79. Basser PJ. Inferring microstructural features and the physiological state of tissues from diffusion-weighted images. NMR Biomed. 1995;8:333–44. https://doi.org/10.1002/nbm.1940080707.
80. Bennett IJ, Madden DJ. Disconnected aging: cerebral white matter integrity and age-related differences in cognition. Neuroscience. 2014;276:187–205. https://doi.org/10.1016/j.neuroscience.2013.11.026.
81. Kraus MF, Susmaras T, Caughlin BP, Walker CJ, Sweeney JA, Little DM. White matter integrity and cognition in chronic traumatic brain injury: a diffusion tensor imaging study. Brain. 2007;130:2508–19. https://doi.org/10.1093/brain/awm216.
82. Morey RA, Haswell CC, Selgrade ES, Massoglia D, Liu C, Weiner J, Marx CE, Cernak I, McCarthy G. Effects of chronic mild traumatic brain injury on white matter integrity in Iraq

and Afghanistan war veterans. Hum Brain Mapp. 2012;34:2986–99. https://doi.org/10.1002/hbm.22117.
83. Shenton ME, Hamoda HM, Schneiderman JS, Bouix S, Pasternak O, Rathi Y, Vu M-A, Purohit MP, Helmer K, Koerte I, Lin AP, Westin C-F, Kikinis R, Kubicki M, Stern RA, Zafonte R. A review of magnetic resonance imaging and diffusion tensor imaging findings in mild traumatic brain injury. Brain Imaging Behav. 2012;6:137–92. https://doi.org/10.1007/s11682-012-9156-5.
84. Yeh P-H, SaraM L, Brickell TA, Ollinger J, French LM, Lange RT. Longitudinal changes of white matter microstructure following traumatic brain injury in U.S. military service members. Brain Commun. 2022;4:fcac132. https://doi.org/10.1093/braincomms/fcac132.
85. Bouchard HC, Sun D, Dennis EL, Newsome MR, Disner SG, Elman J, Silva A, Velez C, Irimia A, Davenport ND, Sponheim SR, Franz CE, Kremen WS, Coleman MJ, Williams MW, Geuze E, Koerte IK, Shenton ME, Adamson MM, Coimbra R, Grant G, Shutter L, George MS, Zafonte RD, McAllister TW, Stein MB, Thompson PM, Wilde EA, Tate DF, Sotiras A, Morey RA. Age-dependent white matter disruptions after military traumatic brain injury: multivariate analysis results from ENIGMA brain injury. Hum Brain Mapp. 2022;43:2653–67. https://doi.org/10.1002/hbm.25811.
86. Tremblay S, Desjardins M, Bermudez P, Iturria-Medina Y, Evans AC, Jolicoeur P, De Beaumont L. Mild traumatic brain injury: the effect of age at trauma onset on brain structure integrity. NeuroImage Clin. 2019;23:101907. https://doi.org/10.1016/j.nicl.2019.101907.
87. Mak CHK, Wong SKH, Wong GK, Ng S, Wang KKW, Lam PK, Poon WS. Traumatic brain injury in the elderly: is it as bad as we think? Curr Transl Geriatr Exp Gerontol Rep. 2012;1:171–8. https://doi.org/10.1007/s13670-012-0017-2.
88. Vakhtin AA, Zhang Y, Wintermark M, Ashford JW, Furst AJ. Distant histories of mild traumatic brain injury exacerbate age-related differences in white matter properties. Neurobiol Aging. 2021;107:30–41. https://doi.org/10.1016/j.neurobiolaging.2021.07.002.
89. Gan S, Shi W, Wang S, Sun Y, Yin B, Bai G, Jia X, Sun C, Niu X, Wang Z, Jiang X, Liu J, Zhang M, Bai L. Accelerated brain aging in mild traumatic brain injury: longitudinal pattern recognition with white matter integrity. J Neurotrauma. 2021;38:2549. https://doi.org/10.1089/neu.2020.7551.
90. Hayes JP, Bigler ED, Verfaellie M. Traumatic brain injury as a disorder of brain connectivity. J Int Neuropsychol Soc JINS. 2016;22:120–37. https://doi.org/10.1017/S1355617715000740.
91. Venkatesan UM, Dennis NA, Hillary FG. Chronology and chronicity of altered resting-state functional connectivity after traumatic brain injury. J Neurotrauma. 2015;32:252–64. https://doi.org/10.1089/neu.2013.3318.
92. de Souza NL, Parker R, Gonzalez CS, Ryan JD, Esopenko C. Effect of age at time of injury on long-term changes in intrinsic functional connectivity in traumatic brain injury. Brain Inj. 2020;34:1646–54. https://doi.org/10.1080/02699052.2020.1832257.
93. Powell JR, Hopfinger JB, Giovanello KS, Walton SR, DeLellis SM, Kane SF, Means GE, Mihalik JP. Mild traumatic brain injury history is associated with lower brain network resilience in soldiers. Brain Commun. 2023;5:fcad201. https://doi.org/10.1093/braincomms/fcad201.
94. Bullmore E, Sporns O. Complex brain networks: graph theoretical analysis of structural and functional systems. Nat Rev Neurosci. 2009;10:186–98. https://doi.org/10.1038/nrn2575.
95. Stocchetti N, Zanier ER. Chronic impact of traumatic brain injury on outcome and quality of life: a narrative review. Crit Care. 2016;20:148. https://doi.org/10.1186/s13054-016-1318-1.
96. McMillan TM, Teasdale GM, Stewart E. Disability in young people and adults after head injury: 12-14 year follow-up of a prospective cohort. J Neurol Neurosurg Psychiatry. 2012;83:1086–91. https://doi.org/10.1136/jnnp-2012-302746.
97. Rabinowitz AR, Kumar RG, Sima A, Venkatesan UM, Juengst SB, O'Neil-Pirozzi TM, Watanabe TK, Goldin Y, Hammond FM, Dreer LE. Aging with traumatic brain injury: deleterious effects of injury chronicity are Most pronounced in later life. J Neurotrauma. 2021;38:2706–13. https://doi.org/10.1089/neu.2021.0038.

98. Rauen K, Reichelt L, Probst P, Schäpers B, Müller F, Jahn K, Plesnila N. Quality of life up to 10 years after traumatic brain injury: a cross-sectional analysis. Health Qual Life Outcomes. 2020;18:166. https://doi.org/10.1186/s12955-020-01391-3.
99. Li LM, Carson A, Dams-O'Connor K. Psychiatric sequelae of traumatic brain injury—future directions in research. Nat Rev Neurol. 2023;19:556–71. https://doi.org/10.1038/s41582-023-00853-8.
100. An KY, Monette MCE. Cognitive profiles of older adults with a prior traumatic brain injury versus healthy controls: a meta-analysis. Brain Inj. 2018;32:832–42. https://doi.org/10.1080/02699052.2018.1463104.
101. Dikmen SS, Corrigan JD, Levin HS, Machamer J, Stiers W, Weisskopf MG. Cognitive outcome following traumatic brain injury. J Head Trauma Rehabil. 2009;24:430. https://doi.org/10.1097/HTR.0b013e3181c133e9.
102. Chanti-Ketterl M, Pieper CF, Yaffe K, Plassman BL. Associations between traumatic brain injury and cognitive decline among older male veterans: a twin study. Neurology. 2023;101:e1761–70. https://doi.org/10.1212/WNL.0000000000207819.
103. Lennon MJ, Brooker H, Creese B, Thayanandan T, Rigney G, Aarsland D, Hampshire A, Ballard C, Corbett A, Raymont V. Lifetime traumatic brain injury and cognitive domain deficits in late life: the PROTECT-TBI cohort study. J Neurotrauma. 2023;40:1423–35. https://doi.org/10.1089/neu.2022.0360.
104. Li W, Risacher SL, McAllister TW, Saykin AJ. Traumatic brain injury and age at onset of cognitive impairment in older adults. J Neurol. 2016;263:1280–5. https://doi.org/10.1007/s00415-016-8093-4.

Part II
Preclinical Perspectives of Traumatic Brain Injuries

Chapter 7
Pediatric Traumatic Brain Injury: Models, Therapeutics, and Outcomes

Eleni H. Moschonas, Ellen M. Annas, Jonathan Zamudio-Flores, Jessica M. Jarvis, Naima Lajud, Corina O. Bondi, and Anthony E. Kline

Abstract Pediatric traumatic brain injury (TBI) is a significant healthcare issue, but potential treatments are absent despite robust investigation in several clinical trials. Factors attributed to clinical TBI, such as heterogeneity of injury and single-dose pharmacological treatments as well as timing of administration, may be reasons for the negative studies. Preclinical models of TBI can reduce some of the impediments by highlighting differences in injury depending on injury severity and location and by conducting dose response studies, thus providing better therapeutic targets and pharmacological profiles for clinical use. In this chapter, there were sufficient reports to make comparisons between the models in terms of pathophysiology, behavioral dysfunction, and the efficacy of therapeutic interventions. The models used to date include controlled cortical impact (CCI), weight drop, fluid percussion, and abusive head trauma. Several therapeutics were identified after CCI

E. H. Moschonas · E. M. Annas · J. M. Jarvis
Department of Physical Medicine & Rehabilitation, and Safar Center for Resuscitation Research, University of Pittsburgh, Pittsburgh, PA, USA
e-mail: ehm18@pitt.edu; ema75@pitt.edu; jmjarvis@pitt.edu

J. Zamudio-Flores
Centro de Investigación Biomédica de Michoacán – Instituto Mexicano del Seguro Social, Morelia, Mexico

N. Lajud
Instituto de Investigaciones sobre los Recursos Naturales – Universidad Michoacana de San Nicolás de Hidalgo, Morelia, Michoacan, Mexico

C. O. Bondi
Departments of Physical Medicine & Rehabilitation and Neurobiology, Center for Neuroscience, and Safar Center for Resuscitation Research, John G. Rangos Research Center, University of Pittsburgh School of Medicine, Pittsburgh, PA, USA
e-mail: bondico@upmc.edu

A. E. Kline (✉)
Departments of Physical Medicine & Rehabilitation, Critical Care Medicine, and Psychology, and Safar Center for Resuscitation Research, University of Pittsburgh School of Medicine, John G. Rangos Research Center, Pittsburgh, PA, USA
e-mail: klineae@upmc.edu

L. J. Noble-Haeusslein, D. M. Schnyer (eds.), *Traumatic Brain Injuries*, Advances in Neurobiology 42, https://doi.org/10.1007/978-3-031-69832-3_7

injury but none in the other models, which underscores the need for studies evaluating the therapies reported after CCI injury as well as novel potential approaches.

Keywords Cognition · Controlled cortical impact · Fluid percussion · Morris water maze · Recovery · Traumatic brain injury

Abbreviations

AHT	Abusive head trauma
CCI	Controlled cortical impact
FPI	Fluid percussion injury
HPA	Hypothalamic-pituitary-adrenal
IL	Interleukin
ICP	Intracranial pressure
MS	Maternal separation
PND	Postnatal day
PFC	Prefrontal cortex
SBS	Shaken baby syndrome
TBI	Traumatic brain injuries

7.1 Introduction

Traumatic brain injury (TBI) is a significant and worldwide public health issue that places enormous burden on patients and their families every year [48]. Broadly defined as an acute injury, often with alterations in consciousness resulting from mechanical energy to the head from external forces, TBI is the leading cause of death and impairment among children in the United States [11]. Determining the exact prevalence is difficult, but it is estimated that 2.8 million TBIs occur yearly in the United States. Among those, very young children (0–4 years) and adolescents experience the highest proportion of TBIs compared to other age groups. This is particularly concerning as the pediatric brain undergoes rapid developmental changes that may be halted or hindered by the TBI [37]. If left unresolved, the sequelae of TBI can lead to greater developmental difficulties across the lifespan [10, 38].

Clinical care and research for children with TBI is complex due to the heterogeneity in their mechanism, severity, and impact [44]. Although most children survive, they are likely to experience impairments in physical, cognitive, communicative, and emotional domains immediately after TBI and long-term if the insult is severe [29, 55]. A TBI can thus be conceptualized as both acute and chronic conditions. While there are limited longitudinal studies on children who survive TBI, existing research suggests childhood TBI increases the risk of adverse outcomes in adulthood, such as lower educational attainment, reduced employment opportunities, and limited social interactions [10]. As a result, a destructive cycle may ensue and lead to a disproportionate number of TBI survivors experiencing homelessness and incarceration later in life [43].

Despite the significance of this disease, high-level recommendations for pediatric TBI are not available in the current pediatric guidelines due to a lack of high-quality successful clinical studies and randomized clinical trials [40]. Unfortunately, clinical trials for TBI have failed to provide significant benefits [7]. Perhaps one of the most salient reasons for the lack of clinical success is the heterogeneity of TBI, with varying degrees of severity and locations of injury. Additionally, the complex pathophysiology of TBI, which encompasses multiple mechanisms of injury and secondary injury from neurochemical and histological changes, makes it difficult to focus on a particular therapeutic target. The heterogeneity in individual characteristics, including age, prior health, and social factors that are known to drive disparities in outcomes, is also a factor that may influence clinical results [22, 31]. Ethical considerations, safety concerns, and poor adherence to intervention protocols can further complicate knowledge acquisition through clinical research [24].

Albeit not without limitations, there are many potential benefits to using rigorous preclinical models of TBI, which may reduce or eliminate some of the factors proposed as limiting clinical success and advance the evidence base for pediatric care and recovery [61]. Preclinical models afford the opportunity to control the heterogeneity in injury severity, location, and mechanism as well as eliminating the influence of social determinants of health. Moreover, these changes can be done without undergoing lengthy enrollment periods or sacrificing sample size. Further, preclinical models provide the opportunity to gather data only possible through tissue samples that cannot be collected from living children.

7.2 Considerations When Modeling Pediatric TBI: Age and Type of Injury

During childhood, brain development has differential milestones and time-windows that could change the dynamics of the effects of TBI. For example, young infants are more susceptible to shaking injuries due to the ratio of head-to-body size, while adolescence is a critical period for impulse control and social behavior that depend on appropriate synaptogenesis and pruning processes in the prefrontal cortex (PFC). Therefore, age plays a significant role when modeling pediatric TBI. Rodents have faster postnatal development but greater central nervous system immaturity at birth compared to humans. Understanding these developmental differences is crucial for accurate TBI modeling. The fundamental processes involving the birth, migration, and death of neurons occur in a similar timeframe. Although these processes are essentially complete in most brain regions at birth, they may continue for up to 18 months in the slower maturing human PFC. The proliferation of neuronal processes, synapse formation, and maturation of neurotransmitter systems occur during early childhood, leading to a substantial increase in synapses that require pruning to refine neural circuits. This process is partly influenced by experience and extends into young adulthood. PFC maturity, including full synaptogenesis and myelination, is achieved at approximately 17–25 years of age in humans, contrasting with postnatal day (PND) 90 in rats.

Another important consideration is that pediatric TBI can result from various causes, each presenting unique features. Falls are often characterized by a single high-impact event. In contrast, shaken baby syndrome (SBS) involves recurrent concussive injuries caused by violent shaking in infants, while sports-related TBIs and automobile accidents in older children have different characteristics due to the involvement of acceleration and deceleration forces. Moreover, abusive head trauma encompasses not only the features associated with TBI but also its interaction with previous exposure to stress commonly associated with abuse. Therefore, diverse TBI models have been developed, or are continually being developed, to mimic these different pathophysiological components of brain trauma (Fig. 7.1).

7.3 Preclinical Models of TBI

Preclinical models of pediatric TBI utilize key developmental processes, encompassing central nervous system milestones such as neurogenesis, synaptogenesis, and myelination, as well as age- and sex-dependent molecular and behavioral alterations to align the maturation of rodents with that of pediatric-aged humans. Therefore, to model pediatric TBI, studies utilize rodents aged between postnatal days 7–11 and 17–21, mirroring the clinical age ranges of an "infant" and "toddler," respectively. In this chapter, our focus centers on preclinical models of pediatric TBI, encompassing studies conducted across the PND 7–21 age range. To this end, not all facets of pediatric development and brain injury can be recapitulated using preclinical TBI models. Therefore, only certain elements of brain maturation relevant to secondary injury and neurobehavioral outcomes can be fully assessed.

Originally developed for adult rodents, several well-established experimental models of TBI have been adapted and scaled for pediatric-aged rodents. Models like the controlled cortical impact (CCI), weight drop (closed/open skull), and fluid percussion (FP) have been further modified to introduce a range in injury severities from mild to severe. In addition, repeated insult frequency has been characterized to resemble repeated concussive-like injuries. These models produce secondary injury milieu, including neuroinflammation, oxidative stress, disturbances in neurochemical signaling, and functional/behavioral deficits commonly observed in adult TBI. Albeit limited, several studies have shown sex differences after pediatric TBI, including physiological response, pathogenesis, behavioral outcomes, and therapeutic effectiveness. Therefore, similar histological, molecular, and behavioral assays are utilized to assess the age-, time-, and sex-dependent outcome measures of pediatric injury. The subsequent sections will discuss the various pediatric TBI models, delve into the molecular and cellular mechanisms driving acute and chronic injury-related pathophysiology, investigate TBI-induced deficits in behavioral outcomes, and discuss the potential therapeutic targets for TBI in the developing brain. Also, because the majority of pediatric TBI cases resulting in death are associated

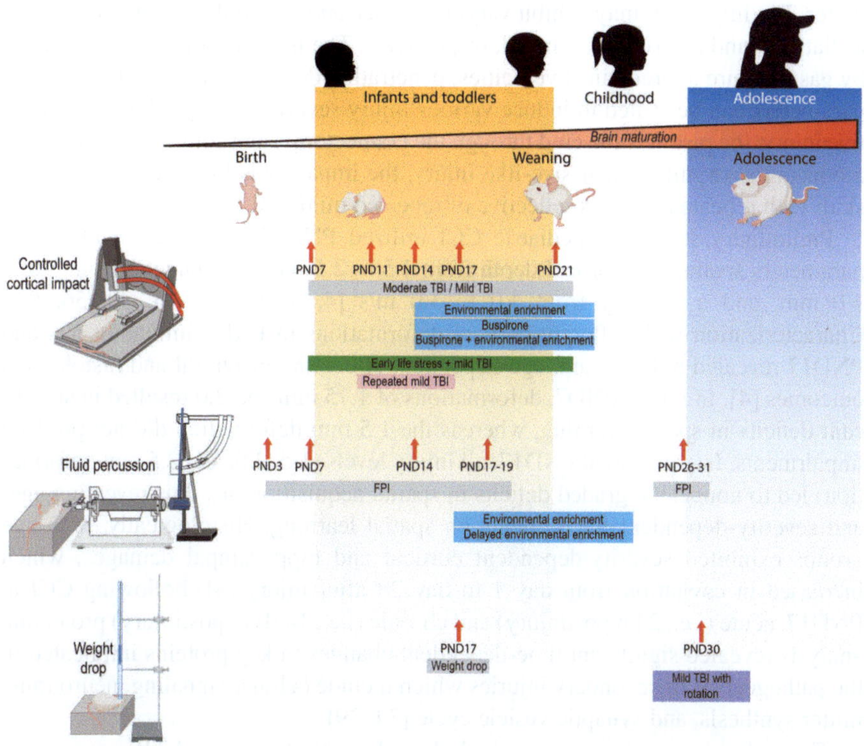

Fig. 7.1 Schematic representing a timeline of human brain development and its corresponding stage in rodents, along with descriptions of the different models employed to simulate various types of injuries. Controlled cortical impact (CCI) has been utilized in both single and repeated injury models to mimic injuries in infants and toddlers (rats from PND7 to PND21). CCI has also been employed to create a model of abusive head trauma by combining it with an early stress model. Fluid percussion has been used to model injuries in infants (PND3–PND19) as well as in adolescents, while the weight drop model has primarily been used at PND17. An additional weight drop model modified to include rotational movements has been used in adolescents to simulate sports-related injuries. These models have been instrumental in evaluating various therapeutic strategies

with child abuse [50], we will briefly discuss a model of early life adversity combined with TBI from our laboratories (see Chap. 9 for a more complete discussion of stressors plus TBI).

7.3.1 *Controlled Cortical Impact*

The most widely employed model for pediatric TBI is the CCI that was originally designed for adult rats [16]. The CCI model consists of an affixed pneumatic impactor positioned through a craniectomy either perpendicular or vertical to the dura mater. The impact tip may exhibit varying surface and material characteristics, such as flat or round and rubber or metal, respectively. The impactor tip is then propelled by gas pressure at predefined velocities, penetration depths, and dwell times. These parameters can be scaled to induce various injury severities. To produce a penetrative injury, the piston is lowered through the craniectomy, and conversely to produce a noncontusive mild concussive-like injury, the impactor is lowered onto a closed skull with repeated impacts reflective of repeated mild TBI (mTBI).

Preliminary studies in pediatric CCI utilized PND17–21 rodents with injury parameters spanning a cortical depth from 1.5 to 2.5 mm, tip diameter from 3.0 to 6.0 mm, and a velocity from 4.0 to 5.0 m/s [4, 8, 32, 34, 35, 39, 56, 63]. Characterization with different cortical deformations in PND7 immature rats and PND17 revealed a depth- and age-dependent effect on behavioral and histological outcomes [4]. In rats at PND7, deformations of 1.75 mm and 2.0 resulted in significant deficits in spatial learning, whereas the 1.5 mm deformation did not produce impairments. In contrast, at PND17, all injury levels (1.5, 2.0, and 2.5 mm deformation) led to noticeable graded deficits in spatial acquisition [4], thus revealing age- and severity-dependent impairments on spatial learning. Histologically, both age groups exhibited severity-dependent cortical and hippocampal damages, which increased in cavitation from day 1 to day 28 after injury [4]. Following CCI at PND17, acute (i.e., 24 h postinjury) and chronic (i.e., 14 days postinjury) proteomic analysis revealed significant time-dependent changes in key proteins implicated in the pathogenesis of secondary injuries which include cellular signaling, neurotransmitter synthesis, and synaptic vesicle cycle [34, 39].

The pediatric population is particularly vulnerable to repeated TBIs over one's lifetime, especially in cases of abusive head trauma and sports-related injuries. As such, it is critical to elucidate the neurobehavioral outcomes following repeated TBIs. The adaptability of the CCI device allows for manipulation in severity and frequency to impart a mild or a mild repeated impact. Expanding on prior research using a metal impacting tip, Huh and colleagues [32] modified the impact by incorporating a silicone tip. They subsequently administered 1–3 mild noncontusive impacts to simulate mild repeated TBIs in PND11 male and female pups. The repeated injuries resulted in a graded histological response such that increased reactive astrocytosis, white matter atrophy, enlarged ventricles, and calpain activation were prominent in the rats that received two or three impacts [32]. Despite histopathological markers across injuries, neither one nor two impacts led to spatial learning deficits [32]. Supporting previous findings, evaluation of rmTBI on short-term histological outcomes demonstrated that pediatric rats (i.e., PND17) subjected to one or two mTBIs exhibited significant axonal damage within the external capsule at the impact site and the corpus callosum [19]. After two impacts, there was an exacerbation in colabeled inducible nitric oxide synthase (iNOS)-positive microglia, signaling the presence of a proinflammatory microglial subtype. Furthermore, rmTBI resulted in persistent deficits in associative fear learning over an extended duration. A sustained presence of axonal silver degeneration was evident at the 21-day assessment, but this subsided by postinjury day 92 [19].

Albeit limited, several studies have also found evidence that sex and sex-related hormones play a role in the outcomes of pediatric TBI. Semple et al. [58] demonstrated that male mice displayed greater impairments in social recognition and reduced dendritic complexity in hippocampal and mPFC neurons [58]. Conversely, brain-injured females had preserved dendritic complexity but a higher sociosexual avoidance. These results suggest that sex-specific neuronal changes may contribute to psychosocial deficits after pediatric TBI [58]. Studies have also examined the efficacy of progesterone, a neuroactive steroid hormone implicated in several neuroprotective pathways, following pediatric TBI and have revealed its sex-dependent therapeutic efficacy [23, 56] (see Therapies, Sect. 7.4).

7.3.2 *Weight Drop*

The weight drop model is a contusion injury model, originally developed for spinal cord injury by Allen and colleagues [6] and later adapted for sensorimotor cortex injury by Feeney et al. [18] to produce a contusion injury that imparts progressive histopathological and behavioral deficits. Briefly, a tube with a specified length guides a falling weight onto a footplate that is placed upon the intact dura [18]. The force of the impact can be altered by changing the mass of the weight and the distance dropped, thereby producing graded injury severities. More recently, studies have investigated the effect of noncontusive close-head injury and frequency of impacts using this model. Grundl et al. [27] adjusted the weight drop model for use in early adolescent rats (PND 26–31) by scaling the impact to the hemispheric mass of the immature brain. Results demonstrated that injury increased edema and acute blood flow, which was postulated to be due to several factors, including age-dependent differences in blood brain permeability, susceptibility to ischemia, and inflammatory states [27]. Several studies further characterized this model for use in pediatric-aged rats demonstrating a weight-dependent effect on injury severity that imparts graded functional, cognitive, and histological deficits. Modifications to the weight drop, including closed-head and/or repeated impacts, have modeled mild and repeated mild type injuries. Specifically, anesthetized rats are placed directly under a hollow guide tube, and a subsequent impact is delivered by a released weight at a specified distance resulting in rotational acceleration of the head.

Early investigations utilizing light (i.e., 100 g) and heavy (i.e., 150 g) weights kept at a constant drop distance produced a moderate-severe and severe injury, respectively, in PND17 rats. Results demonstrated that acutely following injury, the severe impact significantly decreased motor and spatial learning and memory [3]. In a subsequent study, the long-term functional and cognitive abilities of immature rats were assessed using the previously examined severe injury parameters (150 g/2 m) [2]. Assessment of motor performance via beam balance and inclined plane revealed acute but reversible motor deficits 10 days after injury. Assessment of cognition revealed sustained deficits to spatial learning and memory up to 3 months following injury [2]. Postinjury subarachnoid and intraventricular hemorrhage, diffuse axonal injury, and astrogliosis exhibited a time-dependent reactivity, such that acutely following injury these were markedly increased in PND17 rats [5]. Following impact, axonal injury occurring in the long ascending and descending fiber tracts of the brain stem was maximal at 3 days albeit, present until 7 days postinjury [5]. There were no significant histopathological abnormalities 90 days after TBI. Although histopathological assessments did not persist following impact, previous studies demonstrated persistent cognitive deficits suggesting that additional underlying mechanisms, such as synaptic dysfunction or neurotransmitter hypofunction, may also contribute to chronic cognitive impairments [2, 5].

Alternative mechanisms underlying observed cognitive deficits may be due to injury-induced secondary pathogenesis, namely, increased reactive oxidative species and inflammation. Immature rats, aged 3.5–4.5 weeks, subjected to weight drop injury exhibited a gradual increase in the production of iNOS, the enzyme responsible for generating nitric oxide (NO) [13]. Specifically, 2 h after the injury, iNOS was localized to the peritrauma area in the cerebrovascular region. However, at 24 and 48 h postinjury, there was a significant increase in iNOS expression in vascular smooth muscle cells, and it was found to be colocalized with infiltrated neutrophils, indicating a clear association between iNOS and inflammation. By day 7, iNOS expression had subsided. These findings suggest that the expression of inflammation and NO-mediated cerebrovascular disturbances after pediatric brain trauma is both location- and time-dependent [13].

Additionally, weight drop injury in immature rats increased cellular apoptosis and neurodegeneration. Peak neuronal cell death observed at 12–24 h post-injury coincided with a significant increase in mRNA and protein levels for caspase-1, interleukin (IL)-1β, and IL-18 within 2–12 h following injury in PND7 rats [59]. Caspase-1 levels remained elevated for 72 h, while IL-1β decreased at 48 h. IL-18 remained elevated for 3 days but returned to normal levels by day 7 after the injury. These findings suggest a dynamic time-dependent relationship between cytokines (IL-1β and IL-18) and caspase-1 in trauma-induced neuronal cell death in the developing brain [59]. When postinjury apoptotic neurodegeneration was assessed as a function of age (i.e., PND3, 7, 14, and 30), findings revealed an age-dependent pattern in apoptotic severity, with PND3 and PND7 rats displaying the highest levels of apoptotic cell death, followed by a gradual decline with increasing age [9]. Together these studies suggest that time and age impact postinjury function and cognitive and histopathological outcome following moderate-to-severe injury.

Further modifications to the weight drop, including closed-head and/or repeated impacts, has modeled mild and repeated mild type injuries. Specifically, anesthetized rats are positioned on delicate holders (e.g., Kimwipe or aluminum foil) directly under a hollow guide tube, and as previously described, an impact is delivered by a released weight at a specified distance resulting in rotational acceleration of the head as it is pushed through the thin placement material. Assessment of rmTBI on PND30 male and female rats demonstrated time- and sex-dependent outcomes [21]. Results showed that rmTBI negatively impacted motor skills, cognitive abilities, and exploratory behavior as assessed using the beam-walk, elevated plus maze, and novel context mismatch tasks, respectively, in both sexes. Additionally, increased inflammatory markers (CXCL1, IL-5, VEGFa) were found in both blood plasma and the cerebellum shortly after injury. Microglia/macrophage activation markers (Iba1 and CX3CR1) remained elevated, suggesting immune cell activation in the cerebellum. At 72 h postinjury, cerebellar levels of CXCL1, CCL5, and VEGFa correlated with time spent in the open arms of the elevated plus maze, hinting at a link between anxiety-like behavior and increased cytokines and growth factors after mTBI in males but not females [21].

7.3.3 *Fluid Percussion*

Fluid percussion injury (FPI) is a model of experimental TBI producing diffuse brain injury, adapted from other animals for use in rats to produce graded injury in a behaviorally and histologically well-characterized species [17]. Lateral FPI is most common, and in a pediatric model of TBI, FPI can be induced as early as PND17, the youngest time at which pups have the minimum necessary skull thickness to support the FPI cap [54]. The ages for pups undergoing this injury paradigm typically range from PND17 to PND28. For this procedure, the rats are fully anesthetized and placed into a stereotaxic frame. A craniotomy 3–4.8 mm in diameter is created using a high-speed drill. To do this, the plastic injury cap is attached using silicone adhesive, cyanoacrylate, and dental cement. When the cement hardens, the injury cap is filled with saline, anesthesia is briefly discontinued, and the animal is attached to the FPI device by the injury cap. The injury is induced by lifting and releasing a metal pendulum that strikes the piston end of the tube from a predetermined height depending on desired injury severity, creating a pulse of intracranial pressure [47]. Apnea time is measured by the resumption of spontaneous respiration and unconscious time by return of the hind limb reflex. Upon return of hind limb withdrawal reflex, anesthesia is continued, and the injury cap is removed before closure of the surgical opening [25, 26, 28, 33, 54]. Injury severity can be assessed through unconscious time post-FPI, as it has been shown to correlate with fluid pulse pressure and functional outcome in adults [17].

FPI yields many neurobehavioral effects that correspond with injury severity. Subsequent measures such as mortality, intracranial pressure (ICP), and mean arterial blood pressure (MABP) provide insight into age-dependent effects of FPI. For 30 min after FPI in PND17 rats across mild, moderate, and severe injury levels, ICP and MABP increased with injury severity, as did mortality rate. Developing rats showed pronounced hypotension compared to adults after injury, following the trend of greater mortality among the younger rats as well [54].

FPI is a diffuse model of TBI, causing immediate and indiscriminate depolarization and glutamate and potassium efflux from neurons into the extracellular space [25, 26]. This release is followed by an acute increase of cerebral glucose use, before several days of metabolic depression and a prolonged influx of calcium in adult rats. However, an age-related difference is observed in injured pups, who show a nearly indetectable reduction in glycolysis following the acute increase in cerebral glucose uptake [25]. Additionally, comparable injury severity between adult rats and pups does not yield the same morphological damage. In PND17–19-injured pups, the acute cell death following lateral FPI and reduction in quantity of cortical or hippocampal neurons is far less than that seen in adults with the same injury severity [28]. Histologically, FPI did not significantly affect the quantity of neurons in the ipsilateral cortex or CA3 region in pediatric rats, no matter the injury severity. However, the injured rats showed severity-independent deficits in behavior as seen through greater escape latency and trials to criterion in the Morris water maze (MWM), albeit there was no observed deficit in storage and recall as seen in the probe trial. These results indicate that while there is no observed significant neuronal death following FPI in pups, neuronal dysfunction may be the culprit underlying the mild behavioral deficits [28].

Despite the lack of injury-induced cell death in pediatric rats, there are still observed deficits in neuronal plasticity, as seen through injured rats not responding to an enriched environment (EE) as robustly as noninjured rats [33]. While early EE exposure following lateral FPI in adult rats decreases injury-induced behavioral deficits, it has little effect in improving neuronal plasticity in PND17–19 counterparts [33]. The efficacy of EE in pups following lateral FPI has been shown to be time-dependent, with pups receiving delayed EE as opposed to early EE seeing reduced impairments in learning and memory as measured by the acquisition of spatial learning [25]. These findings show that pediatric TBI produces neurobehavioral deficits in experience-dependent plasticity, with time-dependent improvements provided by EE.

Characterization of FPI in pediatric rats has shown that this diffuse model of injury produces age- and severity-dependent differences in physiology underlying neurobehavioral symptoms and plasticity. Continuing experimentation with this model may further elucidate the molecular pathways underlying these disparities. Establishing sex differences following TBI with the FPI model will also allow for increased translatability to the clinical setting, leading to more targeted treatments to reduce the neurobehavioral and cognitive deficits resulting from diffuse injury.

7.3.4 Abusive Head Trauma

Abusive head trauma (AHT) accounts for approximately 12% of physical injuries resulting from child abuse, yet the majority of pediatric TBI cases that result in fatalities are associated with child abuse [50, 60]. Survivors of AHT show higher probabilities of adverse outcomes compared to those who have experienced accidental brain injuries [1], and they represent 90% of post-TBI disability cases [66]. While SBS has been the prevailing definition for brain injuries related to maltreatment, AHT is now considered more accurate [12]. This shift is because SBS exclusively pertains to brain injuries resulting from shaking in infants, whereas AHT encompasses a broader range of nonaccidental traumatic injuries to a child's head [45, 51].

Child abuse, neglect, and violence toward children are forms of early life stress (ELS) that affect brain development and cause a chronic activation of the stress response [30, 62]. Activation of the hypothalamic-pituitary-adrenal (HPA) axis is an essential mechanism for coping with environmental threats that disrupt homeostasis [36, 46]. Acute activation of HPA axis is a highly adaptive response that increases the chances of survival. However, chronic stress exposure can lead to excessive release of glucocorticoids and contribute to psychopathology. This is especially important in the case of child abuse, as the areas responsible for HPA axis negative feedback are still developing during childhood and are highly vulnerable to stress.

Maternal separation (MS) is the most widely used rodent model of early life stress [53, 65]. MS induces a periodic activation of the HPA axis response during the first weeks of life [41] that is accompanied by a chronic elevation in corticotropin-releasing hormone (CRH) expression in the hippocampus and hypothalamus [57]. Thus, to develop an appropriate model capable of accurately replicating the relationship between abuse-induced chronic stress and pediatric TBI, a few studies have examined the combined effects of MS and TBI. Diaz Chávez and colleagues [15] performed a mild CCI in young rats that were previously subjected to MS. The results showed that although MS and mTBI did not affect cognitive performance when presented alone, the combination of stress and TBI negatively impacted spatial learning [15]; however, this additive effect was not observed when evaluating cognitive flexibility in the attentional set shifting test. Additionally, MS increased HPA axis markers such as hypothalamic CRH and corticosterone levels, regardless of injury, which suggests a region-specific effect of stress on the hippocampal formation [42].

It has been proposed that the local immunological response, specifically microglial activity, plays a significant role in the pathophysiology of both child abuse and pediatric TBI [14, 20]. Like the stress system, the activation of the immune response during the acute phase of TBI is a homeostatic mechanism that facilitates the removal of cellular debris and tissue repair. However, if this immune response is not efficiently resolved, it can lead to chronic inflammation and persistent damage [52]. In the chronic phase of pediatric TBI, there is an increase in the proportion of activated microglial cells on the ipsilateral, but not on the contralateral hemisphere, with no significant effect on cytokine expression. However, in the AHT model that combines both MS and mTBI, this heightened proportion of activated microglial cells is accompanied by a higher expression of IL1B, indicative of a primed phenotype. Additionally, there is an increase in the proportion of microglial cells in the contralateral CA1 region [42]. These results suggest that ELS-induced priming of hippocampal microglia could play an important role on the increased vulnerability observed in victims of AHT.

7.4 Therapies

Considering the studies that mechanistically link increased oxidative stress and mitochondrial dysfunction to pediatric CCI, several research investigations have explored potential therapies aimed at scavenging and neutralizing reactive oxygen species and free radicals. The administration of XJB-5-131, a novel pharmacologic that targets and scavenges reactive oxygen specific and free radicals within the mitochondria, reduced CL oxygenation and cortical lesion volume and improved spatial learning and associative memory [35]. Pharmacologic inhibition of COX-2 via SC8125 significantly reduced the CCI-induced COX-2 product, prostaglandin E2 (PGE2), and improved spatial learning and memory in PND17 rats. Alternatively, transgenic overexpression of glutathione peroxidase (GPx), the enzyme that metabolizes free radicals and reduces oxidative stress, exhibited an acute reduction in markers of oxidative stress following CCI in PND21 mice [64]. At 3 months postinjury, the GPx-injured mice exhibited significantly reduced cortical lesion volume and improved spatial memory, albeit worsening sensorimotor and nonspatial memory after injury, indicating a partial mitigation of TBI-induced histopathology and behavioral functioning through GPx overexpression [64]. Once daily postinjury administration of buspirone, a serotonin$_{1A}$ receptor agonist, following CCI showed benefits in spatial learning and a decrease in cortical lesion volume. Notably, when buspirone was combined with environmental enrichment (EE), a preclinical model of neurorehabilitation, an additive therapeutic benefit was achieved [49]. Progesterone has been reported to preserve mitochondrial function and prevent the loss of mitochondrial glutathione in injured male rats. However, in female rats, progesterone did not significantly improve mitochondrial function but did show greater tissue sparing [56].

7.5 Conclusion

In summary, the preclinical models described have been instrumental in advancing our knowledge of experimental pediatric TBI, offering valuable insights into the neurobehavioral mechanisms underlying injury-, temporal-, age, and sex-dependent responses. These findings provide some clarity into the complex injury process and highlight potential therapeutic intervention targets that may lead to increased clinical translation. Potential therapies, such as antioxidants, progesterone, and the serotonin$_{1A}$ receptor agonist buspirone alone or combined with EE, were described in studies utilizing the CCI injury model. Unfortunately, no treatments were evaluated in the other TBI models, which underscore the urgent need for such studies.

Acknowledgments This work was supported, in part, by NIH grants NS084967, NS121037 (AEK), and NS110609 (COB). EHM is a PhD candidate and is supported in part by a Brain Injury Association of America Fund as well as a Research Advisory Committee grant supported by the Children's Hospital of Pittsburgh of the UPMC Health System. JZF is a graduate student in the

Programa de Maestría en Ciencias en Ecología Integrativa—Universidad Michoacana de San Nicolás de Hidalgo, supported by CONACYT (CVU 1187161).

Disclosures None.

References

1. Adamo KB, Prince SA, Tricco AC, Connor-Gorber S, Tremblay M. A comparison of indirect versus direct measures for assessing physical activity in the pediatric population: a systematic review. Int J Pediatr Obes. 2009;4(1):2–27. https://doi.org/10.1080/17477160802315010. PMID: 18720173.
2. Adelson PD, Dixon CE, Kochanek PM. Long-term dysfunction following diffuse traumatic brain injury in the immature rat. J Neurotrauma. 2000;17(4):273–82. https://doi.org/10.1089/neu.2000.17.273. PMID: 10776912.
3. Adelson PD, Dixon CE, Robichaud P, Kochanek PM. Motor and cognitive functional deficits following diffuse traumatic brain injury in the immature rat. J Neurotrauma. 1997;14(2):99–108. https://doi.org/10.1089/neu.1997.14.99. PMID: 9069441.
4. Adelson PD, Fellows-Mayle W, Kochanek PM, Dixon CE. Morris water maze function and histologic characterization of two age-at-injury experimental models of controlled cortical impact in the immature rat. Childs Nerv Syst. 2013;29(1):43–53. https://doi.org/10.1007/s00381-012-1932-4. Epub 2012 Oct 23. PMID: 23089934.
5. Adelson PD, Jenkins LW, Hamilton RL, Robichaud P, Tran MP, Kochanek PM. Histopathologic response of the immature rat to diffuse traumatic brain injury. J Neurotrauma. 2001;18(10):967–76. https://doi.org/10.1089/08977150152693674. PMID: 11686497.
6. Allen AR. Surgery of experimental lesion of spinal cord equivalent to crush injury of fracture dislocation of Spinal Column. JAMA. 1911;LVII:878. https://doi.org/10.1001/jama.1911.04260090100008.
7. Appavu B, Foldes ST, Adelson PD. Clinical trials for pediatric traumatic brain injury: definition of insanity? J Neurosurg Pediatr. 2019;23(6):661–9. https://doi.org/10.3171/2019.2.PEDS18384. PMID: 31153150.
8. Bayir H, Tyurin VA, Tyurina YY, Viner R, Ritov V, Amoscato AA, Zhao Q, Zhang XJ, Janesko-Feldman KL, Alexander H, Basova LV, Clark RS, Kochanek PM, Kagan VE. Selective early cardiolipin peroxidation after traumatic brain injury: an oxidative lipidomics analysis. Ann Neurol. 2007;62(2):154–69. https://doi.org/10.1002/ana.21168. PMID: 17685468.
9. Bittigau P, Sifringer M, Felderhoff-Mueser U, Hansen HH, Ikonomidou C. Neuropathological and biochemical features of traumatic injury in the developing brain. Neurotox Res. 2003;5(7):475–90. https://doi.org/10.1007/BF03033158. PMID: 14715432.
10. Blackwell LS, Grell R. Pediatric traumatic brain injury: impact on the developing brain. Pediatr Neurol. 2023:S0887-8994(23)00191-1. https://doi.org/10.1016/j.pediatrneurol.2023.06.019. Epub ahead of print. PMID: 37652817.
11. Blennow K, Brody DL, Kochanek PM, Levin H, McKee A, Ribbers GM, Yaffe K, Zetterberg H. Traumatic brain injuries. Nat Rev Dis Primers. 2016;2:16084. https://doi.org/10.1038/nrdp.2016.84. PMID: 27853132.
12. Christian CW, Block R. Committee on Child Abuse and Neglect; American Academy of Pediatrics. Abusive head trauma in infants and children. Pediatrics. 2009;123(5):1409–11. https://doi.org/10.1542/peds.2009-0408. PMID: 19403508.
13. Clark RS, Kochanek PM, Schwarz MA, Schiding JK, Turner DS, Chen M, Carlos TM, Watkins SC. Inducible nitric oxide synthase expression in cerebrovascular smooth muscle and neutrophils after traumatic brain injury in immature rats. Pediatr Res. 1996;39(5):784–90. https://doi.org/10.1203/00006450-199605000-00007. PMID: 8726229.

14. Desplats P, Gutierrez AM, Antonelli MC, Frasch MG. Microglial memory of early life stress and inflammation: susceptibility to neurodegeneration in adulthood. Neurosci Biobehav Rev. 2020;117:232–42. https://doi.org/10.1016/j.neubiorev.2019.10.013. Epub 2019 Nov 5. PMID: 31703966; PMCID: PMC7198341.
15. Diaz-Chávez A, Lajud N, Roque A, Cheng JP, Meléndez-Herrera E, Valdéz-Alarcón JJ, Bondi CO, Kline AE. Early life stress increases vulnerability to the sequelae of pediatric mild traumatic brain injury. Exp Neurol. 2020;329:113318. https://doi.org/10.1016/j.expneurol.2020.113318. Epub 2020 Apr 16. PMID: 32305419; PMCID: PMC7245477.
16. Dixon CE, Clifton GL, Lighthall JW, Yaghmai AA, Hayes RL. A controlled cortical impact model of traumatic brain injury in the rat. J Neurosci Methods. 1991;39(3):253–62. https://doi.org/10.1016/0165-0270(91)90104-8. PMID: 1787745.
17. Dixon CE, Lyeth BG, Povlishock JT, Findling RL, Hamm RJ, Marmarou A, Young HF, Hayes RL. A fluid percussion model of experimental brain injury in the rat. J Neurosurg. 1987;67(1):110–9. https://doi.org/10.3171/jns.1987.67.1.0110. PMID: 3598659.
18. Feeney DM, Boyeson MG, Linn RT, Murray HM, Dail WG. Responses to cortical injury: I. Methodology and local effects of contusions in the rat. Brain Res. 1981;211(1):67–77. https://doi.org/10.1016/0006-8993(81)90067-6. PMID: 7225844.
19. Fidan E, Lewis J, Kline AE, Garman RH, Alexander H, Cheng JP, Bondi CO, Clark RS, Dezfulian C, Kochanek PM, Kagan VE, Bayır H. Repetitive mild traumatic brain injury in the developing brain: effects on long-term functional outcome and neuropathology. J Neurotrauma. 2016;33(7):641–51. https://doi.org/10.1089/neu.2015.3958. Epub 2015 Dec 1. PMID: 26214116; PMCID: PMC4827290.
20. Fraunberger E, Esser MJ. Neuro-inflammation in pediatric traumatic brain injury-from mechanisms to inflammatory networks. Brain Sci. 2019;9(11):319. https://doi.org/10.3390/brainsci9110319. PMID: 31717597; PMCID: PMC6895990.
21. Fraunberger EA, DeJesus P, Zanier ER, Shutt TE, Esser MJ. Acute and persistent alterations of cerebellar inflammatory networks and glial activation in a rat model of pediatric mild traumatic brain injury. J Neurotrauma. 2020;37(11):1315–30. https://doi.org/10.1089/neu.2019.6714. Epub 2020 Feb 25. PMID: 31808365.
22. Garg A, Lobner K, Song J, Mitchell R, Egbunine A, Kudchadkar SR. Social determinants of health in pediatric rehabilitation for children with traumatic injury: a systematic review. J Pediatr. 2023;259:113459. https://doi.org/10.1016/j.jpeds.2023.113459. Epub 2023 May 11. PMID: 37172806; PMCID: PMC10524504.
23. Geddes RI, Peterson BL, Stein DG, Sayeed I. Progesterone treatment shows benefit in female rats in a pediatric model of controlled cortical impact injury. PLoS One. 2016;11(1):e0146419. https://doi.org/10.1371/journal.pone.0146419. PMID: 26799561; PMCID: PMC4723082.
24. Gilholm P, Ergetu E, Gelbart B, Raman S, Festa M, Schlapbach LJ, Long D, Gibbons KS, Australian and New Zealand Intensive Care Society Paediatric Study Group. Adaptive clinical trials in pediatric critical care: a systematic review. Pediatr Crit Care Med. 2023;24(9):738–49. https://doi.org/10.1097/PCC.0000000000003273. Epub 2023 May 17. PMID: 37195182.
25. Giza CC, Griesbach GS, Hovda DA. Experience-dependent behavioral plasticity is disturbed following traumatic injury to the immature brain. Behav Brain Res. 2005;157(1):11–22. https://doi.org/10.1016/j.bbr.2004.06.003. PMID: 15617766.
26. Giza CC, Prins ML, Hovda DA, Herschman HR, Feldman JD. Genes preferentially induced by depolarization after concussive brain injury: effects of age and injury severity. J Neurotrauma. 2002;19(4):387–402. https://doi.org/10.1089/08977150252932352. PMID: 11990346.
27. Grundl PD, Biagas KV, Kochanek PM, Schiding JK, Barmada MA, Nemoto EM. Early cerebrovascular response to head injury in immature and mature rats. J Neurotrauma. 1994;11(2):135–48. https://doi.org/10.1089/neu.1994.11.135. PMID: 7932794.
28. Gurkoff GG, Giza CC, Hovda DA. Lateral fluid percussion injury in the developing rat causes an acute, mild behavioral dysfunction in the absence of significant cell death. Brain Res. 2006;1077(1):24–36. https://doi.org/10.1016/j.brainres.2006.01.011. Epub 2006 Feb 20. PMID: 16490184.

29. Hartman ME, Williams CN, Hall TA, Bosworth CC, Piantino JA. Post-intensive-care syndrome for the pediatric neurologist. Pediatr Neurol. 2020;108:47–53. https://doi.org/10.1016/j.pediatrneurol.2020.02.003. Epub 2020 Feb 20. PMID: 32299742; PMCID: PMC7306429.
30. Heim C, Nemeroff CB. The role of childhood trauma in the neurobiology of mood and anxiety disorders: preclinical and clinical studies. Biol Psychiatry. 2001;49(12):1023–39. https://doi.org/10.1016/s0006-3223(01)01157-x. PMID: 11430844.
31. Houtrow AJ, Jarvis JM, Fink E. Children deserve to be protected from traumatic injuries and receive high-quality equitable rehabilitative care when injuries occur. J Pediatr. 2023;261:113586. https://doi.org/10.1016/j.jpeds.2023.113586. Epub 2023 Jun 28. PMID: 37385336.
32. Huh JW, Franklin MA, Widing AG, Raghupathi R. Regionally distinct patterns of calpain activation and traumatic axonal injury following contusive brain injury in immature rats. Dev Neurosci. 2006;28(4–5):466–76. https://doi.org/10.1159/000094172. PMID: 16943669.
33. Ip EY, Giza CC, Griesbach GS, Hovda DA. Effects of enriched environment and fluid percussion injury on dendritic arborization within the cerebral cortex of the developing rat. J Neurotrauma. 2002;19(5):573–85. https://doi.org/10.1089/089771502753754055. PMID: 12042093.
34. Jenkins LW, Peters GW, Dixon CE, Zhang X, Clark RS, Skinner JC, Marion DW, Adelson PD, Kochanek PM. Conventional and functional proteomics using large format two-dimensional gel electrophoresis 24 hours after controlled cortical impact in postnatal day 17 rats. J Neurotrauma. 2002;19(6):715–40. https://doi.org/10.1089/08977150260139101. PMID: 12165133.
35. Ji J, Kline AE, Amoscato A, Samhan-Arias AK, Sparvero LJ, Tyurin VA, Tyurina YY, Fink B, Manole MD, Puccio AM, Okonkwo DO, Cheng JP, Alexander H, Clark RS, Kochanek PM, Wipf P, Kagan VE, Bayır H. Lipidomics identifies cardiolipin oxidation as a mitochondrial target for redox therapy of brain injury. Nat Neurosci. 2012;15(10):1407–13. https://doi.org/10.1038/nn.3195. Epub 2012 Aug 26. PMID: 22922784; PMCID: PMC3697869.
36. Joëls M, Baram TZ. The neuro-symphony of stress. Nat Rev Neurosci. 2009;10(6):459–66. https://doi.org/10.1038/nrn2632. PMID: 19339973; PMCID: PMC2844123.
37. Keenan HT, Clark A, Holubkov R, Ewing-Cobbs L. Longitudinal developmental outcomes of infants and toddlers with traumatic brain injury. JAMA Netw Open. 2023;6(1):e2251195. https://doi.org/10.1001/jamanetworkopen.2022.51195. PMID: 36648943; PMCID: PMC9856699.
38. Keenan HT, Presson AP, Clark AE, Cox CS, Ewing-Cobbs L. Longitudinal developmental outcomes after traumatic brain injury in young children: are infants more vulnerable than toddlers? J Neurotrauma. 2019;36(2):282–92. https://doi.org/10.1089/neu.2018.5687. Epub 2018 Sep 6. PMID: 30019631; PMCID: PMC6338576.
39. Kochanek AR, Kline AE, Gao WM, Chadha M, Lai Y, Clark RS, Dixon CE, Jenkins LW. Gel-based hippocampal proteomic analysis 2 weeks following traumatic brain injury to immature rats using controlled cortical impact. Dev Neurosci. 2006;28(4–5):410–9. https://doi.org/10.1159/000094167. PMID: 16943664; PMCID: PMC2721469.
40. Kochanek PM, Tasker RC, Carney N, Totten AM, Adelson PD, Selden NR, Davis-O'Reilly C, Hart EL, Bell MJ, Bratton SL, Grant GA, Kissoon N, Reuter-Rice KE, Vavilala MS, Wainwright MS. Guidelines for the management of pediatric severe traumatic brain injury, third edition: update of the brain trauma foundation guidelines. Pediatr Crit Care Med. 2019;20(3S Suppl 1):S1–S82. https://doi.org/10.1097/PCC.0000000000001735. Erratum in: Pediatr Crit Care Med. 2019 Apr;20(4):404. PMID: 30829890.
41. Lajud N, Roque A, Cajero M, Gutiérrez-Ospina G, Torner L. Periodic maternal separation decreases hippocampal neurogenesis without affecting basal corticosterone during the stress hyporesponsive period but alters HPA axis and coping behavior in adulthood. Psychoneuroendocrinology. 2012;37(3):410–20. https://doi.org/10.1016/j.psyneuen.2011.07.011. Epub 2011 Aug 20. PMID: 21862224.

42. Lajud N, Roque A, Cheng JP, Bondi CO, Kline AE. Early life stress preceding mild pediatric traumatic brain injury increases neuroinflammation but does not exacerbate impairment of cognitive flexibility during adolescence. J Neurotrauma. 2021;38(4):411–21. https://doi.org/10.1089/neu.2020.7354. Epub 2020 Nov 6. PMID: 33040677; PMCID: PMC7875608.
43. Mackelprang JL, Harpin SB, Grubenhoff JA, Rivara FP. Adverse outcomes among homeless adolescents and young adults who report a history of traumatic brain injury. Am J Public Health. 2014;104(10):1986–92. https://doi.org/10.2105/AJPH.2014.302087. Epub 2014 Aug 14. PMID: 25122029; PMCID: PMC4167112.
44. Madurski C, Jarvis JM, Beers SR, Houtrow AJ, Wagner AK, Fabio A, Wang C, Smith CM, Doughty L, Janesko-Feldman K, Rubin P, Pollon D, Treble-Barna A, Kochanek PM, Fink EL, PICU-Rehabilitation Study Group. Serum biomarkers of regeneration and plasticity are associated with functional outcome in pediatric neurocritical illness: an exploratory study. Neurocrit Care. 2021;35(2):457–67. https://doi.org/10.1007/s12028-021-01199-z. Epub 2021 Mar 4. PMID: 33665769; PMCID: PMC9344468.
45. Maiese A, Iannaccone F, Scatena A, et al. Pediatric abusive head trauma: a systematic review. Diagnostics. 2021;11:734. https://doi.org/10.3390/diagnostics11040734.
46. McEwen BS. Brain on stress: how the social environment gets under the skin. Proc Natl Acad Sci U S A. 2012;109(Suppl 2(Suppl 2)):17180–5. https://doi.org/10.1073/pnas.1121254109. Epub 2012 Oct 8. Erratum in: Proc Natl Acad Sci U S A. 2013 Jan 22;110(4):1561. PMID: 23045648; PMCID: PMC3477378.
47. McIntosh TK, Vink R, Noble L, Yamakami I, Fernyak S, Soares H, Faden AL. Traumatic brain injury in the rat: characterization of a lateral fluid-percussion model. Neuroscience. 1989;28(1):233–44. https://doi.org/10.1016/0306-4522(89)90247-9. PMID: 2761692.
48. Miley AE, Fisher AP, Moscato EL, Culp A, Mitchell MJ, Hindert KC, Makoroff KL, Rhine TD, Wade SL. A mixed-methods analysis examining child and family needs following early brain injury. Disabil Rehabil. 2022;44(14):3566–76. https://doi.org/10.1080/09638288.2020.1870757.
49. Monaco CM, Gebhardt KM, Chlebowski SM, Shaw KE, Cheng JP, Henchir JJ, Zupa MF, Kline AE. A combined therapeutic regimen of buspirone and environmental enrichment is more efficacious than either alone in enhancing spatial learning in brain-injured pediatric rats. J Neurotrauma. 2014;31(23):1934–41. https://doi.org/10.1089/neu.2014.3541. Epub 2014 Sep 29. PMID: 25050595; PMCID: PMC4238252.
50. Niederkrotenthaler T, Xu L, Parks SE, Sugerman DE. Descriptive factors of abusive head trauma in young children--United States, 2000–2009. Child Abuse Negl. 2013;37(7):446–55. https://doi.org/10.1016/j.chiabu.2013.02.002. Epub 2013 Mar 25. PMID: 23535075.
51. Parks S, Sugerman D, Xu L, Coronado V. Characteristics of non-fatal abusive head trauma among children in the USA, 2003–2008: application of the CDC operational case definition to national hospital inpatient data. Inj Prev. 2012;18(6):392–8. https://doi.org/10.1136/injuryprev-2011-040234. Epub 2012 Feb 10. PMID: 22328632; PMCID: PMC4772141.
52. Patterson ZR, Holahan MR. Understanding the neuroinflammatory response following concussion to develop treatment strategies. Front Cell Neurosci. 2012;6:58. https://doi.org/10.3389/fncel.2012.00058. PMID: 23248582; PMCID: PMC3520152.
53. Plotsky PM, Meaney MJ. Early, postnatal experience alters hypothalamic corticotropin-releasing factor (CRF) mRNA, median eminence CRF content and stress-induced release in adult rats. Brain Res Mol Brain Res. 1993;18(3):195–200. https://doi.org/10.1016/0169-328x(93)90189-v. PMID: 8497182.
54. Prins ML, Lee SM, Cheng CL, Becker DP, Hovda DA. Fluid percussion brain injury in the developing and adult rat: a comparative study of mortality, morphology, intracranial pressure and mean arterial blood pressure. Brain Res Dev Brain Res. 1996;95(2):272–82. https://doi.org/10.1016/0165-3806(96)00098-3. PMID: 8874903.
55. Rivara FP, Koepsell TD, Wang J, Temkin N, Dorsch A, Vavilala MS, Durbin D, Jaffe KM. Incidence of disability among children 12 months after traumatic brain injury. Am J

Public Health. 2012;102(11):2074–9. https://doi.org/10.2105/AJPH.2012.300696. Epub 2012 Sep 20. PMID: 22994196; PMCID: PMC3477965.

56. Robertson CL, Saraswati M. Progesterone protects mitochondrial function in a rat model of pediatric traumatic brain injury. J Bioenerg Biomembr. 2015;47(1–2):43–51. https://doi.org/10.1007/s10863-014-9585-5. Epub 2014 Oct 28. PMID: 25348484.
57. Roque A, Valles Méndez KM, Ruiz R, Pineda E, Lajud N. Early life stress induces a transient increase in hippocampal corticotropin-releasing hormone in rat neonates that precedes the effects on hypothalamic neuropeptides. Eur J Neurosci. 2022;55(9–10):2108–21. https://doi.org/10.1111/ejn.15193. Epub 2021 Mar 26. PMID: 33745155.
58. Semple BD, Dixit S, Shultz SR, Boon WC, O'Brien TJ. Sex-dependent changes in neuronal morphology and psychosocial behaviors after pediatric brain injury. Behav Brain Res. 2017;15(319):48–62. https://doi.org/10.1016/j.bbr.2016.10.045. Epub 2016 Nov 6. PMID: 27829127.
59. Sifringer M, Stefovska V, Endesfelder S, Stahel PF, Genz K, Dzietko M, Ikonomidou C, Felderhoff-Mueser U. Activation of caspase-,1 dependent interleukins in developmental brain trauma. Neurobiol Dis. 2007;25(3):614–22. https://doi.org/10.1016/j.nbd.2006.11.003. Epub 2006 Dec 22. PMID: 17188500.
60. Sieswerda-Hoogendoorn T, Boos S, Spivack B, Bilo RA, van Rijn RR. Educational paper: abusive head trauma part I. Clinical aspects. Eur J Pediatr. 2012;171(3):415–23. https://doi.org/10.1007/s00431-011-1598-z. Epub 2011 Oct 28. PMID: 22033697; PMCID: PMC3284669.
61. Smith DH, Kochanek PM, Rosi S, Meyer R, Ferland-Beckham C, Prager EM, Ahlers ST, Crawford F. Roadmap for advancing pre-clinical science in traumatic brain injury. J Neurotrauma. 2021;38(23):3204–21. https://doi.org/10.1089/neu.2021.0094. Epub 2021 Aug 13. PMID: 34210174; PMCID: PMC8820284.
62. Teicher MH, Andersen SL, Polcari A, Anderson CM, Navalta CP, Kim DM. The neurobiological consequences of early stress and childhood maltreatment. Neurosci Biobehav Rev. 2003;27(1–2):33–44. https://doi.org/10.1016/s0149-7634(03)00007-1. PMID: 12732221.
63. Tong W, Igarashi T, Ferriero DM, Noble LJ. Traumatic brain injury in the immature mouse brain: characterization of regional vulnerability. Exp Neurol. 2002;176(1):105–16. https://doi.org/10.1006/exnr.2002.7941. PMID: 12093087.
64. Tsuru-Aoyagi K, Potts MB, Trivedi A, Pfankuch T, Raber J, Wendland M, Claus CP, Koh SE, Ferriero D, Noble-Haeusslein LJ. Glutathione peroxidase activity modulates recovery in the injured immature brain. Ann Neurol. 2009;65(5):540–9. https://doi.org/10.1002/ana.21600. PMID: 19475669; PMCID: PMC2690612.
65. van Bodegom M, Homberg JR, Henckens MJAG. Modulation of the hypothalamic-pituitary-adrenal axis by early life stress exposure. Front Cell Neurosci. 2017;11:87. https://doi.org/10.3389/fncel.2017.00087. PMID: 28469557; PMCID: PMC5395581.
66. Vinchon M, Defoort-Dhellemmes S, Desurmont M, Dhellemmes P. Accidental and nonaccidental head injuries in infants: a prospective study. J Neurosurg. 2005;102(4 Suppl):380–4. https://doi.org/10.3171/ped.2005.102.4.0380. PMID: 15926388.

Chapter 8
The Emerging Landscape of the Cerebellum after a Pediatric Traumatic Brain Injury: From Diaschisis to Sociality

Michael Donovan, Morgan Frentz, Anakaren Romero Lozano, Shripriya Rao, Madeline Rodriguez, and Linda J. Noble-Haeusslein

Abstract There is an expanding interest in the cerebellum in the context of focal and diffuse traumatic injuries to the cerebral cortex. In the adult brain, preclinical studies have revealed acute as well as progressive loss of Purkinje cells in the cerebellum coincident with microglial activation. This pathogenesis, remote to the site of the primary injury, is termed "diaschisis." Here we consider traumatic injuries to the developing brain, where the cerebellum likewise undergoes neurodegeneration. As injury is superimposed on a young brain, long-term adverse consequences may reflect diaschisis that is compounded by disruption of brain development.

Keywords Pediatric traumatic brain injuries · Cerebellum · Diaschisis · Sociality
Abbreviations

M. Donovan · M. Frentz · S. Rao
Department of Neurology, Dell Medical School, The University of Texas at Austin, Austin, TX, USA
e-mail: michael.donovan@austin.utexas.edu; morgan.frentz@austin.utexas.edu; ssr2454@utexas.edu

A. R. Lozano
Department of Biomedical Engineering, Cockrell School of Engineering, The University of Texas at Austin, Austin, TX, USA
e-mail: anakarenr@austin.utexas.edu

M. Rodriguez
Texas Tech University Health Sciences Center, Lubbock, TX, USA
e-mail: madeline.rodriguez@ttuhsc.edu

L. J. Noble-Haeusslein (✉)
Departments of Neurology and Psychology, Dell Medical School and the College of Liberal Arts, The University of Texas at Austin, Austin, TX, USA
e-mail: linda.noble@austin.utexas.edu

L. J. Noble-Haeusslein, D. M. Schnyer (eds.), *Traumatic Brain Injuries*, Advances in Neurobiology 42, https://doi.org/10.1007/978-3-031-69832-3_8

ENIGMA	Enhancing NeuroImaging Genetics through Meta-Analysis
ICP	Intracranial pressure
TBIs	Traumatic brain injuries

8.1 Introduction

From a global perspective, pediatric traumatic brain injuries assume a bimodal age distribution involving very young children (0–2 years) and adolescents (15–18 years) [1]. After 3 years of age, males have a higher incidence of traumatic brain injuries than females with 80% of these injuries classified as mild and only 10% requiring surgical interventions [1]. In the United States, there is a growing awareness of brain injuries in children, in particular those resulting from mild injuries including concussions [2]. The Centers for Disease Control and Prevention has been actively engaged in efforts to reduce both incidence and severity of traumatic brain injuries (https://www.cdc.gov/headsup/) and have supported evidence-based guidelines on diagnosis and management of pediatric traumatic brain injuries (https://www.cdc.gov/injury/pdfs/fundedprograms/Ped-mTBI-Guideline_ Response-to-Public-Comments-508.pdf).

While pediatric traumatic brain injuries have received considerable attention by the medical/research communities, we have yet to understand how a traumatic brain injury influences those areas of the brain that are remote from the primary insult. To address this issue, we present both preclinical and clinical data that support the vulnerability of the cerebellum after a traumatic brain injury. We provide an overview of its anatomy and connectivity, consider the emerging role of diaschisis in cerebellar pathogenesis, and highlight advances in our understanding of cerebellar function beyond its traditional roles in motor control.

8.2 The Cerebellum

In this overview, we address the general anatomy of the cerebellum, including its meninges, with an emphasis on how unique properties of these structures may influence regional vulnerability after a traumatic brain injury.

There are several distinguishing factors when comparing the cerebral and cerebellar cortices. While both the human cortex and cerebellum are gyrencephalic, the latter is distinguished by a much tighter folding pattern. As such, its surface area represents 78% of that of the cerebral cortex [3]. Remarkedly, the cerebellum is rich in neurons, being home to 80% of the brain's total neurons [4, 5], most of which are cerebellar granule cells [6].

Early preclinical studies have revealed a significant loss of neurons in the cerebellum after a focal injury to the adult cerebral cortex [6], with finite element modeling predicting risk of damage to the cerebellum. More recent studies have likewise demonstrated ongoing inflammatory processes in the cerebellum after an early age traumatic brain injury [7]. These collective findings highlight the remote effects of a traumatic brain injury on both the mature and developing cerebellum and raise questions regarding the mechanisms underlying its vulnerability.

8.2.1 The Meninges: A Species-Specific Perspective

Like other regions of the brain, the cerebellum is covered by the meninges (see Review, [8]). However, there are species-specific differences in in the meninges associated with the cerebellum (see Review, [9]), and these may further compound the debate regarding accuracy of preclinical models of pathogenesis in human brain injuries [10]. This is exemplified in studies of human and nonhuman primates where the head is positioned above the cervical spine [8]. The dura forms the tentorium cerebelli, a tent-like structure that serves as the roof of the posterior cranial fossa. Finite element modeling studies of the human brain suggest that the tentorium of the cerebellum serves to reduce its movement while increasing strain in the adjacent brainstem in response to head rotation in both the coronal and transverse rotations [11]. Such findings are intriguing as they may, at least in part, explain vulnerability of the brainstem to diffuse axonal injury [11].

The tentorium, which separates the supratentorial fossa from infratentorial fossa, likely plays an important role in response to an increase in intracranial pressure (ICP), a key feature of traumatic brain injuries. However, few studies have considered ICP in the supra- and infratentorial compartments, where it may not be uniformly distributed [12]. A recent study measured ICP after a stroke in each of these compartments [13]. The findings revealed a significant transtentorial pressure gradient with higher ICP in the infratentorial compartment. Moreover, elevated ICP was associated with poor neurological outcomes, irrespective of ICP findings in the supratentorial compartment.

The tentorium differs across species, and this is most notable when considering the position of the head relative to the cervical spinal and its structure [8]. In humans and nonhuman primates, the rostral-caudal axis of the brain is at near right angles to the cervical spine. In contrast, this axis in rodents, the most commonly studied preclinical models, is in line with the spine. Additionally, the tentorium in rodents is a more delicate partition, compared to that of humans, and is a separated, rather than fused, structure. Such distinctions may contribute to differences in pathophysiology that have been reported in the injured human brain. This is exemplified in a recent review that addressed preclinical models of strokes [14]. Similar to humans, there is an abnormal increase in ICP in those large animal models that, like humans, have a well-developed tentorium. It is thought that the tentorium serves to confine an increase in ICP to the supratentorial compartment. Importantly, this abnormal increase may lead to loss of consciousness and cerebral herniation. Similar findings are not observed in rodent models of stroke, which raises a cautionary note against overreliance on similar model species in the study of human traumatic brain injuries [10].

8.2.2 General Anatomy: From Outside to Inside

The cerebellum, from the Latin phrase "little brain," is located within the posterior fossa and posterior to the fourth ventricle, pons, and medulla oblongata. This structure has a unique anatomy, from its distinct external features to the intricacies of its folia (Fig. 8.1, see Review, [15]). From a mediolateral position, the cerebellum is divided into the cerebrocerebellum, representing the two lateral hemispheres, and the more medial spinocerebellum and vestibulocerebellum, each of which includes the midline vermis. From an anterior/posterior direction, the cerebellum consists of three lobes, namely, the anterior, posterior, and flocculo-nodular lobes. The human

cerebellar cortex is gyrencephalic with folds referred to as folia (Fig. 8.1a). These folia, consisting of both gray and white matter, serve to increase the surface area of the cerebellum. Within these folia, the dendritic arborization of Purkinje cells is distributed in one plane that is perpendicular to the longitudinal axis of these folds. Thus, their profile may appear either linear or fan-like, depending on the plane of view (Fig. 8.1b).

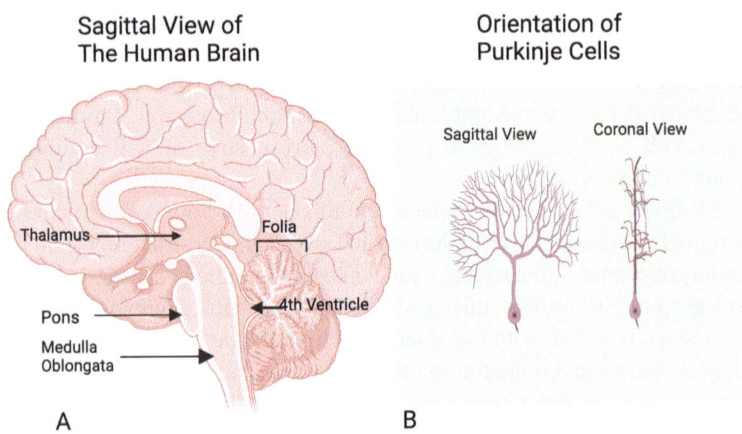

Fig. 8.1 (**a**)This midsagittal section of the human brain demonstrates the position of the cerebellum in the posterior fossa. The vermis of the cerebellum (unlabeled) lies immediately adjacent to the fourth ventricle. This region of the cerebellum has been shown to be vulnerable to both and diffuse and focal traumatic brain injuries [16–18], raising the possibility that fluid pressure changes in the adjacent cerebrospinal compartment may damage the vermis. (**b**) The broadly distributed dendritic arborization of Purkinje cells is only apparent in the sagittal plane. Rotating the cerebellum by 90 degrees reveals a more linear orientation of these dendrites in the coronal plane. (Created by BioRender.com)

8.2.3 Microcircuitry and Connectivity

The microcircuitry of the cerebellum is distributed within the cerebellar cortex and the deep cerebellar nuclei (see Review, [19]). The cortex consists of three regions—the outermost molecular layer, the middle Purkinje layer, and the granular layer (Fig. 8.2, [20]). The molecular layer consists of dendritic arborizations of Purkinje cells that reside in the middle layer and parallel fibers that arise from the innermost granular layer. Neurons in the deep cerebellar nuclei are central to the circuitry of the cerebellum. These neurons receive input from cerebellar afferents and Purkinje cells, and their axons are a major output for the cerebellum.

From the standpoint of connectivity, climbing and mossy fibers, the two major afferents to the cerebellum, provide excitatory input to the molecular and granular layers, respectively (Fig. 8.1b). Climbing fibers project solely from the inferior

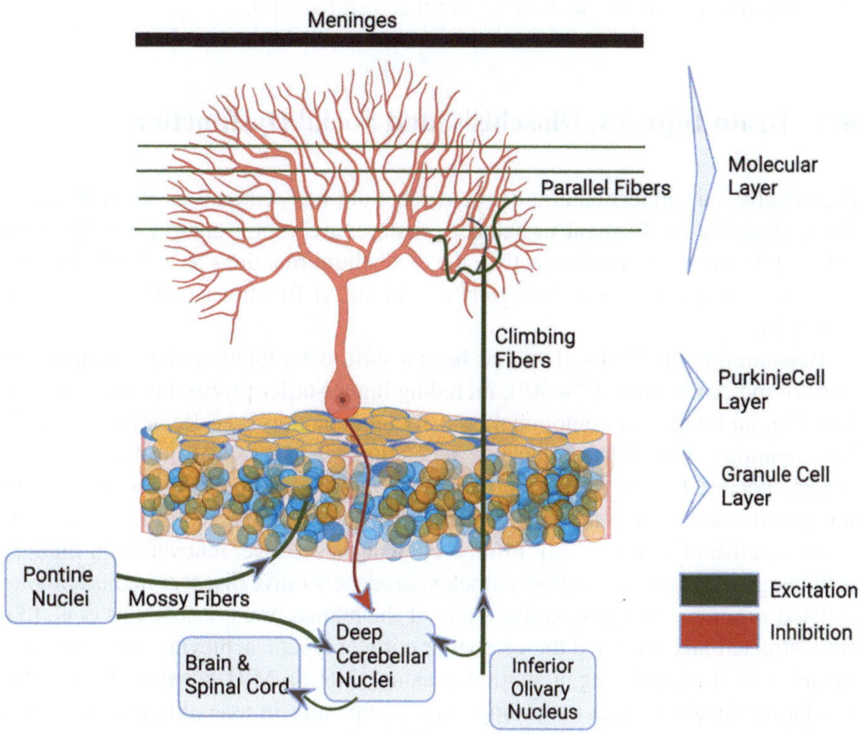

Fig. 8.2 An overview of the circuitry of the adult murine cerebellum. The cerebellum consists of the cortex and deep cerebellar nuclei. The cortex is divided into three layers—the outermost layer representing the dendritic arborization of Purkinje cells, the adjacent cell bodies of the Purkinje cell layer, and the granule cell layer. From a more generalized perspective, inputs to the cerebellum include mossy fibers, projecting solely from the inferior olivary nucleus to the granule cell layer and deep cerebellar nuclei and climbing fibers that originate solely from the inferior olive. Created by BioRender.com

olivary nucleus to the molecular layer and deep cerebellar nuclei, conveying sensorimotor information that is relevant to motor control and learning [21, 22]. Their axons synapse with Purkinje cells, which are the exclusive output of the cerebellar cortex to the deep cerebellar nuclei [22]. Importantly, inputs from these axons are organized into microzones. These are functional units that coordinate activity, support plasticity, and modulate cerebellar output through the deep cerebellar nuclei (see Review [23]).

Mossy fibers originate from multiple sources, the largest of which arises from the cerebral cortex and the pontocerebellar pathway. Signals, entering the cerebellum through mossy fibers, are processed in the granular layer and transmitted to Purkinje cells, while a collateral pathway inhibits the deep cerebellar nuclei. These nuclei play a key role in integrating information from the brain stem, inferior olive, and spinal cord [24]. Regardless of their input, all cerebellar cortical output is derived from the Purkinje neurons, almost all of which project to one of the four deep cerebellar nuclei [fastigial, interposed (emboliform and globose) and dentate] which in turn project to targets throughout the brain and spinal cord.

8.3 Brain Injuries, Diaschisis, and Social Dysfunction

Historically, our understanding of the cerebellum has centered on its motor functions, given its involvement in balance, posture, and motor control (see Reviews, [25, 26]. From a current perspective, the cerebellum functions as a "hub" that supports diverse sensory, motor, cognitive, and social functions (see Review, [27]) (Fig. 8.3).

Beginning in the 1990s, there has been a shift to exploration of nonmotor functions of the cerebellum ([28–30]), including higher-order processing of autonomic function, and cognitive, emotional, and social behaviors (see Reviews, [27, 31]). Investigations, fueled by innovative technologies, have spear-headed discovery across species. For example, the advent of high-resolution two-photon calcium imaging has led to the findings that cerebellar granule cells are both responsive to and predictive of rewards [32]. From a clinical perspective, innovation in imaging technology has made it possible to track neurodegenerative diseases in this complex foliated structure. In a prospective study of the human brain, Priovoulos et al. [33] reconstructed and unfolded the cerebellar cortical layers, achieving a resolution of up to 0.19 mm, utilizing a motion-resistant 7-tesla MRI scanner. Such high-resolution imaging serves as an important comparator in assessing progression of neurodegeneration, a common feature of TBIs [34].

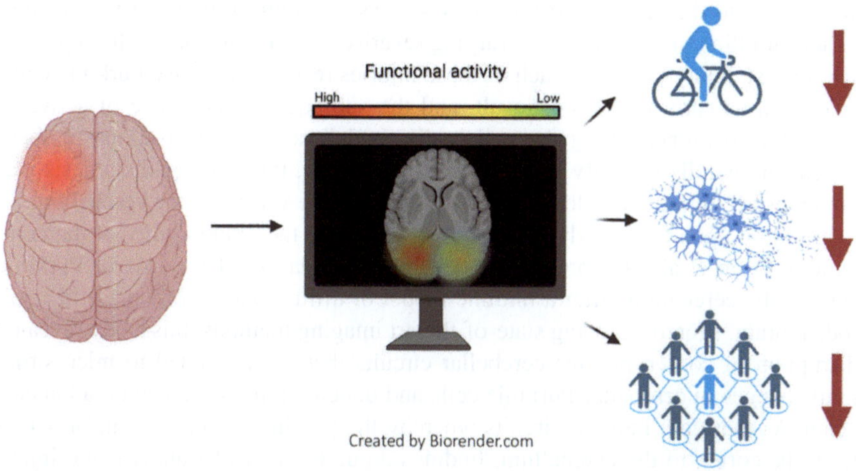

Fig. 8.3 A traumatic injury to the cerebral cortex may result in connectional diaschisis in the cerebellum, characterized by neurodegeneration and neurological deficits ranging from motor dysfunction to deficits in executive function including sociality

8.3.1 Diaschisis and the Cerebellum

Age at time of injury is a critical factor in recovery from TBI, with vulnerability most evident in infants/young children. The loss of function in the injured young brain is dictated, in part, by the nature of the injury (focal and/or diffuse), biomechanical properties (lack of rigidity of the skull, a large head/body ratio coupled with weaker neck musculature), incomplete myelination that influences the extent of energy absorption, and inadequate antioxidant reserves [35–40].

Although less studied, diaschisis, from the Greek meaning "split throughout," is also a determinant of neurological recovery after a TBI. This term, introduced by Monakow in 1914, is defined by a focal injury to the brain concomitant with a loss of function and connectivity in a region that is remote to the primary insult (see Review [41]) (Fig. 8.2). With advances in technology including imaging, the terminology has evolved to "focal" and "nonfocal" diaschisis, with the former addressing focal decreases or abnormalities in metabolism and neuronal activity and the latter reflecting selective coupling between two nodes of a network or changes in the structural and functional connectome.

8.3.2 Preclinical Models

Preclinical models have primarily focused on structures rostral to the cerebellar tentorium. However, beginning in the late 1960s, investigations of the cerebellum in adult rodent models revealed regionally specific vulnerability of the cerebellum after injury to the cerebral cortex. Damage was apparent in the adult vermis after either a midline focal injury, with varying severity of injury or a lateral fluid percussive diffuse injury [17, 18]. Each of these injuries resulted in injury Purkinje cells, evidenced as early at 3 h postinsult, and the subsequent emergence of activated microglia positioned along the cell bodies and dendritic arborizations of these degenerating cells [18]. By several weeks postinjury, this early pathogenesis was predictive of Purkinje cell loss in discrete zones in the vermis of the cerebellum.

Building upon these early findings that addressed the vulnerability of the cerebellum, Ozen et al. [42] provided independent validation of the loss of Purkinje cells in the cerebellum after a midline model of diffuse axonal injury to the adult rodent brain. Moreover, using state-of-the-art imaging methods, this study revealed disruption of cortico-pontine-cerebellar circuits that corresponded to microstructural changes in cerebellar Purkinje cells and deficits in performance on a balance beam. As these are indirect circuits whereby the pontine nuclei relay information from the cortex to the cerebellum, findings argue in favor of transsynaptic injury mechanisms, initiated by mechanical injury to the cerebral cortex.

While the above examples speak to the putative involvement of diaschisis in subcortical pathogenesis, there remains the possibility that the cerebellum may be directly injured as a result of a focal cerebral cortical impact. This possibility is exemplified in a study that measured intracranial tissue strain along the neuroaxis in an adult rodent focal injury model [6]. Finite element modeling predicted a maximum principal strain of 10% in the cerebellum after a severe injury. Thus, injury of Purkinje cells within the first several hours postinjury may have resulted from a mechanical insult. Most recently, Wang and Hamill [43] identified Piezo2 channels, low threshold pressure sensors, that are expressed in cerebellar Purkinje cells. These channels may serve as an intrinsic mechanism that responds to intracranial pressure pulses. Such findings add a new dimension to our understanding of how the cerebellum responds to a cerebral cortical injury, where vulnerability may reflect diaschisis and mechanotransductive mechanisms that influence the more distant regions of the neural axis [44].

Few preclinical studies have examined the cerebellum after traumatic injury to the developing brain, where injury is superimposed upon brain development. One of the earliest studies examined a discrete needle-induced injury to the rodent cerebral cortex. The findings of cerebellar apoptosis were surprising, given the mild nature of the injury and its impact on the more remote cerebellum [45]. Most recently, Fraunberger and colleagues [7] examined inflammatory networks in the cerebellum after mild injury by "weight drop" to the intact skull in adolescent rodents. These investigators reported a persistent activation of glia in the cerebellum in the absence of behavioral changes. While the mild degree of injury may not have reached a threshold for behavioral modifications, it may render the brain more susceptible to a subsequent mild traumatic brain injury. This scenario of repeated mild traumatic brain injuries has been commonly associated with sports-related concussions in youths [46]. Thus, there is a clear need to further explore this line of research.

8.3.3 Clinical Studies

Recent studies have likewise evaluated the cerebellum after a pediatric traumatic brain injury. Key findings are exemplified in several recent studies that address injury, subsequent brain development, and the intersection between structural changes and executive function. An early age injury results in increased gyrification in those regions of the brain that mature later in brain development, a finding that suggests a compensatory response to the initial injury [47]. Most recently, a large-scale, multiinstitutional clinical study addressed changes in the cerebellum after accidental mild traumatic brains injuries in children and adolescence. Findings were based upon longitudinal assessments and three-dimensional T1-weighted MR images that were processed using the ENIGMA (Enhancing NeuroImaging Genetics through Meta-Analysis) pipeline [48]. Volume growth of the cerebellum which was smaller than controls was most prominent in the posterior lobe, a region that matures more slowly than the anterior lobe. Moreover, cerebellar volumes corresponded to poor executive function as measured by multiple parameters. While the mechanisms underlying these changes are unclear, the authors postulate connectional diaschisis, a process whereby there are changes in structural and functional connectivity between the cerebellum and the cerebral cortical lesion.

8.3.4 Cerebellum and Sociality

Given the pathological findings in the cerebellum after a traumatic brain injury, a key question relates to the extent to which degenerative changes in the cerebellum may contribute to disruption of normal behaviors. The challenge of studying the consequences of a traumatic brain injury is that various regions of the neuroaxis are commonly affected. Moreover, while a traumatic brain injury may result in degeneration of the cerebellum, it is noteworthy that the cerebellum also has the capacity for compensation and restoration after injury, a feature that is referred to as "reserve" [49]. Beyond multiple forms of synaptic plasticity, cerebellar reserve is attributed to mossy fibers that innervate numerous microzones, which convey redundant information. Thus, if neurodegeneration is confined to a limited area, there may be compensation from other areas.

Despite the above limitations, insights regarding the relationship between the cerebellum and sociality have been gained in children diagnosed with brain tumors in the posterior fossa and those exposed to early age cerebellar hemorrhages. Surgical interventions to remove a tumor in the posterior fossa, a region that houses the brainstem and cerebellum, results in posterior fossa syndrome, a condition that presents with transient mutism, emotional lability, and changes in social behaviors (see Review, [50]). Changes in social behaviors are further manifested at adulthood, where there is increased risk of severe social cognitive impairments in those who survived pediatric CNS tumors [51]. Beyond early age brain tumors, preterm infants that have sustained cerebellar hemorrhages or have congenital malformations of the cerebellum may likewise present with deficits in social cognition during their subsequent maturation. Collectively, these findings, based upon different forms of injury (cerebellar tumor removal and hemorrhage), reinforce the social consequences of an injured cerebellum.

Preclinical studies have further reinforced the intersection between traumatic brain injuries and disruption of the cerebellum. Focal injuries to the developing rodent brain result in deficits in sociosexual behaviors and communication, as evidenced by reduced interactions with a novel rodent and abnormal mounting behavior [52, 53]. It is noteworthy that these changes are not evident at adolescence but rather emerge at adulthood. There are several possible explanations for this delay in onset. The delay may be related to the timeline for maturation of skills associated with sociality [52]. Although not mutually exclusive, it is also conceivable that the delay reflects diaschisis—the timeline for neurodegeneration in the more remotely located cerebellum.

8.4 The Next Steps

Early age traumatic injuries to the cerebral cortex result in changes to both the structure and function of the cerebellum during subsequent maturation of the brain. Given this progressive pathogenesis during brain maturation, it is conceivable that interventions could either protect those regions that are initially intact after injury and/or support reparative processes. Potential avenues include environmental enrichment [54], targeting neurogenesis [55], and identification of new candidate therapeutics [56] including electrical stimulation [57]. A challenge, unique to the developing brain, is that candidate approaches are superimposed upon a developing terrain. As such, the window of opportunity for therapeutic intervention may need to be specifically tailored to developmental age at time of injury.

Acknowledgments This research was supported by start-up funds for LJN, provided by the Dell Medical School, University of Texas at Austin.

Disclosures None.

References

1. Dewan MC, Mummareddy N, Wellons JC 3rd, Bonfield CM. Epidemiology of global pediatric traumatic brain injury: qualitative review. World Neurosurg. 2016;91:497–509. e491.
2. Podolak OE, Chaudhary S, Haarbauer-Krupa J, Metzger KB, Curry AE, Kessler RS, Pfeiffer MR, Breiding MJ, Master CL, Arbogast KB. Characteristics of diagnosed concussions in children aged 0 to 4 years presenting to a large pediatric healthcare network. Pediatr Emerg Care. 2021;37:e1652–7.

3. Sereno MI, Diedrichsen J, Tachrount M, Testa-Silva G, d'Arceuil H, De Zeeuw C. The human cerebellum has almost 80% of the surface area of the neocortex. Proc Natl Acad Sci USA. 2020;117:19538–43.
4. Azevedo FA, Carvalho LR, Grinberg LT, Farfel JM, Ferretti RE, Leite RE, Jacob Filho W, Lent R, Herculano-Houzel S. Equal numbers of neuronal and nonneuronal cells make the human brain an isometrically scaled-up primate brain. J Comp Neurol. 2009;513:532–41.
5. Van Essen DC, Donahue CJ, Glasser MF. Development and evolution of cerebral and cerebellar cortex. Brain Behav Evol. 2018;91:158–69.
6. Mao H, Jin X, Zhang L, Yang KH, Igarashi T, Noble-Haeusslein LJ, King AI. Finite element analysis of controlled cortical impact-induced cell loss. J Neurotrauma. 2010;27:877–88.
7. Fraunberger EA, DeJesus P, Zanier ER, Shutt TE, Esser MJ. Acute and persistent alterations of cerebellar inflammatory networks and glial activation in a rat model of pediatric mild traumatic brain injury. J Neurotrauma. 2020;37:1315–30.
8. Rai R, Iwanaga J, Shokouhi G, Oskouian RJ, Tubbs RS. The tentorium cerebelli: a comprehensive review including its anatomy, embryology, and surgical techniques. Cureus. 2018;10:e3079.
9. Klintworth GK. The comparative anatomy and phylogeny of the tentorium cerebelli. Anat Rec. 1968;160:635–42.
10. Smith DH, Kochanek PM, Rosi S, Meyer R, Ferland-Beckham C, Prager EM, Ahlers ST, Crawford F. Roadmap for advancing pre-clinical science in traumatic brain injury. J Neurotrauma. 2021;38:3204–21.
11. Ho J, Zhou Z, Li X, Kleiven S. The peculiar properties of the falx and tentorium in brain injury biomechanics. J Biomech. 2017;60:243–7.
12. Wolfla CE, Luerssen TG, Bowman RM, Putty TK. Brain tissue pressure gradients created by expanding frontal epidural mass lesion. J Neurosurg. 1996;84:642–7.
13. Petr O, Ho WM, Petutschnigg T, Krigers A, Treichl SA, Preuss-Hernandez C, Brawanski K, Helbok R, Thome C. Link between both infratentorial and supratentorial intracranial pressure burdens and final outcome in patients with infratentorial brain injury. J Neurosurg. 2023;139:1430–8.
14. Kaiser EE, West FD. Large animal ischemic stroke models: replicating human stroke pathophysiology. Neural Regen Res. 2020;15:1377–87.
15. Dekeyzer S, Vanden Bossche S, De Cocker L. Anything but little: a pictorial review on anatomy and pathology of the cerebellum. Clin Neuroradiol. 2023;33:907–29.
16. Fukuda K, Aihara N, Sagar SM, Sharp FR, Pitts LH, Honkaniemi J, Noble LJ. Purkinje cell vulnerability to mild traumatic brain injury. J Neurotrauma. 1996;13:255–66.
17. Igarashi T, Potts MB, Noble-Haeusslein LJ. Injury severity determines Purkinje cell loss and microglial activation in the cerebellum after cortical contusion injury. Exp Neurol. 2007;203:258–68.
18. Mautes AE, Fukuda K, Noble LJ. Cellular response in the cerebellum after midline traumatic brain injury in the rat. Neurosci Lett. 1996;214:95–8.
19. Prestori F, Mapelli L, D'Angelo E. Diverse neuron properties and complex network dynamics in the cerebellar cortical inhibitory circuit. Front Mol Neurosci. 2019;12:267.
20. Maxwell DL, Orian JM. Cerebellar pathology in multiple sclerosis and experimental autoimmune encephalomyelitis: current status and future directions. J Cent Nerv Syst Dis. 2023;15:11795735231211508.
21. Ikezoe K, Hidaka N, Manita S, Murakami M, Tsutsumi S, Isomura Y, Kano M, Kitamura K. Cerebellar climbing fibers multiplex movement and reward signals during a voluntary movement task in mice. Commun Biol. 2023;6:924.
22. Zang Y, De Schutter E. Climbing fibers provide graded error signals in cerebellar learning. Front Syst Neurosci. 2019;13:46.
23. Kostadinov D, Hausser M. Reward signals in the cerebellum: origins, targets, and functional implications. Neuron. 2022;110:1290–303.

24. Uusisaari M, Obata K, Knopfel T. Morphological and electrophysiological properties of GABAergic and non-GABAergic cells in the deep cerebellar nuclei. J Neurophysiol. 2007;97:901–11.
25. Timmann D, Drepper J, Frings M, Maschke M, Richter S, Gerwig M, Kolb FP. The human cerebellum contributes to motor, emotional and cognitive associative learning. A review. Cortex. 2010;46:845–57.
26. Amore G, Spoto G, Ieni A, Vetri L, Quatrosi G, Di Rosa G, Nicotera AG. A focus on the cerebellum: from embryogenesis to an age-related clinical perspective. Front Syst Neurosci. 2021;15:646052.
27. Rudolph S, Badura A, Lutzu S, Pathak SS, Thieme A, Verpeut JL, Wagner MJ, Yang YM, Fioravante D. Cognitive-affective functions of the cerebellum. J Neurosci. 2023;43:7554–64.
28. Schmahmann JD. An emerging concept. The cerebellar contribution to higher function. Arch Neurol. 1991;48:1178–87.
29. Schmahmann JD. The role of the cerebellum in cognition and emotion: personal reflections since 1982 on the dysmetria of thought hypothesis, and its historical evolution from theory to therapy. Neuropsychol Rev. 2010;20:236–60.
30. Schmahmann JD. The cerebellum and cognition. Neurosci Lett. 2019;688:62–75.
31. Beuriat PA, Cohen-Zimerman S, Smith GNL, Krueger F, Gordon B, Grafman J. Evidence of the role of the cerebellum in cognitive theory of mind using voxel-based lesion mapping. Sci Rep. 2022;12:4999.
32. Wagner MJ, Kim TH, Savall J, Schnitzer MJ, Luo L. Cerebellar granule cells encode the expectation of reward. Nature. 2017;544:96–100.
33. Priovoulos N, Andersen M, Dumoulin SO, Boer VO, van der Zwaag W. High-resolution motion-corrected 7.0-T MRI to derive morphologic measures from the human cerebellum in vivo. Radiology. 2023;307:e220989.
34. Brett BL, Gardner RC, Godbout J, Dams-O'Connor K, Keene CD. Traumatic brain injury and risk of neurodegenerative disorder. Biol Psychiatry. 2022;91:498–507.
35. Araki T, Yokota H, Morita A. Pediatric traumatic brain injury: characteristic features, diagnosis, and management. Neurol Med Chir (Tokyo). 2017;57:82–93.
36. Bayir H, Kagan VE, Tyurina YY, Tyurin V, Ruppel RA, Adelson PD, Graham SH, Janesko K, Clark RS, Kochanek PM. Assessment of antioxidant reserves and oxidative stress in cerebrospinal fluid after severe traumatic brain injury in infants and children. Pediatr Res. 2002;51:571–8.
37. Bayir H, Kochanek PM, Kagan VE. Oxidative stress in immature brain after traumatic brain injury. Dev Neurosci. 2006;28:420–31.
38. Corrigan JD, Bogner J, Mellick D, Bushnik T, Dams-O'Connor K, Hammond FM, Hart T, Kolakowsky-Hayner S. Prior history of traumatic brain injury among persons in the Traumatic Brain Injury Model Systems National Database. Arch Phys Med Rehabil. 2013;94:1940–50.
39. Tsuru-Aoyagi K, Potts MB, Trivedi A, Pfankuch T, Raber J, Wendland M, Claus CP, Koh SE, Ferriero D, Noble-Haeusslein LJ. Glutathione peroxidase activity modulates recovery in the injured immature brain. Ann Neurol. 2009;65:540–9.
40. Potts MB, Koh SE, Whetstone WD, Walker BA, Yoneyama T, Claus CP, Manvelyan HM, Noble-Haeusslein LJ. Traumatic injury to the immature brain: inflammation, oxidative injury, and iron-mediated damage as potential therapeutic targets. NeuroRx. 2006;3:143–53.
41. Carrera E, Tononi G. Diaschisis: past, present, future. Brain. 2014;137:2408–22.
42. Ozen I, Mai H, De Maio A, Ruscher K, Michalettos G, Clausen F, Gottschalk M, Ansar S, Arkan S, Erturk A, Marklund N. Purkinje cell vulnerability induced by diffuse traumatic brain injury is linked to disruption of long-range neuronal circuits. Acta Neuropathol Commun. 2022;10:129.
43. Wang J, Hamill OP. Piezo2-peripheral baroreceptor channel expressed in select neurons of the mouse brain: a putative mechanism for synchronizing neural networks by transducing intracranial pressure pulses. J Integr Neurosci. 2021;20:825–37.

44. Keating CE, Cullen DK. Mechanosensation in traumatic brain injury. Neurobiol Dis. 2021;148:105210.
45. Taylor DL, Joashi UC, Sarraf C, Edwards AD, Mehmet H. Consequential apoptosis in the cerebellum following injury to the developing rat forebrain. Brain Pathol. 2006;16:195–201.
46. Greco T, Ferguson L, Giza C, Prins ML. Mechanisms underlying vulnerabilities after repeat mild traumatic brain injuries. Exp Neurol. 2019;317:206–13.
47. Wilde EA, Merkley TL, Lindsey HM, Bigler ED, Hunter JV, Ewing-Cobbs L, Aitken ME, MacLeod MC, Hanten G, Chu ZD, Abildskov TJ, Noble-Haeusslein LJ, Levin HS. Developmental alterations in cortical organization and socialization in adolescents who sustained a traumatic brain injury in early childhood. J Neurotrauma. 2021;38:133–43.
48. Keleher F, Lindsey HM, Kerestes R, Amiri H, Asarnow RF, Babikian T, Bartnik-Olson B, Bigler ED, Caeyenberghs K, Esopenko C, Ewing-Cobbs L, Giza CC, Goodrich-Hunsaker NJ, Hodges CB, Hoskinson KR, Irimia A, Konigs M, Max JE, Newsome MR, Olsen A, Ryan NP, Schmidt AT, Stein DJ, Suskauer SJ, Ware AL, Wheeler AL, Zielinski BA, Thompson PM, Harding IH, Tate DF, Wilde EA, Dennis EL. Multimodal analysis of secondary cerebellar alterations after pediatric traumatic brain injury. JAMA Netw Open. 2023;6:e2343410.
49. Mitoma H, Kakei S, Yamaguchi K, Manto M. Physiology of cerebellar reserve: redundancy and plasticity of a modular machine. Int J Mol Sci. 2021;22:477.
50. Olson IR, Hoffman LJ, Jobson KR, Popal HS, Wang Y. Little brain, little minds: the big role of the cerebellum in social development. Dev Cogn Neurosci. 2023;60:101238.
51. Papini C, Willard VW, Gajjar A, Merchant TE, Srivastava D, Armstrong GT, Hudson MM, Krull KR, Brinkman TM. Social cognition and adjustment in adult survivors of pediatric central nervous system tumors. Cancer. 2023;129:3064–75.
52. Semple BD, Canchola SA, Noble-Haeusslein LJ. Deficits in social behavior emerge during development after pediatric traumatic brain injury in mice. J Neurotrauma. 2012;29:2672–83.
53. Semple BD, Noble-Haeusslein LJ, Jun Kwon Y, Sam PN, Gibson AM, Grissom S, Brown S, Adahman Z, Hollingsworth CA, Kwakye A, Gimlin K, Wilde EA, Hanten G, Levin HS, Schenk AK. Sociosexual and communication deficits after traumatic injury to the developing murine brain. PLoS One. 2014;9:e103386.
54. Bozkurt S, Lannin NA, Mychasiuk R, Semple BD. Environmental modifications to rehabilitate social behavior deficits after acquired brain injury: what is the evidence? Neurosci Biobehav Rev. 2023;152:105278.
55. Rizk M, Vu J, Zhang Z. Impact of pediatric traumatic brain injury on hippocampal neurogenesis. Neural Regen Res. 2021;16:926–33.
56. Bielanin JP, Metwally SAH, Paruchuri SS, Sun D. An overview of mild traumatic brain injuries and emerging therapeutic targets. Neurochem Int. 2024;172:105655.
57. Ziesel D, Nowakowska M, Scheruebel S, Kornmueller K, Schafer U, Schindl R, Baumgartner C, Ucal M, Rienmuller T. Electrical stimulation methods and protocols for the treatment of traumatic brain injury: a critical review of preclinical research. J Neuroeng Rehabil. 2023;20:51.

Chapter 9
Chronic Stress in Early Development and Effects on Traumatic Brain Injury Outcome

Aditi S. Gorthy, Alyssa F. Balleste, Fabiola Placeres-Uray, and Coleen M. Atkins

Abstract In recent years, significant advances have been made in the study of mild traumatic brain injury (mTBI). Complete recovery from mTBI normally requires days to weeks, yet a subset of the population suffers from symptoms for weeks to months after injury. The risk factors for these prolonged symptoms have not yet been fully understood. In this chapter, we address one proposed risk factor, early life stress (ELS) and its influence on mTBI recovery. To study the effects of ELS on mTBI recovery, accepted animal models of ELS, including maternal separation, limited bedding and nesting, and chronic unpredictable stress, have been implemented. Combining these ELS models with standardized mTBI models, such as fluid percussion injury or controlled cortical impact, has allowed for a deeper understanding of the neuronal, hormonal, and cognitive changes that occur after mTBI following ELS. These preclinical findings are being used to understand how adverse childhood experiences may predispose a subset of individuals to poorer recovery after mTBI.

Keywords Traumatic brain injury · Chronic stress · Microglial activation

Abbreviations

ACEs	Adverse childhood experiences
CCI	Controlled cortical impact
CUS	Chronic unpredictable stress
ELS	Early life stress
FPI	Fluid percussion injury

A. S. Gorthy · A. F. Balleste · F. Placeres-Uray · C. M. Atkins (✉)
Department of Neurological Surgery, University of Miami Miller School of Medicine, Miami, FL, USA
e-mail: axg5752@miami.edu; afl43@med.miami.edu; fxp226@miami.edu; catkins@miami.edu

L. J. Noble-Haeusslein, D. M. Schnyer (eds.), *Traumatic Brain Injuries*, Advances in Neurobiology 42, https://doi.org/10.1007/978-3-031-69832-3_9

GRs	Glucocorticoid receptors
LBN	Limited bedding and nesting
MS	Maternal separation
mTBI	Mild traumatic brain injury

9.1 Introduction

Each year approximately 2.8 million Americans experience a traumatic brain injury (TBI), which is defined by the CDC as a disruption in the normal functioning of the brain caused by a bump, blow, or jolt to the head, a penetrating head injury, or an explosive blast to the head [1]. Among these cases, 80% are classified as mild traumatic brain injuries (mTBIs) and are colloquially known as concussions. While recovery from mTBI typically takes 1–2 weeks, 15–30% of individuals report persistent neurological symptoms lasting for weeks to months after injury [2, 3]. Although some risk factors related to poor recovery have been identified, a mechanistic understanding of how these risk factors alter the consequences of mTBI is still needed. Known risk factors for poor outcomes after mTBI include comorbid PTSD, a previous TBI, drug/alcohol abuse, low education level, low socioeconomic status, low preinjury resilience, and psychiatric problems such as depression and anxiety [4, 5]. Many of these comorbid risk factors have an underlying relationship with a history of chronic stressful life experiences [6]. Stressful life events that occur during childhood are known as adverse childhood events (ACEs) clinically and as early life stress (ELS) in animal models. Common examples of ACEs are neglect, physical abuse, emotional abuse, and sexual abuse during early development. During this time, a child's central nervous system undergoes rapid and programmed development. Consequently, ACEs can severely disrupt normal brain development and increase risk of developing neuropsychiatric disorders and neurodegenerative diseases later in life. This chapter will review how ELS is modeled in rodents and how these models have furthered our understanding of the effects ELS on mTBI recovery.

9.2 Epidemiological Studies

ACEs are highly prevalent across the world with 38.8% of adults reporting at least one ACE before the age of 18 [7]. In the USA, 61% of adults report experiencing at least one ACE, and 16% of adults have reported experiencing four or more ACEs [8]. The most common type of ACE in the US population is childhood abuse with approximately 600,000 cases reported annually. Of the total number of child abuse cases, 76% are related to neglect, 16% are due to physical abuse, and 10% are caused by sexual abuse [9]. Some populations have a higher prevalence of ACEs; for example, US veterans are more likely to have experienced at least one ACE in their life compared to civilians [10]. The prevalence of ACEs as measured by reported cases only gives a glimpse into the actual total number of cases [11].

Experiencing ACEs increases the risk of experiencing a TBI. Individuals who reported at least one ACE were 2.6 times more likely to have suffered a TBI and experiencing four or more ACEs increased this risk by 3.5-fold [12]. Clinical studies have investigated the coincidence of stressful life events and mTBI outcome. In a study of pediatric mTBI patients, a history of prior life stress was associated with prolonged postconcussive symptoms after mTBI [13]. Similarly, in a study of adult mTBI patients admitted to an emergency room, a history of stressful life events was associated with worsened physical and mental health quality of life at 3 months postinjury [14]. These findings were replicated in a retrospective study of 165 mTBI US military service members, which found that exposure to chronic stressful life experiences positively correlated with persistent postconcussion symptoms [4]. The role of ACEs in mTBI recovery is significant, especially as the coincidence of these events is greater than the incidence of either event alone. Thus, in individuals with a high probability of acquiring brain trauma, understanding the mechanism by which ACEs may affect mTBI recovery is of utmost importance.

9.3 Modeling Childhood Adversity in Animals

To model the effects of childhood stress on mTBI recovery in animals, the animal must exhibit measurable behavioral, cognitive, and pathological changes in response to stress and neurotrauma. Rodents, such as mice and rats, have well-established patterns of behavior for anxiety and depression, learning and memory, and sensorimotor functioning that can be measured using accepted, validated tasks. Several models of ELS and mTBI have been developed and documented in rodentsv Rodents are also advantageous for these types of studies because of their relatively short reproductive life cycle of only 2 months, allowing for expedited experiments compared to other animal models. Therefore, rodents are ideal animals to conduct preclinical studies regarding the effects of childhood stress on mTBI recovery.

While the rodent model must have measurable, reliable changes in response to ELS, the ELS models employed must also reflect the real-life ACEs that humans may experience. In humans, ACEs often involve mistreatment during rearing, such as neglect, physical abuse, or emotional and psychological abuse [9]. Therefore, rodent models of ELS often involve manipulating maternal care as a means to model rearing mistreatment [15]. Manipulating maternal care in rodents most commonly involves separation of dams and pups or providing limited nesting and bedding materials. These two models of ELS implement a controlled, scheduled series of stressors. However, ACEs in humans often include unpredictably intense and randomly timed stressors. Therefore, the chronic unpredictable stress model was designed to model the randomness of human ACEs in rodents. While other models of ELS exist, including early weaning, early exposure to threats, and social isolation, the most common models of ELS in rodents are maternal separation, limited bedding/nesting, and chronic unpredictable stress [16].

The face validity of these ELS models depends upon how similarly the effects of the ELS models on rodent development mimic the effects of ACEs on human development. As we proceed to assess the different models of ELS, we must consider their measurable consequences in rodent behavior to assure their translation to human behavior.

9.3.1 Maternal Separation

The maternal separation model employs our understanding of maternal care during development [17]. In humans, an increase in affectionate touch during infancy results in subsequent psychosocial and emotional development [18]. The importance of touch is also seen in rodent studies, whereby pup touching reinforces maternal behavior and results in decreased anxious and depressive behaviors in the pups [19, 20]. Rodents, as well as humans, also undergo physiological changes that further strengthen maternal care behaviors [21]. To induce ELS, this normal maternal care is disturbed through physical separation of dams and pups. This model of stress is known as maternal separation (MS).

The main types of MS are repetitive MS and single exposure MS. Repetitive MS is performed by separating dams and pups for 3 h daily during postnatal days (P) 2–14 [15]. Single exposure MS involves separation of dams and pups for 24 h on P9 or P12 [22, 23]. In both forms of the MS model, the pups are deprived of maternal care during an early developmental period (Fig. 9.1). The absence of the dam implies that pups cannot benefit from her touch or feed during this time [24, 25]. Both forms of MS, single and repetitive exposure, induces lasting neurobiological and behavioral changes in the offspring, causing profound behavioral and cognitive consequences that are observable in adulthood and aging.

9.3.2 Maternal Separation Behavioral Consequences

The consequences of the various ELS models depend on the pup's stage of development, stress duration, and severity of the separation experience [26]. The separation of pups from their dam during early development with the MS model induces

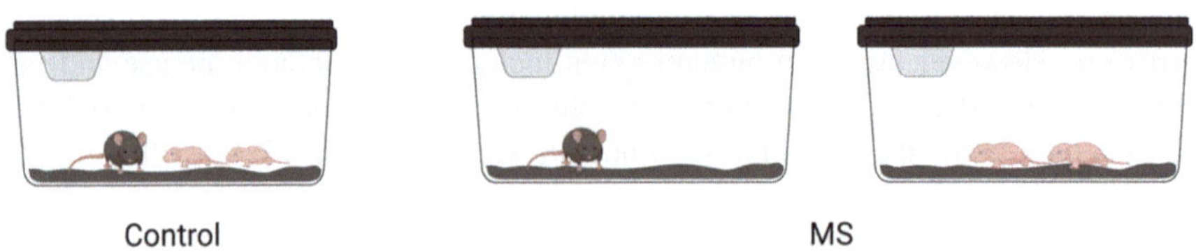

Fig. 9.1 Maternal separation as a model of ELS. Control: Standard housing of the dam with free access to the pups for touch and with normal bedding available for nesting. Maternal separation: Separate housing of the dam and pups for a predetermined time [15]. (Created in BioRender)

long-lasting depressive and anxious behaviors as the pups progress to adulthood [27, 28]. While normally reared rodents have low hypothalamic-pituitary-adrenal (HPA) axis activity during the first days after birth, rodents exposed to early MS have more dramatic stress responses with increased HPA axis activity [29, 30]. The HPA axis leads to glucocorticoid production, which is an essential component of the hormonal stress response in humans and rodents with respective secretion of cortisol and corticosterone [31]. The premature and prolonged activation of the HPA axis with the MS model and the associated increase in corticosterone have lasting cognitive and behavioral effects, which may be assessed with several accepted behavioral tests in rodents. Mice that underwent MS at P1–P21 and then testing in the Barnes maze during young adulthood (P53) had a significant increase in distance-to-escape, indicating that the stressed mice had deficits in spatial memory [32]. Rats exposed to MS at P2–P15 and assessed in the forced swim test at 3 months old had a shorter latency time to first immobility and a longer duration of immobility, suggesting increased depressive behavior [33]. Using infrared tracking to assess interindividual distances among cage mates, rats exposed to MS during P1–P14 exhibited greater social distancing from cage mates at 3 months of age as compared to normally reared animals, demonstrating a decrease in sociability [27]. In a systematic review analyzing results of MS rats who completed the elevated plus maze (EPM), MS rats spent significantly more time in the closed arm of the apparatus, indicating increased anxiety-like behaviors [27, 28]. Therefore, MS results in expected cognitive and behavioral changes in rodents similar to those associated with ACEs in humans.

9.3.3 Maternal Separation Challenges and Caveats

While there are many behavioral and cognitive consequences seen after MS that support face validity, some caveats must be addressed. Immediately after MS, there is a compensatory increase in maternal care. The fluctuations in maternal care from periods of low care during MS followed by periods of heightened care during reunion may contribute to the behavioral consequences of the model, in addition to the separation itself [34]. The behavioral and cognitive consequences of MS may also be due to the fluctuations in nutrition that pups receive during periods of separation from their dams. MS decreases weight gain in developing pups [35]. For these reasons, the use of MS requires awareness of possible confounding variables inherent to the model, as well as incorporation of controls and analyses to isolate the specific effects of ELS [36].

9.3.4 Limited Bedding and Nesting

Limited bedding and nesting (LBN) is another model for ELS used to mimic the experience of an impoverished environment during childhood. This stress paradigm involves providing dams with limited materials to construct proper nesting areas

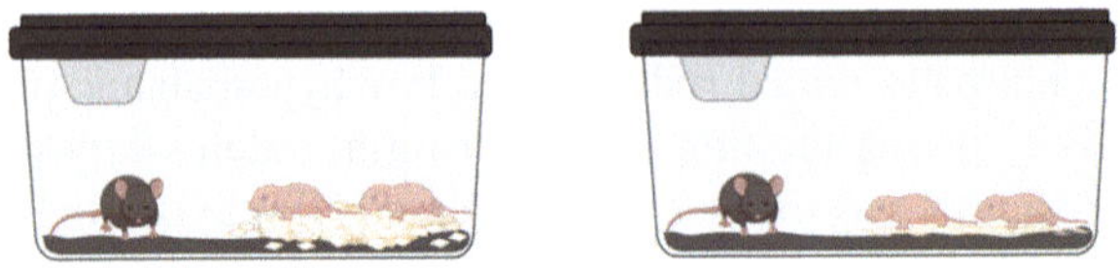

Fig. 9.2 Limited bedding and nesting as a model of ELS. Control: A dam housed with her pups in the postnatal period with enough bedding to create an adequate nest. LBN: Bedding is significantly diminished during the early postnatal period, contributing to stress for both the dam and pups [37]. (Created in BioRender)

during the early postnatal period (Fig. 9.2). While dams usually receive 500 mL of bedding during the postnatal period, the LBN model only allows for 100 mL of bedding during this time. LBN is typically implemented during the first 2 weeks of development (P2–P14). Studies that limit bedding as early as P2–9 still result in effects on both dam and pup behavior, indicating the robustness of this model [37]. Limiting bedding during this developmental period results in cognitive and behavioral deficits in both dams and offspring [38].

9.3.5 Limited Bedding and Nesting Behavioral Consequences

Both dams and pups exhibit measurable behavioral deficits in response to the LBN model. The dam has increased anxiety-like behaviors, such as decreased licking and increased isolation, both of which result in fragmented and abnormal interactions between the dam and her pups [39]. During the EPM, dams exhibit an increase in anxiety following LBN [40]. Pups experience a similar increase in anxiety as well as depression-like behaviors following LBN [41, 42]. Moreover, pups exhibit learning and memory deficits in adulthood following rearing under the LBN model. At 12 months of age, rats exposed to LBN during rearing had decreased spatial memory function during the Morris water maze (MWM) test and impaired novel object recognition memory [43]. In the contextual fear conditioning task, mice housed with LBN at P2–P9 had decreased freezing time during recall in adulthood at 6 months of age, suggesting that stressed rodents had decreased associative fear-based memory [44]. These findings suggest that the LBN and MS models of ELS are similar in rodents because both elicit measurable behavioral and cognitive deficits comparable to those seen in humans exposed to ACEs.

9.3.6 Limited Bedding and Nesting Challenges and Caveats

Unlike the MS model, there are no major challenges that have been made against LBN in the literature, making LBN one of the stronger models of ELS. The results of LBN studies are reproducible, and the deficits seen in rodents reflect the human experience. While there are no major caveats, this model has been shown to result in prominent sex-dependent differences. A systematic review of the LBN in rats revealed that males generally exhibited more depressive and anxious behaviors compared to females across behavioral tasks, including the open field and EPM [45]. Therefore, dimorphic differences must be considered when using LBN and analyzing its results.

9.3.7 Chronic Unpredictable Stress

The chronic unpredictable stress (CUS) protocol is used to more effectively model the unpredictability and randomness of ACEs in humans. In the CUS model, rodents are exposed to a variety of stressors at random times throughout development (Fig. 9.3). These protocols can span 4–8 weeks in total, and the stressors may include food/water deprivation, social defeat, exposure to a hot air stream, overnight illumination, and others [46]. The duration of stress exposure in CUS is greater than that of the MS and LBN models. Moreover, the stressors occur randomly and sporadically in CUS, while they are scheduled regularly during MS and LBN. Despite the increased duration and unpredictability of the stress, the CUS model has a similar intensity of stress as other models and comparable changes in cognition and behavior [47]. The greatest benefit of the CUS model is face validity with ACEs in humans, which may span several developmental stages and include a diverse array of stressful experiences.

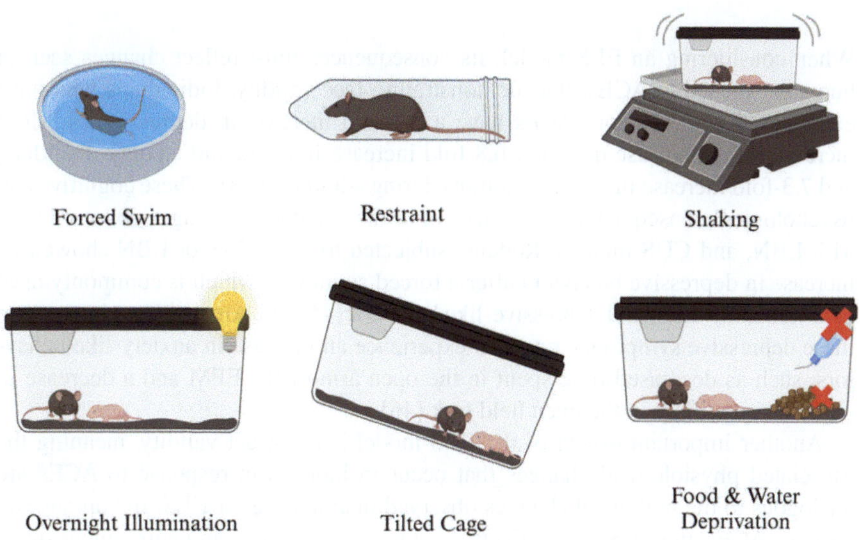

Fig. 9.3 Chronic unpredictable stress as a model of ELS. Stressors commonly implemented in the CUS model include forced swimming, physical restraint, constant shaking of the cage, bright illumination overnight, tilting the cage, and food and water deprivation [46]. (Created in BioRender)

9.3.8 Chronic Unpredictable Stress Behavioral Consequences

Rats exposed to CUS during P1–P42 exhibit cognitive deficits, increased anxiety, and worsened depression when tested in adolescence at P43 [48]. For example, the forced swim test revealed that CUS exposure in early development significantly increased immobility time. CUS-exposed rats assessed in the EPM spent significantly more time in the closed arm as compared to normally reared rats. In the MWM test, rats exposed to CUS had deficits in spatial learning and memory. These behavioral changes are congruent with similar learning and memory deficits reported with the other ELS models.

9.3.9 Chronic Unpredictable Stress Challenges and Caveats

While the CUS model leads to cognitive and behavioral impacts that are comparable to those seen in humans after ACEs, there are some challenges to using it in an experimental design. As there are multiple stressors involved, even a slight change in the design, timepoint, and environment can vary the impact of the stressors and influence the long-term effects. Therefore, reproducibility can be difficult when using this model [49].

9.4 ELS Model Validity

When considering an ELS model, its consequences must reflect changes seen in humans exposed to ACEs, thus demonstrating face validity. Individuals who have experienced four or more ACEs have a 3.2-fold increase in depression, 4.3-fold increase in alcohol use disorder, 6.8-fold increase in panic and anxiety disorders, and 7.3-fold increase in suicide attempts during adulthood [50]. These cognitive and psychological consequences are reflective of the behavioral changes observed with MS, LBN, and CUS models. Rodents subjected to either MS or LBN showed an increase in depressive behaviors after a forced swim test, which is commonly used to assess motivation and depressive-like behavior [45, 51]. Alongside an increase in these depressive symptoms, rats also experience an increase in anxiety-like behaviors, such as decreased time spent in the open arm of the EPM and a decrease in exploratory activity in the open field task [46].

Another important aspect of the ELS model is construct validity, meaning the associated physiological changes that occur in humans in response to ACEs are analogous to the biological changes observed in animals after ELS. In humans and rodents, ACEs and ELS, respectively, lead to major changes in neurological structures (Table 9.1). A predominant set of structures affected after ACEs and ELS is the corticolimbic system. The corticolimbic system consists of the prefrontal cortices, amygdala, and hippocampus. This system is related to cognition and emotion, making its function a critical component for higher-order processing of stimuli, threat assessment, and emotion regulation [52]. The hippocampus and amygdala have reduced volumes in humans who have experienced ACEs [53]. Furthermore, previously institutionalized youth have decreased connectivity between the amygdala and medial prefrontal cortex (mPFC) [54]. In male rats exposed to either MS or CUS, several regions of the corticolimbic system, including the mPFC, CA1, and CA3 regions of the hippocampus, and amygdala, exhibited comparable reductions in volume and dendritic arborization [55, 56]. MS also induced a decrease in dendritic spine density and length of mPFC neurons [56]. These changes to the corticolimbic system, seen in both humans exposed to ACEs and rodent models of ELS, reinforce the validity and applicability of these models.

Alongside the corticolimbic system, the HPA axis also undergoes changes after ACEs and ELS. The primary glucocorticoid product of the HPA axis, cortisol in humans and corticosterone in rodents, becomes dysregulated after the chronic stress. In children exposed to maltreatment, cortisol levels have significantly more daily variability when compared to children who have not experienced maltreatment [57]. This variability in cortisol manifests as more unpredictable emotional and behavioral dysregulation among children experiencing maltreatment. Cortisol levels are altered in a time-dependent manner during and after the stressor. During the stressor, individuals experience higher basal cortisol levels and then subsequent blunting of circadian fluctuations in cortisol release [58, 59]. In adulthood, exposure to a stressor causes basal levels of both adrenocorticotropin hormone (ACTH) and cortisol to fall below normal ranges [59, 60]. These fluctuations in the HPA-mediated stress response may, in part, contribute to the increased risk of depression and anxiety in adults with a history of childhood maltreatment [61].

The changes in the HPA axis exhibited in humans in response to ACEs are reflected in rodents exposed to ELS. For example, rats reared with the LBN model of ELS had significantly increased corticosterone levels in early development as compared to their normally reared counterparts [62]. Moreover, pups exposed to LBN had blunted ACTH and corticosterone increases in response to a stressor on P10 [63]. Finally, rats and mice exposed to MS during P1–14 and P1–21, respectively, had similar blunting of corticosterone levels in response to stressors in adulthood [64, 65]. Therefore, similar changes to the HPA axis are exhibited in humans and rodents exposed to respective ACEs and ELS, reinforcing the construct validity of the rodent models of ELS.

The glucocorticoids released as part of the HPA-mediated stress response modulate the immune system in both humans and rodents. The hippocampus has high expression of glucocorticoid and mineralocorticoid receptors in several cell types, including microglia [66, 67]. ELS increases the number of activated microglia in the hippocampus, including the CA3 region [68, 69]. ELS primes microglia by increasing synthesis of the NOD-like receptor protein 3 (NLRP3) [70, 71]. NLRP3 inflammasome recruits procaspase 1, which undergoes autoproteolysis and leads to the cleavage and maturation of interleukin-1β (IL-1β) and interleukin-18 (IL-18). These cytokines result in neuroinflammation and cell death via pyroptosis [71, 72].

Table 9.1 Face and construct validity of ELS models

		Type of ACE/ELS	Subject	Result	Citation
Structural changes	*Corticolimbic structures and circuitry dysregulation*	Physical abuse, early neglect, low socioeconomic status	Humans	Reduced amygdala and hippocampus volumes	[53]
		Previous institutionalization	Humans	Reduced PFC and hippocampus volumes	[91]
		MS	Humans	Mature connectivity (negative amygdala—mPFC coupling)	[54]
		Previous institutionalization	Humans	Nucleus accumbens (NAcc) hypoactivation during adolescence	[92]
		Early life adversity	Humans	Decrease in cortical thickness of the PFC and anterior cingulate cortex; increase in glial density	[93]
		CUS	Male Wistar rats	Atrophy of bipolar neurons of the basolateral complex of the amygdala (BLA)	[55]
		MS	Sprague-Dawley rats	Decrease in dendritic spine density and length in PFC, CA1, and NAcc	[56]
		MS	Male Wistar rats	Decrease in neuronal activation in amygdala, NAcc, and mPFC upon social novelty exposure	[94]
		LBN	Long-Evans rats	Decrease in neuronal activation in amygdala, NAcc, and mPFC upon social novelty exposure	[42]

(continued)

Table 9.1 (continued)

		Type of ACE/ELS	Subject	Result	Citation
		MS	C57BL/6N mice	Increase in GABAergic neurons in dentate gyrus, CA3, presubiculum, and parasubiculum	[95]
	Dysregulation of striatum circuitry	Reactive attachment disorder	Humans	Decrease in caudate and NAcc activity	[96]
		Threat trauma	Humans	Decrease in functional connectivity between ventral tegmental area (VTA) and hippocampus	[97]
		MS	Wistar rats	Decrease in glial cells and proliferation in VTA and substantia nigra pars compacta (SNc)	[98]
		CUS	Sprague-Dawley rats	Decrease in dopaminergic neuron population activity	[99]
		MS	C57BL/6N mice	Hyperexcitability of PFC-projecting neurons in the VTA	[100]
Behavioral Consequences	*Anxiety*	Childhood trauma	Humans	Increase in reported anxiety	[101]
		ACEs	Humans	Increase in health anxiety	[102]
		Emotional abuse	Humans	Increase in risk of social anxiety and panic disorder	[103]
		MS	C57BL/6N mice	Decrease in time spent in the open arm of EPM	[104]
		LBN	129S2/Sv, C57BL/6J mice	Increase in exploratory behavior and anxiety measured by EPM	[105]

(continued)

Table 9.1 (continued)

		Type of ACE/ELS	Subject	Result	Citation
		CUS	Sprague Dawley rats	Decrease in time spent in the open arm of EPM	[106]
	Depression	Emotional abuse, sexual abuse, severe family conflict	Humans	Increase in adult major depressive disorder	[107]
		Childhood trauma	Humans	Increase in depressive symptoms	[108]
		MS	Sprague Dawley rats	Inability to develop appropriate biases in response to rewards	[109]
		LBN	C57BL/6N mice	Increase in depressive-like behaviors (sucrose preference test, forced swim test)	[110]
		CUS	C57BL/6N mice	Increase in depressive-like behaviors (sucrose preference test and forced swim test)	[111]
	Learning and memory deficits	Physical neglect	Humans	Impaired spatial working memory and pattern recognition memory	[112]
		Childhood trauma	Humans	Impaired speed, attention, and executive functioning	[113]
		MS	Wistar rats	Impaired spatial memory	[32]
		MS	Wistar rats	Impaired aversive and recognition memory	[114]
		LBN	C57BL/6N mice	Impaired context memory	[115]
		LBN	C57BL/6N mice	Impaired recognition memory	[116]

(continued)

Table 9.1 (continued)

		Type of ACE/ELS	Subject	Result	Citation
		CUS	C57BL/6N mice	Impaired recognition, short-term spatial, long-term spatial, and reference memory	[111]

Proinflammatory cytokines such as IL-1β and IL-18 are elevated in both rodents exposed to ELS and humans exposed to ACEs [69, 73]. Glucocorticoid receptor activation of microglia and increased expression of proinflammatory cytokines are seen in both rodents and humans with ELS and ACEs, strengthening the validity of these models.

Other physiologic changes that further validate the use of ELS animal models include epigenetic modifications in response to stress [74]. Epigenetic modifications involve covalent bonding such as methylation or acetylation of the DNA or histone proteins that subsequently alter gene expression patterns [75]. Gene pathways that are epigenetically modified in response to stress in both humans and rodents include genes of the HPA axis [76]. For example, humans exhibit methylation to the glucocorticoid receptor gene known as nuclear receptor 3 subfamily c member 1 (*Nr3c1*) [77]. *Nr3c1* normally regulates the HPA axis by initiating a negative feedback loop to reduce glucocorticoid actions [78]. *Nr3c1* downregulation in response to stress may result in HPA axis dysfunction and increased anxiety and depression [79–81]. Similarly, rodent models of ELS experience modification of the homologous *Nr3c1* gene [82]. Consequences of *Nr3c1* methylation in rodents reflect those seen in humans, including a decrease in NR3C1 protein levels and an increase in depressive-like behaviors [83].

Another gene regulated by stress is brain-derived neurotrophic factor (*Bdnf*). BDNF has plays a large role in hippocampal-dependent memory processes [84]. In response to stress, humans exhibit increased methylation of the *Bdnf* promoter region, leading to downregulated expression [85, 86]. Decreased production of BDNF is associated with cognitive impairments, such as learning and memory deficits, and white matter atrophy seen in neurodegenerative diseases and aging [87, 88]. Similarly, rodents that experience ELS have increased methylation of the analogous *Bdnf* gene, leading to a comparable downregulation of its expression and cognitive changes, such as a decrease in sucrose preference, reduced sociability, and short-term memory deficits in response to stress [89, 90]. The similar nature of stress-induced epigenetic modifications in humans and rodents after stress further supports the validity of the use of these ELS models to understand the effects of ACEs in humans (Table 9.1).

9.4.1 Changes Occurring After Both ELS and mTBI

Since ACEs are a risk factor for poor prognosis following mTBI, animal models of ELS have been combined with mTBI to understand the relationship of a prior exposure to ELS on mTBI recovery. Modeling mTBI in preclinical models is complicated by the heterogeneity of human TBI, the biomechanical differences between rodent and human brains, and the difficulty of modeling the psychological consequences of mTBI [117]. Despite these challenges, the implementation of well-accepted, rigorous ELS and mTBI preclinical models has yielded significant results in uncovering the role of ELS on mTBI recovery.

9.4.2 ELS and TBI: Structural Changes

Cerebral atrophy is a volumetric loss of brain tissue [118]. Cortical thickness, a proxy measurement for cortical atrophy, is associated with worsened outcomes after mTBI [119]. Childhood stress and mTBI independently lead to cerebral atrophy in the months to years following the exposure in both humans and rodents [120, 121]. The combination of these two experiences leads to a synergistic loss of volume in particular brain regions [122, 123]. For example, male rats that experienced MS followed by a pediatric mild controlled cortical impact (CCI) had a significant decrease in the volume of the dentate gyrus, compared to either exposure alone [13]. Therefore, the pathogenesis and extent of brain atrophy after ELS and mTBI may account for some of the behavioral consequences in this combinatorial injury.

Cerebral atrophy is largely dependent on reductions in dendritic arborization [123]. Rats exhibit decreased dendritic length in adulthood after MS [124]. Moreover, mice exposed to LBN also have reduced apical dendritic length and complexity of CA3 neurons during adulthood [125]. These morphological changes to dendrites contribute to the loss of brain volume with ELS. When the ELS-exposed brain later receives an insult from a mTBI, further dendritic loss may result. This is unsurprising given that mTBI alone has shown to decrease the neuronal projections from pyramidal neurons in layers II/III of the infralimbic cortex in rats and mice subacutely after injury [126]. Therefore, the decrease in brain volume after a mTBI is in part attributable to dendritic dearborization. When ELS precedes a mTBI in adulthood, a further predictable increase in cortical and hippocampal atrophy has been observed [122].

In addition to causing changes in neuronal morphology, ELS and mTBI may also decrease neurogenesis. As the developing brain reaches adulthood, two primary sites of neurogenesis remain: the subventricular zone of the lateral ventricles and the subgranular zone of the dentate gyrus [127]. The neurons derived from the subgranular zone of the dentate gyrus have been specifically implicated in cognition and spatial memory formation [127, 128]. The dentate gyrus exhibits well-documented reductions in adulthood neurogenesis after exposure to ELS alone [127, 129, 130]. It is less well-established whether exposure to mTBI alone leads to changes in neurogenesis; these discrepant results are likely due to variations in the timepoints measured and the stage of neurogenesis classification. Some reports reveal that mild fluid percussion injury (FPI) in mice triggers an initial increase in neurogenesis of the ipsilateral dentate gyrus [131, 132]. Other reports document a reduction in neurogenesis in mice following mild CCI [133]. Some have even reported no changes in neurogenesis in mice following mild closed head injury [134]. Therefore, a more comprehensive meta-analysis of these findings is required to understand the effects of mTBI alone on neurogenesis.

When ELS precedes a later mTBI, the effects on neurogenesis are more established. In rats exposed to a combination of MS and subsequent pediatric mild CCI, the subgranular zone of the ipsilateral dentate gyrus exhibited significant neuronal depletion compared to either exposure alone [13]. This decrease in neurogenic cells in the hippocampus suggests an impairment in either the proliferation, migration, or differentiation of neurons following the combination of ELS and mTBI [135]. Interestingly, the timing of the stressor relative to the mTBI affects the measured changes to neurogenesis. For example, adult mice exposed to unpredictable foot shock and predator odor stressors at the time of mild CCI did not exhibit any changes to neurogenesis within the subgranular zone of the dentate gyrus [134]. Therefore, careful consideration must be given of the timing of the stressors when understanding their effects on neurogenesis after mTBI. Neurogenesis is largely influenced by microglial functioning in the brain. Microglia are the primary immune cells of the CNS. These specialized immune cells may become activated into a proinflammatory phenotype by a variety of stressors [136]. Activated microglia secrete proinflammatory cytokines to launch the local immune response, regulate blood-brain barrier permeability, and promote wound healing [136, 137]. However, proinflammatory cytokines such as interleukin-6 (IL-6) and IL-1β may also induce cellular damage. IL-6 and IL-1β trigger apoptosis of transient amplifying progenitor cells that would have ultimately differentiated into neurons in the subgranular zone of the dentate gyrus in adult brains [138, 139]. Therefore, activation of the proinflammatory microglia phenotype poses a threat to neurogenesis.

Microglial activation and neuroinflammation are induced by both ELS and mTBI. Mice exposed to MS during P2–P14 had increased microglial activation in the dentate gyrus, CA1, and CA3 regions of the hippocampus during adolescence [140]. In another study, rats that experienced MS on P2 had increased serum levels of NLRP3 in the PFC during adulthood [141]. Similarly, mice exposed to LBN during P2–P19 exhibited proinflammatory transcriptional changes in adulthood (P200), including upregulation of the proinflammatory genes *Ripk*, *Gas6*, *Akirin1*, and *C1qbp*. The protein products of these genes modulate tumor necrosis factor production, increase chemotaxis of leukocytes, and stimulate cellular adhesion, all of which are integral parts of inflammation [142]. The upregulation of proinflammatory pathways is also seen following mTBI. Rats that received mild lateral FPI in adulthood had increased colocalization of activated microglia and proinflammatory cytokines in the cortex starting 6 h postinjury and persisting for at least 18 days postinjury [143]. Moreover, adult male rats who underwent mild CCI had elevated levels of proinflammatory microglia shortly after injury that persisted over 42 days postinjury; this cohort of rats also had a transient increase in antiinflammatory microglia postinjury that returned to baseline at 42 days postinjury [144]. These findings suggest that proinflammatory microglia have more permanent effects in the brain than the antiinflammatory phenotype initiated for endogenous repair.

When ELS precedes mTBI, the inflammatory response after the injury is exacerbated. Rats exposed to ELS on P1–P20 with a pediatric mild CCI on P21 had significantly increased numbers of activated microglia surrounding the contusion site [13]. Moreover, ELS and mTBI rats also had significantly increased levels of IL-1β in the ipsilateral hippocampus [139]. Interestingly, animals with ELS preceding mTBI also had a significant reduction of the volume of the subgranular zone in the dentate gyrus [13]. These results are consistent with the current understanding that ELS before mTBI exacerbates microglial activation, which in turn launches a proinflammatory response that inhibits neurogenesis.

9.4.3 ELS and TBI: HPA Axis

Given that microglial activation plays an integral role in the stress response, we must consider how the combination of ELS and mTBI alters the HPA axis to potentially influence injury recovery. The final product of the HPA axis in rodents is corticosterone. Corticosterone binds to glucocorticoid receptors (GRs) to propagate its effects. In the neurons and astrocytes of the pituitary gland and hippocampus, GRs modulate the negative feedback response of the HPA axis [67, 145]. In microglia, however, the effects of GR activation by corticosterone range from maintaining microglial homeostasis during steady state to inducing expression of proinflammatory genes and their protein products during a stressed state [122, 146]. Repetitive activation of GRs with chronic corticosterone secretion leads to the downregulation of GRs and ultimately allows inflammation to continue without appropriate negative feedback [147].

When mTBI follows ELS, the HPA axis at baseline, as measured by corticosterone levels under nonstressful conditions, is relatively unchanged [122, 139]. However, the HPA axis undergoes prolonged activation following subsequent stressors. For example, rats that experienced MS and then mild FPI at P60 had more prolonged responses to restraint stress at 8 weeks postinjury. Corticosterone levels at 90 min after the cessation of restraint stress were significantly higher in the ELS and mTBI group than the ELS-only or mTBI-only groups [122]. These findings are consistent with the hypothesis that downregulation of GRs after ELS and mTBI dampens the negative feedback loop on the HPA axis and prolongs the stress response. Therefore, the combination of ELS and mTBI influences how the HPA axis perceives and responds to future stressors, rather than how the axis acts at baseline.

9.4.4 ELS and TBI: Behavioral and Cognitive Changes

The ultimate manifestations of the cellular and molecular changes in the brain after ELS and mTBI are behavioral and cognitive deficits. Both independent stressors induce anxiety and depression [148–150]. Unsurprisingly, the combination of ELS and mTBI significantly exacerbates cognitive deficits. Male rats exposed to MS during P1–P20 and a subsequent mild CCI in adolescence had significantly worse spatial learning and memory deficits, as indicated by a longer pathlength in the MWM test [13]. In another study, rats exposed to MS on P2–P14 and later subjected to mild FPI in adulthood had significant deficits in fear-based contextual learning and memory, spatial learning, and spatial working memory as compared to rats that experienced either stressor alone. Some deficits, such as fear-based contextual memory, persisted for over 1 month after mTBI [122]. In a third study, male rats that experienced CUS during adolescence and moderate CCI in adulthood exhibited deficits in spatial learning and memory as measured in the Barnes maze [151]. Therefore, the cognitive deficits seen after the combination of ELS and mTBI are measurable and greater than those observed with either insult alone.

9.4.5 ELS and TBI: Experimental Limitations

There are many factors that influence the effects of ELS on recovery after mTBI. One important variable is the time point at which the ELS and mTBI are experienced. The time of onset and duration of ELS across various developmental periods plays a role in the intensity of symptoms and outcomes [152]. In a study with human participants across four different developmental stages, stress during the perinatal period was found to negatively impact sensory integration and self-regulation, while stress during infancy and childhood negatively impacted cognitive development [153]. In rodents, stress during the postweaning/prepuberty and adulthood periods decreased spatial learning and memory as compared to rodents with stress exposure in adulthood [154]. Therefore, the variety of outcomes from different ELS onsets and durations poses a challenge when interpreting the results of preclinical models.

Similarly, the timepoint of the mTBI acquisition is important to understand how an individual may recover from the mTBI. Individuals who sustained a mTBI after age 40 had greater cognitive decline compared to those who sustained a mTBI before age 40 [155]. Older patients also required more inpatient rehabilitation following a TBI and have an increased risk of death [156]. Similar implications of age are observed in rodent models of TBI. Rats that received a TBI during middle age (>1 year of age) had worsened cognitive function and decreased cortical volume as compared to rats that were injured during young adulthood at 10 weeks of age [157]. Given such variations in the outcomes from ELS and mTBI at various timepoints, it is difficult to generalize results of studies of ELS and mTBI when such studies implement different stressor timelines.

Another confounding variable when conducting and interpreting clinical and preclinical studies in rodents is sex-based differences. The effects of ELS are dependent on sex. In a study that implemented the MS model of ELS, male rats had higher corticosterone levels and reduced social play behaviors as compared to females [158]. In a study of mice exposed to the LBN model of ELS, there was an increase in depressive-like behaviors in adult female but not male mice [110, 159]. The CUS model of ELS led to higher rates of frontal-limbic hyperconnectivity in males but not females [160]. Given the sex-based variability in the biochemical and behavioral responses to ELS in rodents, interpreting the results from male-only studies is limited. Moreover, mechanisms of recovery may vary between the sexes, which may necessitate tailored interventions for each sex.

9.5 Future Directions

In this chapter, we have outlined and validated three commonly implemented ELS models in rodents, as well as the preclinical findings that have helped further our understanding as to how ACEs contribute to a worse recovery from mTBI in humans. Although further studies are needed, multiple studies have demonstrated how exposure to ELS impairs mTBI recovery through a multimodal combination of HPA activation, microglial priming, neuroinflammation, and cerebral atrophy. However, these studies have provoked greater questions that warrant answering. For example, future studies may aim to understand the longevity of ELS-induced microglial activation and proliferation. Understanding how the microglia phenotypic changes with time may help predict recovery from mTBI at subsequent time points. Moreover, investigating compounds that prevent microglia activation and proliferation may promote neurogenesis after the combination of ELS and mTBI. Other neuroprotective interventions during or after ELS may also preserve dendritic arborization and prevent the development of cerebral atrophy that is worsened by a subsequent mTBI. Ultimately, these future endeavors will provide insight into therapeutics to ameliorate the cognitive and behavioral deficits associated with ELS followed by a mTBI.

Acknowledgments This work was supported by the Miami Project to Cure Paralysis, Brain Injury Association of America's Brain Research Fund, and Department of Veterans Affairs, Rehabilitation Research and Development Service (I21RX003017, I01RX003782).

Disclosures The authors have no conflicts of interest to declare that are relevant to the content of this chapter.

References

1. Taylor CA, Bell JM, Breiding MJ, Xu L. Traumatic brain injury-related emergency department visits, hospitalizations, and deaths—United States, 2007 and 2013. MMWR. 2017;66:1–16.
2. Maas AIR, Menon DK, Adelson PD, Andelic N, Bell MJ, Belli A, et al. Traumatic brain injury: integrated approaches to improve prevention, clinical care, and research. Lancet Neurol. 2017;16:987–1048.
3. McInnes K, Friesen CL, MacKenzie DE, Westwood DA, Boe SG. Mild traumatic brain Injury (mTBI) and chronic cognitive impairment: a scoping review. PLoS One. 2017;12:e0174847.
4. Reid MW, Cooper DB, Lu LH, Iverson GL, Kennedy JE. Adversity and resilience are associated with outcome after mild traumatic brain injury in military service members. J Neurotrauma. 2018;35:1146–55.

5. Ramanathan-Elion DM, Baydoun HA, Johnstone B. Psychological predictors of functional outcomes in service members with traumatic brain injury. Brain Inj. 2020;34:1183–92.
6. Norman RE, Byambaa M, De R, Butchart A, Scott J, Vos T. The long-term health consequences of child physical abuse, emotional abuse, and neglect: a systematic review and meta-analysis. PLoS Med. 2012;9:e1001349.
7. Kessler RC, McLaughlin KA, Green JG, Gruber MJ, Sampson NA, Zaslavsky AM, et al. Childhood adversities and adult psychopathology in the WHO world mental health surveys. Br J Psychiatry. 2010;197:378–85.
8. Merrick MT, Ford DC, Ports KA, Guinn AS, Chen J, Klevens J, et al. Vital Signs: estimated proportion of adult health problems attributable to adverse childhood experiences and implications for prevention—25 States, 2015–2017. MMWR. 2019;68:999–1005.
9. Child Maltreatment 2021. Children's Bureau, Administration on Children, Youth and Families, Administration for Children and Families, US Department of Health and Human Services. 2022.
10. McGuinness TM, Waldrop JR. Adverse childhood experiences and the mental health of Veterans. J Psychosoc Nurs Ment Health Serv. 2015;53:23–6.
11. Petersen RC, Caracciolo B, Brayne C, Gauthier S, Jelic V, Fratiglioni L. Mild cognitive impairment: a concept in evolution. J Intern Med. 2014;275:214–28.
12. Guinn AS, Ports KA, Ford DC, Breiding M, Merrick MT. Associations between adverse childhood experiences and acquired brain injury, including traumatic brain injuries, among adults: 2014 BRFSS North Carolina. Inj Prev. 2019;25:514–20.
13. Diaz-Chavez A, Lajud N, Roque A, Cheng JP, Melendez-Herrera E, Valdez-Alarcon JJ, et al. Early life stress increases vulnerability to the sequelae of pediatric mild traumatic brain injury. Exp Neurol. 2020;329:113318.
14. Van Veldhoven LM, Sander AM, Struchen MA, Sherer M, Clark AN, Hudnall GE, et al. Predictive ability of preinjury stressful life events and post-traumatic stress symptoms for outcomes following mild traumatic brain injury: analysis in a prospective emergency room sample. J Neurol Neurosurg Psychiatry. 2011;82:782–7.
15. Van Bodegom M, Homberg JR, Henckens M. Modulation of the hypothalamic-pituitary-adrenal axis by early life stress exposure. Front Cell Neurosci. 2017;11:87.
16. Atrooz F, Alkadhi KA, Salim S. Understanding stress: insights from rodent models. Curr Res Neurobiol. 2021;2:100013.
17. Champagne FA, Francis DD, Mar A, Meaney MJ. Variations in maternal care in the rat as a mediating influence for the effects of environment on development. Physiol Behav. 2003;79:359–71.
18. Narvaez D, Wang L, Cheng A, Gleason TR, Woodbury R, Kurth A, et al. The importance of early life touch for psychosocial and moral development. Psych Res Rev. 2019;32:16.
19. Pereira M, Ferreira A. Demanding pups improve maternal behavioral impairments in sensitized and haloperidol-treated lactating female rats. Behav Brain Res. 2006;175:139–48.
20. Weaver IC, Cervoni N, Champagne FA, D'Alessio AC, Sharma S, Seckl JR, et al. Epigenetic programming by maternal behavior. Nat Neurosci. 2004;7:847–54.
21. Mayer AD, Rosenblatt JS. Prepartum changes in maternal responsiveness and nest defense in *Rattus norvegicus*. J Comp Psychol. 1984;98:177–88.
22. Macri S, Laviola G. Single episode of maternal deprivation and adult depressive profile in mice: interaction with cannabinoid exposure during adolescence. Behav Brain Res. 2004;154:231–8.
23. Marco EM, Adriani W, Llorente R, Laviola G, Viveros MP. Detrimental psychophysiological effects of early maternal deprivation in adolescent and adult rodents: altered responses to cannabinoid exposure. Neurosci Biobehav Rev. 2009;33:498–507.
24. Suchecki D, Mozaffarian D, Gross G, Rosenfeld P, Levine S. Effects of maternal deprivation on the ACTH stress response in the infant rat. Neuroendocrinology. 1993;57:204–12.
25. Van Oers HJ, de Kloet ER, Li C, Levine S. The ontogeny of glucocorticoid negative feedback: influence of maternal deprivation. Endocrinology. 1998;139:2838–46.

26. Arborelius L, Eklund MB. Both long and brief maternal separation produces persistent changes in tissue levels of brain monoamines in middle-aged female rats. Neuroscience. 2007;145:738–50.
27. Endo N, Makinodan M, Mannari-Sasagawa T, Horii-Hayashi N, Somayama N, Komori T, et al. The effects of maternal separation on behaviours under social-housing environments in adult male C57BL/6 mice. Sci Rep. 2021;11:527.
28. Wang D, Levine JLS, Avila-Quintero V, Bloch M, Kaffman A. Systematic review and meta-analysis: effects of maternal separation on anxiety-like behavior in rodents. Transl Psychiatry. 2020;10:174.
29. Pihoker C, Owens MJ, Kuhn CM, Schanberg SM, Nemeroff CB. Maternal separation in neonatal rats elicits activation of the hypothalamic-pituitary-adrenocortical axis: a putative role for corticotropin-releasing factor. Psychoneuroendocrinology. 1993;18:485–93.
30. Schmidt MV, Levine S, Alam S, Harbich D, Sterlemann V, Ganea K, et al. Metabolic signals modulate hypothalamic-pituitary-adrenal axis activation during maternal separation of the neonatal mouse. J Neuroendocrinol. 2006;18:865–74.
31. Smith SM, Vale WW. The role of the hypothalamic-pituitary-adrenal axis in neuroendocrine responses to stress. Dialogues Clin Neurosci. 2006;8:383–95.
32. Maghami S, Zardooz H, Khodagholi F, Binayi F, Ranjbar Saber R, Hedayati M, et al. Maternal separation blunted spatial memory formation independent of peripheral and hippocampal insulin content in young adult male rats. PLoS One. 2018;13:e0204731.
33. Roque S, Mesquita AR, Palha JA, Sousa N, Correia-Neves M. The behavioral and immunological impact of maternal separation: a matter of timing. Front Behav Neurosci. 2014;8:192.
34. Huot RL, Gonzalez ME, Ladd CO, Thrivikraman KV, Plotsky PM. Foster litters prevent hypothalamic-pituitary-adrenal axis sensitization mediated by neonatal maternal separation. Psychoneuroendocrinology. 2004;29:279–89.
35. Figueiredo IL, Frota PB, da Cunha DG, da Silva Raposo R, Canuto KM, de Andrade GM, et al. Prolonged maternal separation induces undernutrition and systemic inflammation with disrupted hippocampal development in mice. Nutrition. 2016;32:1019–27.
36. Orso R, Creutzberg KC, Wearick-Silva LE, Wendt Viola T, Tractenberg SG, Benetti F, et al. How early life stress impact maternal care: a systematic review of rodent studies. Front Behav Neurosci. 2019;13:197.
37. Ivy AS, Brunson KL, Sandman C, Baram TZ. Dysfunctional nurturing behavior in rat dams with limited access to nesting material: a clinically relevant model for early-life stress. Neuroscience. 2008;154:1132–42.
38. Davis LK, Bolton JL, Hanson H, Guarraci FA. Modified limited bedding and nesting is a model of early-life stress that affects reproductive physiology and behavior in female and male Long-Evans rats. Physiol Behav. 2020;224:113037.
39. Gallo M, Shleifer DG, Godoy LD, Ofray D, Olaniyan A, Campbell T, et al. Limited bedding and nesting induces maternal behavior resembling both hypervigilance and abuse. Front Behav Neurosci. 2019;13:167.
40. Gifford JJ, Pluchino JR, Della Valle R, Van Weele B, Brezoczky E, Caulfield JI, et al. Effects of limited bedding and nesting on postpartum mood state in rats. J Neuroendocrinol. 2023;35:e13275.
41. Dalle Molle R, Portella AK, Goldani MZ, Kapczinski FP, Leistner-Segal S, Salum GA, et al. Associations between parenting behavior and anxiety in a rodent model and a clinical sample: relationship to peripheral BDNF levels. Transl Psychiatry. 2012;2:e195.
42. Rincon-Cortes M, Sullivan RM. Emergence of social behavior deficit, blunted corticolimbic activity and adult depression-like behavior in a rodent model of maternal maltreatment. Transl Psychiatry. 2016;6:e930.
43. Brunson KL, Kramar E, Lin B, Chen Y, Colgin LL, Yanagihara TK, et al. Mechanisms of late-onset cognitive decline after early-life stress. J Neurosci. 2005;25:9328–38.
44. Lesuis SL, Lucassen PJ, Krugers HJ. Early life stress impairs fear memory and synaptic plasticity; a potential role for GluN2B. Neuropharmacology. 2019;149:195–203.

45. Walker CD, Bath KG, Joels M, Korosi A, Larauche M, Lucassen PJ, et al. Chronic early life stress induced by limited bedding and nesting (LBN) material in rodents: critical considerations of methodology, outcomes and translational potential. Stress. 2017;20:421–48.
46. Monteiro S, Roque S, de Sa-Calcada D, Sousa N, Correia-Neves M, Cerqueira JJ. An efficient chronic unpredictable stress protocol to induce stress-related responses in C57BL/6 mice. Front Psych. 2015;6:6.
47. Zhang Y, Zhang L, Shi B, Huang F, Gao Y, Miao Z, et al. Comparison of the chronic unpredictable mild stress and the maternal separation in mice postpartum depression modeling. Biochem Biophys Res Commun. 2022;632:24–31.
48. Hu C, Luo Y, Wang H, Kuang S, Liang G, Yang Y, et al. Re-evaluation of the interrelationships among the behavioral tests in rats exposed to chronic unpredictable mild stress. PLoS One. 2017;12:e0185129.
49. Markov DD, Novosadova EV. Chronic unpredictable mild stress model of depression: possible sources of poor reproducibility and latent variables. Biology (Basel). 2022;11:1621.
50. Petruccelli K, Davis J, Berman T. Adverse childhood experiences and associated health outcomes: a systematic review and meta-analysis. Child Abuse Negl. 2019;97:104127.
51. Kwak HR, Lee JW, Kwon KJ, Kang CD, Cheong IY, Chun W, et al. Maternal social separation of adolescent rats induces hyperactivity and anxiolytic behavior. Korean J Physiol Pharmacol. 2009;13:79–83.
52. Swartz JR, Monk CS. The role of corticolimbic circuitry in the development of anxiety disorders in children and adolescents. Curr Top Behav Neurosci. 2014;16:133–48.
53. Hanson JL, Nacewicz BM, Sutterer MJ, Cayo AA, Schaefer SM, Rudolph KD, et al. Behavioral problems after early life stress: contributions of the hippocampus and amygdala. Biol Psychiatry. 2015;77:314–23.
54. Gee DG, Gabard-Durnam LJ, Flannery J, Goff B, Humphreys KL, Telzer EH, et al. Early developmental emergence of human amygdala-prefrontal connectivity after maternal deprivation. Proc Natl Acad Sci USA. 2013;110:15638–43.
55. Vyas A, Mitra R, Shankaranarayana Rao BS, Chattarji S. Chronic stress induces contrasting patterns of dendritic remodeling in hippocampal and amygdaloid neurons. J Neurosci. 2002;22:6810–8.
56. Monroy E, Hernandez-Torres E, Flores G. Maternal separation disrupts dendritic morphology of neurons in prefrontal cortex, hippocampus, and nucleus accumbens in male rat offspring. J Chem Neuroanat. 2010;40:93–101.
57. Doom JR, Cicchetti D, Rogosch FA. Longitudinal patterns of cortisol regulation differ in maltreated and nonmaltreated children. J Am Acad Child Adolesc Psychiatry. 2014;53:1206–15.
58. Miller GE, Chen E, Zhou ES. If it goes up, must it come down? Chronic stress and the hypothalamic-pituitary-adrenocortical axis in humans. Psychol Bull. 2007;133:25–45.
59. Carpenter LL, Carvalho JP, Tyrka AR, Wier LM, Mello AF, Mello MF, et al. Decreased adrenocorticotropic hormone and cortisol responses to stress in healthy adults reporting significant childhood maltreatment. Biol Psychiatry. 2007;62:1080–7.
60. Simsek S, Kaplan I, Uysal C, Yuksel T, Alaca R. The levels of cortisol, oxidative stress, and DNA damage in the victims of childhood sexual abuse: a preliminary study. J Child Sex Abus. 2016;25:175–84.
61. McLaughlin KA, Conron KJ, Koenen KC, Gilman SE. Childhood adversity, adult stressful life events, and risk of past-year psychiatric disorder: a test of the stress sensitization hypothesis in a population-based sample of adults. Psychol Med. 2010;40:1647–58.
62. Moussaoui N, Jacobs JP, Larauche M, Biraud M, Million M, Mayer E, et al. Chronic early-life stress in rat pups alters basal corticosterone, intestinal permeability, and fecal microbiota at weaning: influence of sex. J Neurogastroenterol Motil. 2017;23:135–43.
63. McLaughlin RJ, Verlezza S, Gray JM, Hill MN, Walker CD. Inhibition of anandamide hydrolysis dampens the neuroendocrine response to stress in neonatal rats subjected to suboptimal rearing conditions. Stress. 2016;19:114–24.

64. Kim HJ, Lee JH, Choi SH, Lee YS, Jahng JW. Fasting-induced increases of arcuate NPY mRNA and plasma corticosterone are blunted in the rat experienced neonatal maternal separation. Neuropeptides. 2005;39:587–94.
65. Roman E, Gustafsson L, Berg M, Nylander I. Behavioral profiles and stress-induced corticosteroid secretion in male Wistar rats subjected to short and prolonged periods of maternal separation. Horm Behav. 2006;50:736–47.
66. Morimoto M, Morita N, Ozawa H, Yokoyama K, Kawata M. Distribution of glucocorticoid receptor immunoreactivity and mRNA in the rat brain: an immunohistochemical and *in situ* hybridization study. J Neurosci Res. 1996;26:235–69.
67. Sierra A, Gottfried-Blackmore A, Milner TA, McEwen BS, Bulloch K. Steroid hormone receptor expression and function in microglia. Glia. 2008;56:659–74.
68. Roque A, Ochoa-Zarzosa A, Torner L. Maternal separation activates microglial cells and induces an inflammatory response in the hippocampus of male rat pups, independently of hypothalamic and peripheral cytokine levels. Brain Behav Immun. 2016;55:39–48.
69. Warhaftig G, Almeida D, Turecki G. Early life adversity across different cell-types in the brain. Neurosci Biobehav Rev. 2023;148:105113.
70. Su WJ, Zhang Y, Chen Y, Gong H, Lian YJ, Peng W, et al. NLRP3 gene knockout blocks NF-kappaB and MAPK signaling pathway in CUMS-induced depression mouse model. Behav Brain Res. 2017;322:1–8.
71. Frank MG, Hershman SA, Weber MD, Watkins LR, Maier SF. Chronic exposure to exogenous glucocorticoids primes microglia to pro-inflammatory stimuli and induces NLRP3 mRNA in the hippocampus. Psychoneuroendocrinology. 2014;40:191–200.
72. Weber MD, Frank MG, Tracey KJ, Watkins LR, Maier SF. Stress induces the danger-associated molecular pattern HMGB-1 in the hippocampus of male Sprague Dawley rats: a priming stimulus of microglia and the NLRP3 inflammasome. J Neurosci. 2015;35:316–24.
73. Lo Iacono L, Catale C, Martini A, Valzania A, Viscomi MT, Chiurchiu V, et al. From traumatic childhood to cocaine abuse: the critical function of the immune system. Biol Psychiatry. 2018;84:905–16.
74. Gudsnuk K, Champagne FA. Epigenetic influence of stress and the social environment. Instit Lab Animal Res J. 2012;53:279–88.
75. Gibney ER, Nolan CM. Epigenetics and gene expression. Heredity (Edinb). 2010;105:4–13.
76. Juruena MF, Eror F, Cleare AJ, Young AH. The role of early life stress in HPA axis and anxiety. Adv Exp Med Biol. 2020;1191:141–53.
77. Van Der Knaap LJ, Oldehinkel AJ, Verhulst FC, van Oort FV, Riese H. Glucocorticoid receptor gene methylation and HPA-axis regulation in adolescents. The TRAILS study. Psychoneuroendocrinology. 2015;58:46–50.
78. Liu PZ, Nusslock R. How stress gets under the skin: early life adversity and glucocorticoid receptor epigenetic regulation. Curr Genomics. 2018;19:653–64.
79. Derijk RH. Single nucleotide polymorphisms related to HPA axis reactivity. Neuroimmunomodulation. 2009;16:340–52.
80. Cicchetti D, Handley ED. Methylation of the glucocorticoid receptor gene, nuclear receptor subfamily 3, group C, member 1 (*NR3C1*), in maltreated and nonmaltreated children: associations with behavioral undercontrol, emotional lability/negativity, and externalizing and internalizing symptoms. Dev Psychopathol. 2017;29:1795–806.
81. Van Der Knaap LJ, Riese H, Hudziak JJ, Verbiest MM, Verhulst FC, Oldehinkel AJ, et al. Glucocorticoid receptor gene (NR3C1) methylation following stressful events between birth and adolescence. The TRAILS study. Transl Psychiatry. 2014;4:e381.
82. Palma-Gudiel H, Cordova-Palomera A, Eixarch E, Deuschle M, Fananas L. Maternal psychosocial stress during pregnancy alters the epigenetic signature of the glucocorticoid receptor gene promoter in their offspring: a meta-analysis. Epigenetics. 2015;10:893–902.
83. Ozturk KH, Unal GO, Doguc DK, Togay VA, Kosar PA, Sezik M. Hypothalamic NR3C1 DNA methylation in rats exposed to prenatal stress. Mol Biol Rep. 2022;49:7921–8.

84. Miranda M, Morici JF, Zanoni MB, Bekinschtein P. Brain-derived neurotrophic factor: a key molecule for memory in the healthy and the pathological brain. Front Cell Neurosci. 2019;13:363.
85. Weder N, Zhang H, Jensen K, Yang BZ, Simen A, Jackowski A, et al. Child abuse, depression, and methylation in genes involved with stress, neural plasticity, and brain circuitry. J Am Acad Child Adolesc Psychiatry. 2014;53:417–24. e5.
86. Unternaehrer E, Meyer AH, Burkhardt SC, Dempster E, Staehli S, Theill N, et al. Childhood maternal care is associated with DNA methylation of the genes for brain-derived neurotrophic factor (BDNF) and oxytocin receptor (OXTR) in peripheral blood cells in adult men and women. Stress. 2015;18:451–61.
87. Azman KF, Zakaria R. Recent advances on the role of brain-derived neurotrophic factor (BDNF) in neurodegenerative diseases. Int J Mol Sci. 2022;23:6827.
88. Driscoll I, Martin B, An Y, Maudsley S, Ferrucci L, Mattson MP, et al. Plasma BDNF is associated with age-related white matter atrophy but not with cognitive function in older, non-demented adults. PLoS One. 2012;7:e35217.
89. Kundakovic M, Lim S, Gudsnuk K, Champagne FA. Sex-specific and strain-dependent effects of early life adversity on behavioral and epigenetic outcomes. Front Psych. 2013;4:78.
90. Daskalakis NP, De Kloet ER, Yehuda R, Malaspina D, Kranz TM. Early life stress effects on glucocorticoid-BDNF interplay in the hippocampus. Front Mol Neurosci. 2015;8:68.
91. Hodel AS, Hunt RH, Cowell RA, Van Den Heuvel SE, Gunnar MR, Thomas KM. Duration of early adversity and structural brain development in post-institutionalized adolescents. NeuroImage. 2015;105:112–9.
92. Goff B, Gee DG, Telzer EH, Humphreys KL, Gabard-Durnam L, Flannery J, et al. Reduced nucleus accumbens reactivity and adolescent depression following early-life stress. Neuroscience. 2013;249:129–38.
93. Underwood MD, Bakalian MJ, Escobar T, Kassir S, Mann JJ, Arango V. Early-life adversity, but not suicide, is associated with less prefrontal cortex gray matter in adulthood. Int J Neuropsychopharmacol. 2019;22:349–57.
94. Mejia-Chavez S, Venebra-Munoz A, Garcia-Garcia F, Corona-Morales AA, Orozco-Vargas AE. Maternal separation modifies the activity of social processing brain nuclei upon social novelty exposure. Front Behav Neurosci. 2021;15:651263.
95. Kim EG, Chang W, Shin S, Adhikari AS, Seol GH, Song DY, et al. Maternal separation in mice leads to anxiety-like/aggressive behavior and increases immunoreactivity for glutamic acid decarboxylase and parvalbumin in the adolescence ventral hippocampus. Korean J Physiol Pharmacol. 2023;27:113–25.
96. Takiguchi S, Fujisawa TX, Mizushima S, Saito DN, Okamoto Y, Shimada K, et al. Ventral striatum dysfunction in children and adolescents with reactive attachment disorder: functional MRI study. B J Psych Open. 2015;1:121–8.
97. Marusak HA, Hatfield JRB, Thomason ME, Rabinak CA. Reduced ventral tegmental area-hippocampal connectivity in children and adolescents exposed to early threat. Biol Psychiatry Cogn Neurosci Neuroimaging. 2017;2:130–7.
98. Chocyk A, Dudys D, Przyborowska A, Majcher I, Mackowiak M, Wedzony K. Maternal separation affects the number, proliferation and apoptosis of glia cells in the substantia nigra and ventral tegmental area of juvenile rats. Neuroscience. 2011;173:1–18.
99. Chang CH, Grace AA. Amygdala-ventral pallidum pathway decreases dopamine activity after chronic mild stress in rats. Biol Psychiatry. 2014;76:223–30.
100. Oh WC, Rodriguez G, Asede D, Jung K, Hwang IW, Ogelman R, et al. Dysregulation of the mesoprefrontal dopamine circuit mediates an early-life stress-induced synaptic imbalance in the prefrontal cortex. Cell Rep. 2021;35:109074.
101. Kascakova N, Furstova J, Hasto J, Madarasova Geckova A, Tavel P. The unholy trinity: childhood trauma, adulthood anxiety, and long-term pain. Int J Environ Res Public Health. 2020;17:414.

102. Reiser SJ, McMillan KA, Wright KD, Asmundson GJ. Adverse childhood experiences and health anxiety in adulthood. Child Abuse Negl. 2014;38:407–13.
103. Lochner C, Seedat S, Allgulander C, Kidd M, Stein D, Gerdner A. Childhood trauma in adults with social anxiety disorder and panic disorder: a cross-national study. Afr J Psychiatry (Johannesbg). 2010;13:376–81.
104. Romeo RD, Mueller A, Sisti HM, Ogawa S, McEwen BS, Brake WG. Anxiety and fear behaviors in adult male and female C57BL/6 mice are modulated by maternal separation. Horm Behav. 2003;43:561–7.
105. Wang XD, Labermaier C, Holsboer F, Wurst W, Deussing JM, Muller MB, et al. Early-life stress-induced anxiety-related behavior in adult mice partially requires forebrain corticotropin-releasing hormone receptor 1. Eur J Neurosci. 2012;36:2360–7.
106. Bondi CO, Rodriguez G, Gould GG, Frazer A, Morilak DA. Chronic unpredictable stress induces a cognitive deficit and anxiety-like behavior in rats that is prevented by chronic antidepressant drug treatment. Neuropsychopharmacology. 2008;33:320–31.
107. Saleh A, Potter GG, McQuoid DR, Boyd B, Turner R, MacFall JR, et al. Effects of early life stress on depression, cognitive performance and brain morphology. Psychol Med. 2017;47:171–81.
108. Kuzminskaite E, Vinkers CH, Milaneschi Y, Giltay EJ, Penninx B. Childhood trauma and its impact on depressive and anxiety symptomatology in adulthood: a 6-year longitudinal study. J Affect Disord. 2022;312:322–30.
109. Stuart SA, Hinchcliffe JK, Robinson ESJ. Evidence that neuropsychological deficits following early life adversity may underlie vulnerability to depression. Neuropsychopharmacology. 2019;44:1623–30.
110. Goodwill HL, Manzano-Nieves G, Gallo M, Lee HI, Oyerinde E, Serre T, et al. Early life stress leads to sex differences in development of depressive-like outcomes in a mouse model. Neuropsychopharmacology. 2019;44:711–20.
111. Alqurashi GK, Hindi EA, Zayed MA, Abd El-Aziz GS, Alturkistani HA, Ibrahim RF, et al. The impact of chronic unpredictable mild stress-induced depression on spatial, recognition and reference memory tasks in mice: behavioral and histological study. Behav Sci (Basel). 2022;12:166.
112. Majer M, Nater UM, Lin JM, Capuron L, Reeves WC. Association of childhood trauma with cognitive function in healthy adults: a pilot study. BMC Neurol. 2010;10:61.
113. Petkus AJ, Lenze EJ, Butters MA, Twamley EW, Wetherell JL. Childhood trauma is associated with poorer cognitive performance in older adults. J Clin Psychiatry. 2018;79:16m11021.
114. Menezes J, Souto das Neves BH, Goncalves R, Benetti F, Mello-Carpes PB. Maternal deprivation impairs memory and cognitive flexibility, effect that is avoided by environmental enrichment. Behav Brain Res. 2020;381:112468.
115. Manzano-Nieves G, Gaillard M, Gallo M, Bath KG. Early life stress impairs contextual threat expression in female, but not male, mice. Behav Neurosci. 2018;132:247–57.
116. Bath KG, Nitenson AS, Lichtman E, Lopez C, Chen W, Gallo M, et al. Early life stress leads to developmental and sex selective effects on performance in a novel object placement task. Neurobiol Stress. 2017;7:57–67.
117. Hoogenboom WS, Branch CA, Lipton ML. Animal models of closed-skull, repetitive mild traumatic brain injury. Pharmacol Ther. 2019;198:109–22.
118. Harris TC, de Rooij R, Kuhl E. The shrinking brain: cerebral atrophy following traumatic brain injury. Ann Biomed Eng. 2019;47:1941–59.
119. Gimbel SI, Hungerford L, Twamley E, Ettenhoffer M. White matter organization and cortical thickness differ among active duty service members with chronic mild, moderate, and severe traumatic brain injury. J Neurotrauma. 2023;41:818.
120. Wang X, Xie H, Cotton AS, Tamburrino MB, Brickman KR, Lewis TJ, et al. Early cortical thickness change after mild traumatic brain injury following motor vehicle collision. J Neurotrauma. 2015;32:455–63.
121. Duque A, Coman D, Carlyle BC, Bordner KA, George ED, Papademetris X, et al. Neuroanatomical changes in a mouse model of early life neglect. Brain Struct Funct. 2012;217:459–72.

122. Sanchez CM, Titus DJ, Wilson NM, Freund JE, Atkins CM. Early life stress exacerbates outcome after traumatic brain injury. J Neurotrauma. 2021;38:555–65.
123. Wiegand TLT, Sollmann N, Bonke EM, Umeasalugo KE, Sobolewski KR, Plesnila N, et al. Translational neuroimaging in mild traumatic brain injury. J Neurosci Res. 2022;100:1201–17.
124. Rincel M, Lepinay AL, Janthakhin Y, Soudain G, Yvon S, Da Silva S, et al. Maternal high-fat diet and early life stress differentially modulate spine density and dendritic morphology in the medial prefrontal cortex of juvenile and adult rats. Brain Struct Funct. 2018;223:883–95.
125. Liu R, Yang XD, Liao XM, Xie XM, Su YA, Li JT, et al. Early postnatal stress suppresses the developmental trajectory of hippocampal pyramidal neurons: the role of CRHR1. Brain Struct Funct. 2016;221:4525–36.
126. Zhao J, Huynh J, Hylin MJ, O'Malley JJ, Perez A, Moore AN, et al. Mild traumatic brain injury reduces spine density of projection neurons in the medial prefrontal cortex and impairs extinction of contextual fear memory. J Neurotrauma. 2018;35:149–56.
127. Loi M, Koricka S, Lucassen PJ, Joels M. Age- and sex-dependent effects of early life stress on hippocampal neurogenesis. Front Endocrinol. 2014;5:13.
128. Sawada M, Sawamoto K. Mechanisms of neurogenesis in the normal and injured adult brain. Keio J Med. 2013;62:13–28.
129. Korosi A, Naninck EF, Oomen CA, Schouten M, Krugers H, Fitzsimons C, et al. Early-life stress mediated modulation of adult neurogenesis and behavior. Behav Brain Res. 2012;227:400–9.
130. Naninck EF, Hoeijmakers L, Kakava-Georgiadou N, Meesters A, Lazic SE, Lucassen PJ, et al. Chronic early life stress alters developmental and adult neurogenesis and impairs cognitive function in mice. Hippocampus. 2015;25:309–28.
131. Clark LR, Yun S, Acquah NK, Kumar PL, Metheny HE, Paixao RCC, et al. Mild traumatic brain injury induces transient, sequential increases in proliferation, neuroblasts/immature neurons, and cell survival: a time course study in the male mouse dentate gyrus. Front Neurosci. 2020;14:612749.
132. Mohd A, Goswami N, Manda K. Early changes in the hippocampal neurogenesis, neuroinflammation, and behavioral functions following mild traumatic brain injury. IBRO Neurosci Rep. 2019;6:S190.
133. Hood KN, Zhao J, Redell JB, Hylin MJ, Harris B, Perez A, et al. Endoplasmic reticulum stress contributes to the loss of newborn hippocampal neurons after traumatic brain injury. J Neurosci. 2018;38:2372–84.
134. Villasana LE, Westbrook GL, Schnell E. Neurologic impairment following closed head injury predicts post-traumatic neurogenesis. Exp Neurol. 2014;261:156–62.
135. Parker KN, Donovan MH, Smith K, Noble-Haeusslein LJ. Traumatic injury to the developing brain: emerging relationship to early life stress. Front Neurol. 2021;12:708800.
136. Dheen ST, Kaur C, Ling EA. Microglial activation and its implications in the brain diseases. Curr Med Chem. 2007;14:1189–97.
137. Jurga AM, Paleczna M, Kuter KZ. Overview of general and discriminating markers of differential microglia phenotypes. Front Cell Neurosci. 2020;14:198.
138. Perez-Rodriguez DR, Blanco-Luquin I, Mendioroz M. The participation of microglia in neurogenesis: a review. Brain Sci. 2021;11:658.
139. Lajud N, Roque A, Cheng JP, Bondi CO, Kline AE. Early life stress preceding mild pediatric traumatic brain injury increases neuroinflammation but does not exacerbate impairment of cognitive flexibility during adolescence. J Neurotrauma. 2021;38:411–21.
140. Bachiller S, Paulus A, Vazquez-Reyes S, Garcia-Dominguez I, Deierborg T. Maternal separation leads to regional hippocampal microglial activation and alters the behavior in the adolescence in a sex-specific manner. Brain Behav Immun Health. 2020;9:100142.
141. Yurtdas C, Zortul H, Yilmaz B, Aricioglu F. Microglial activation mediates maternal separation-induced depressive-like behavior in rats: a neurodevelopmental depression model. J Affect Disord. 2023;12:100462.
142. Reemst K, Kracht L, Kotah JM, Rahimian R, van Irsen AAS, Congrains Sotomayor G, et al. Early-life stress lastingly impacts microglial transcriptome and function under basal and immune-challenged conditions. Transl Psychiatry. 2022;12:507.

143. Perez-Polo JR, Rea HC, Johnson KM, Parsley MA, Unabia GC, Xu G, et al. Inflammatory consequences in a rodent model of mild traumatic brain injury. J Neurotrauma. 2013;30:727–40.
144. Bai R, Gao H, Han Z, Huang S, Ge X, Chen F, et al. Flow cytometric characterization of T cell subsets and microglia after repetitive mild traumatic brain injury in rats. Neurochem Res. 2017;42:2892–901.
145. Tertil M, Skupio U, Barut J, Dubovyk V, Wawrzczak-Bargiela A, Soltys Z, et al. Glucocorticoid receptor signaling in astrocytes is required for aversive memory formation. Transl Psychiatry. 2018;8:255.
146. Schramm E, Waisman A. Microglia as central protagonists in the chronic stress response. Neurol Neuroimmunol Neuroinflamm. 2022;9:e200023.
147. Cohen S, Janicki-Deverts D, Doyle WJ, Miller GE, Frank E, Rabin BS, et al. Chronic stress, glucocorticoid receptor resistance, inflammation, and disease risk. Proc Natl Acad Sci USA. 2012;109:5995–9.
148. Qin X, Liu XX, Wang Y, Wang D, Song Y, Zou JX, et al. Early life stress induces anxiety-like behavior during adulthood through dysregulation of neuronal plasticity in the basolateral amygdala. Life Sci. 2021;285:119959.
149. Juruena MF. Early-life stress and HPA axis trigger recurrent adulthood depression. Epilep Behav. 2014;38:148–59.
150. Fidan E, Lewis J, Kline AE, Garman RH, Alexander H, Cheng JP, et al. Repetitive mild traumatic brain injury in the developing brain: effects on long-term functional outcome and neuropathology. J Neurotrauma. 2016;33:641–51.
151. Park SJ, Park HJ, Kim B, Kim YM, Lee SW, Kim H. Chronic juvenile stress exacerbates neurobehavioral dysfunction and neuroinflammation following traumatic brain injury in adult mice. Clin Exp Emerg Med. 2023;10:200–12.
152. Lupien SJ, McEwen BS, Gunnar MR, Heim C. Effects of stress throughout the lifespan on the brain, behaviour and cognition. Nat Rev Neuro. 2009;10:434–45.
153. Hambrick EP, Brawner TW, Perry BD. Timing of early-life stress and the development of brain-related capacities. Front Behav Neurosci. 2019;13:183.
154. Avital A, Richter-Levin G. Exposure to juvenile stress exacerbates the behavioural consequences of exposure to stress in the adult rat. Int J Neuropsychopharmacol. 2005;8:163–73.
155. Marquez de la Plata CD, Hart T, Hammond FM, Frol AB, Hudak A, Harper CR, et al. Impact of age on long-term recovery from traumatic brain injury. Arch Phys Med Rehabil. 2008;89:896–903.
156. Mosenthal AC, Livingston DH, Lavery RF, Knudson MM, Lee S, Morabito D, et al. The effect of age on functional outcome in mild traumatic brain injury: 6-month report of a prospective multicenter trial. J Trauma Acute Care Surg. 2004;56:1042–8.
157. Sun M, Brady RD, Casillas-Espinosa PM, Wright DK, Semple BD, Kim HA, et al. Aged rats have an altered immune response and worse outcomes after traumatic brain injury. Brain Behav Immun. 2019;80:536–50.
158. Lundberg S, Martinsson M, Nylander I, Roman E. Altered corticosterone levels and social play behavior after prolonged maternal separation in adolescent male but not female Wistar rats. Horm Behav. 2017;87:137–44.
159. Liu X, Sun YX, Zhang CC, Zhang XQ, Zhang Y, Wang T, et al. Vortioxetine attenuates the effects of early-life stress on depression-like behaviors and monoamine transporters in female mice. Neuropharmacology. 2021;186:108468.
160. White JD, Arefin TM, Pugliese A, Lee CH, Gassen J, Zhang J, et al. Early life stress causes sex-specific changes in adult fronto-limbic connectivity that differentially drive learning. Elife. 2020;9:e58301.

Chapter 10
Sex and Age-at-Injury as Determinants of Social Behavior Outcomes After TBI

Bridgette D. Semple and Richelle Mychasiuk

Abstract While our understanding of long-term disability after traumatic brain injury (TBI) has habitually focused on cognitive and sensorimotor functioning, it is increasingly appreciated that changes in social function for survivors of a brain injury are common and have a profound impact on one's quality of life. In this chapter, we highlight the consequences of TBI on social behavior, taking into account evidence from studies of patient populations as well as from preclinical animal models. After first considering the protracted nature of the development of social behavior across the lifespan, including the neurobiological networks that underlie social functioning, we discuss how TBI results in social behavior impairments and how these manifest. We focus particularly on how age-at-injury influences TBI-induced social impairments, with most of the evidence suggesting age-dependent vulnerability after injury at a younger age. In addition, we explore how biological sex is a key determinant of social behavior impairments after TBI, while gender in humans may also influence the nature and extent of social outcomes. Finally, we identify key knowledge gaps and emphasize the need for further research in the field.

Keywords Pediatric traumatic brain injury · Social behavior · Sex · Age · Brain development · Preclinical

Abbreviations

ADHD	Attention-deficit hyperactivity disorder
MRI	Magnetic resonance imaging
mPFC	Medial prefrontal cortex
SBN	Social brain network
TBI	Traumatic brain injury

B. D. Semple (✉) · R. Mychasiuk
Department of Neuroscience, The School of Translational Medicine, Monash University, Melbourne, VIC, Australia
e-mail: Bridgette.Semple@monash.edu; Richelle.Mychasiuk@monash.edu

L. J. Noble-Haeusslein, D. M. Schnyer (eds.), *Traumatic Brain Injuries*, Advances in Neurobiology 42, https://doi.org/10.1007/978-3-031-69832-3_10

10.1 Introduction

For social species, which includes most mammals and birds, social behaviors permeate all aspects of our ecosystem and are critical for survival. From an evolutionary perspective, interdependence increases a species' overall fitness and allows for the transmission of learned skills [32]. As such, social interactions are particularly important for learning; individuals learn through their own experiences but also by observing the actions, interactions, and responses of others. For children, this often begins with social play, where individuals have the opportunity to develop flexible behavioral strategies to respond to novel and uncertain situations, develop important motor and sensory skills, gain knowledge of social hierarchies, and begin to understand tradeoffs between risk and reward [1]. During adolescence, individuals spend less time with family, become more influenced by social interactions, and cultivate a greater ability to regulate socioemotional responses, all of which depend upon the development of executive function [12]. As we move into adulthood, social behaviors are necessary for work, relationships, family dynamics, and integration within general societal practices. Social interactions then continue to evolve with aging, and later-life changes in sociality are increasingly recognized to contribute to health and well-being in an ageing population [73]. Each social interaction requires that the individual select the correct/appropriate behavior for each particular situation. For example, applause and other displays of group cheering are appropriate and expected at public sporting events but are not acceptable forms of social support during funerals.

Given the behavioral flexibility and comprehension of social intent that is required to engage in optimal social interaction, it is not surprising that a significant and widely distributed network of brain regions—termed the social brain network (SBN)—is dedicated to the regulation and modulation of social behaviors. The SBN describes the group of brain structures that are involved in identifying and interpreting conspecific behaviors and where we fit into that picture [12, 38]. The regions that encompass the SBN include, but are not limited to, the temporoparietal junction, posterior temporal sulcus, cingulate cortex, hippocampus, nucleus accumbens, amygdala, anterior temporal cortex, and prefrontal cortex. Within the SBN, three distinct but interconnected circuits that underlie social processing and interactions have been defined: the *face perception network* (involved in the social processing and perception of others), the *mirroring network* (required for understanding another's intentions and emotions and responding appropriately), and the *mentalizing network* (important for theory of mind or the capacity to make accurate inferences about the mental state of others) [83]. Damage to, or disruption of, typical functioning within various regions of the SBN due to traumatic brain injury (TBI) therefore has the potential to significantly impair a wide range of social functions.

Social behaviors, or the ways in which we interact with and influence other people, can be affected by TBI in various capacities. TBI may affect social skills, social cognitive ability (e.g., perception of social cues, empathy, and understanding of intentions), and social communication (e.g., language competency and nonverbal communication) [48]. Social problems following a TBI typically manifest as a reduction in social participation (withdrawal) and/or inappropriate social or sociosexual behaviors. Other behavioral changes associated with TBI, such as increased irritability, reduced concentration, disinhibition, deficits in emotional regulation, and concurrent anxiety or depression, can also profoundly influence social functioning [63].

Such changes in social functioning are commonly reported and significantly impact a TBI survivor's quality of life, with consequences for one's performance and satisfaction at home as well as, in academic and workplace settings. For example, inappropriate social behaviors may prevent an individual from participating in mainstream school systems or reduce their ability to find and hold down a job. Relationships with caregivers, family members, and friends may be altered, and an individual's social support network typically diminishes. Together, these trickle-down effects of social dysfunction after TBI lead to reduced quality of life for survivors, particularly when TBI is sustained during childhood [19, 87, 88]. A long-term increased risk of antisocial behavior has also been reported after TBI in both children and adults, such as threatening of others, physical aggression, and engaging in illegal activities, leading to an increase in violent criminal convictions being reported in this population [9, 64, 76].

The impact of TBI on an individual's social functioning is dependent on a range of factors, including the nature of the injury (e.g., severity, location, mechanism), the preinjury state (e.g., preexisting cognitive and socioemotional level, socioeconomic status), and postinjury variables (e.g., family and social support networks and access to rehabilitation resources) [89]. Age at the time of injury—or rather the developmental period when an injury is sustained—appears to have a particularly profound impact on social outcomes after TBI.

10.2 Age Differences in Social Outcomes After TBI

The majority of research on TBI-induced social and emotional dysfunction is in the context of pediatric TBI or injuries sustained during infancy and childhood [3, 4, 21]. Specifically, an earlier age-at-injury is associated with increased risk of social behavior problems, including deficits in social cognition and social communication [63]. Vulnerability of the young brain to social deficits after injury is likely attributed to the protracted development of the SBN [12, 28, 63]. Various cortical (e.g., the medial prefrontal cortex, mPFC) and subcortical components of the SBN (e.g., the anterior insula and amygdala) continue to mature through childhood and adolescence into early adulthood [12, 38, 84]. As such, a TBI sustained during different phases of postnatal brain development may disrupt the normal trajectory of maturation of these brain regions and lead to issues with social function [11, 62, 78] (Fig. 10.1).

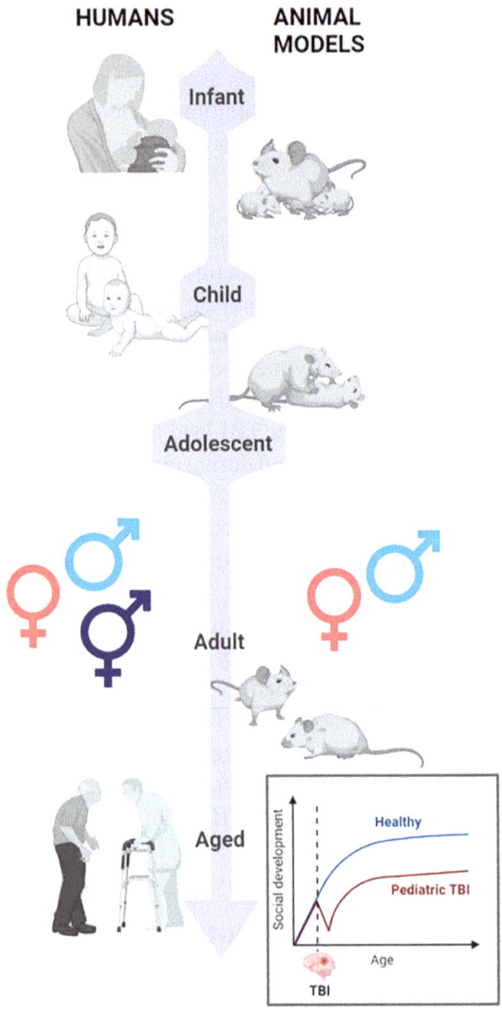

Fig. 10.1 Schematic illustration of the lifespan in humans and animal models such as rodents, from infancy through to old age. Expansion of the central timeline at the infant, child, and adolescent times reflects these as critical periods for the development of social cognition and expansion of social skills. Traumatic brain injury (TBI) sustained during these periods likely has differing consequences for the manifestation of social behavior impairments. Biological sex is another key variable that influences social behavior outcomes after TBI, as supported by evidence comparing male and female rodents in experimental TBI models, as well as clinical studies examining the effects of sex and gender on social outcomes in human populations. Insert depicts the normal trajectory of social development with age in healthy individuals (blue). A TBI during early life (dotted line) can acutely interfere with social behavior as well as alter the subsequent trajectory of development (red). (Figure generated in Biorender.com)

10.2.1 Clinical Evidence That Age-at-Injury Influences Social Function

Several studies have examined different cohorts of TBI survivors to specifically address the hypothesis of age-dependent vulnerability to social dysfunction after TBI. For example, Donders and Warschausky [20] compared children who sustained a TBI between 6 and 12 years of age with those who were injured between ages 16 and 20 (late adolescence). While no differences were found between groups in terms of overall cognitive performance, those who were injured at a younger age exhibited poorer high-level cognitive skills, social integration, and reduced independent living [20]. Similarly, a younger age-at-injury has been associated with greater impairments in social communication including reduced verbal comprehension, vocabulary, and fluency [5, 22].

More recently, Ryan and colleagues explored the concept of age-dependent vulnerability to social deficits after TBI in the context of the SBN, by evaluating social and emotional behavior alongside magnetic resonance imaging (MRI), assessing neuropathology with susceptibility-weighted imaging [62]. Youth with TBI were categorized by the timing of their brain injury ("middle childhood" as 5–9 years, "late childhood" as 10–11 years, or "adolescence" as 12–15 years). While they reported some evidence of a relationship between brain structure and function (e.g., between the extent of diffuse neuropathology and social cognitive impairments in Theory of Mind assessments), this was not a simple linear relationship. Instead, the youngest age group (middle childhood) showed a time-dependent emergence of social cognitive impairment across a 24-month period postinjury, while adolescence was also identified as a critical period for the acquisition and establishment of high-level social cognitive skills [62].

Considering antisocial behaviors after TBI including violent offences, several studies have identified younger age-at-injury as a risk factor and predictor of offending [47]. For example, McKinlay and colleagues reported that individuals injured at 0–5 years of age had the highest relative risk of property and violent offences [46]. Other studies, using an arbitrary cut-off of 12–13 years of age, have found that offenders who sustained a TBI before this time had higher scores for aggression [26] and started their criminal careers earlier [77] compared to those who sustained a TBI after ages 12–13. Further, in adolescents and adults (15–65 years) with moderate and severe TBI, a younger age-at-injury was reported to predict more aggressive and rule-breaking behavior 2–5 years postinjury [25]. Together, these studies support the hypothesis that an earlier age-at-injury increases an individual's risk of antisocial behaviors.

Increasing evidence indicates that severe TBI early in life typically leads to poorer behavioral outcomes more broadly, not merely in terms of social functioning. For example, one study examined the consequences of mild-complicated to severe TBI sustained between 3 and 7 years of age and reported significantly higher levels of parent-reported attention-deficit hyperactivity disorder (ADHD) and anxiety symptoms in younger children compared to those who were older at injury, when followed up ~38 months later [40]. Similarly, Keenan and colleagues reported that children aged 6–11 years had more ADHD symptoms, affective and emotional problems, but it was the younger children (preschoolers aged 2.5–6 years) who displayed an increase in symptom burden over time, from 3 to 12 months postinjury [41]. Further, in children who were tested more than 3 years after moderate-to-severe TBI, age and sex were found to interact, such that younger boys displayed higher levels of oppositional defiant behaviors compared to girls, while older girls displayed more symptoms of ADHD [80].

Another important consideration is the time that has elapsed since the TBI was sustained and how social skills, social cognition, and social brain networks evolve over time as one ages with an injury. While limited, several longitudinal studies have reported aberrant trajectories of neurodevelopment alongside progressive neurodegeneration following mild or moderate-to-severe TBI sustained during childhood and adolescence [17, 86]. Similarly for social functioning after TBI, social problems may resolve, persist, or even emerge with time; and this trajectory is influenced by age-at-injury [62, 63]. For example, a cross-sectional study of TBI survivors stratified by time postinjury (1–30 years) and age-at-injury (categorized as before or after age 30)—while controlling for sex and injury severity—found that increased decades postinjury and an older age-at-injury both predicted declines in societal participation, alongside decreasing cognitive function and reduced functional independence [69].

As noted earlier, most of the literature on social outcomes after TBI has focused on pediatric populations or young adults following injury at an earlier age. However, social dysfunction and social communication difficulties are also a significant and understudied issue for those who sustain a TBI as older adults [72]. In this population, social deficits after TBI need to be considered in the context of changes in cognitive function as a feature of the normal ageing process, even in the absence of any neurological conditions [27]. As older adulthood is a time of high risk for TBI [23, 24], more research is needed in this age period, particularly to account for potential confounds such as age-related health comorbidities and polypharmacy which can affect TBI outcomes [72].

10.2.2 Preclinical Evidence That Age-at-Injury Influences Social Function

The influence of age on the consequences of TBI can also be explored in preclinical animal models, where age-at-injury can be manipulated and outcome measures can be applied across a time course, then considered relative to postmortem analyses, to provide insight regarding the underlying biological mechanisms. Social consequences of TBI have predominantly been examined in rodent models of pediatric and adolescent TBI, with reports of reduced social interest, play behavior, social recognition, and social communication (scent marking and ultrasonic vocalizations) (e.g., Hazari et al. [34], Mychasiuk et al. [52], Runyan et al. [61], Ryan et al. [63], Semple et al. [66], Semple et al. [67], Semple et al. [68]). Social behavior changes have also been reported in adult TBI rodents, in a variety of models and injury severities [7, 42, 43, 56]. Changes in sexual behaviors, including aberrant interactions with the opposite sex, have also been observed after focal pediatric TBI in mice [68], repetitive mild TBI in adolescent rats [31], and impact-acceleration diffuse TBI in adult rats [56]. In adolescent rats, these inappropriate sexual behaviors were associated with reduced testosterone production, erectile dysfunction, and delayed development of reproductive organs through puberty, suggesting that TBI-induced hypopituitarism contributes to this phenotype [30, 31].

Consistent with what has been observed in children after moderate-to-severe TBI, male mice that sustain a focal TBI via the controlled cortical impactor model at 21 days of age (approximating a preschool-aged child) exhibit social behavior deficits that emerge over time [66]. Of note, this TBI mouse model is characterized by progressive regional brain atrophy and a time-dependent trajectory of other neurobehavioral consequences, including age-dependent anxiolysis and hyperactivity [59]—supporting the hypothesis that this neurodegenerative trajectory may underlie the delayed presentation of social deficits. However, progressive neurodegeneration does not seem to be a feature exclusive to injury during early life, as several studies of adult TBI of varying severities (in both humans and rodent models) have also reported temporal evolution of brain damage [8, 33, 44, 57], in association with persistent cognitive deficits [50].

Further exploring age-at-injury, Semple and colleagues compared male mice who received a TBI at 21 days of age versus those injured at 35 days of age (approximating adolescence) [66, 68]. They reported that the earlier injury resulted in more pronounced deficits in sociability, social memory, sociosexual behavior, and social communication. At the other end of the lifespan, experimental TBI in aged rodents (12–20 months of age) exhibit an exacerbated neuroinflammatory response compared to TBI in young adult animals (2–3 months of age), alongside poorer sensorimotor performance, increased anxiety-like behavior, and poorer cognitive function [37, 60, 75]. However, how an older age-at-injury influences social behavior outcomes has not been examined in such models to date.

10.2.3 Sex Differences in Social Outcomes After TBI

Just as age or developmental stage at the time of TBI is a significant moderator of outcomes so is biological sex and gender. Although we will discuss them as a single concept for the remainder of this chapter, it is important to consider the interplay and differences that exist between sex and gender. Sex is a biological construct based on chromosomal complement (XX/XY), reproductive anatomy, and gonadal hormones [13]. Gender, on the other hand, is a complex and multidimensional social construct of male or female identity based on societal norms, roles, and behaviors [13, 49]. Although preclinical research is wholly based upon biological sex, and there is no way to ascribe gender to nonhuman animals, gender-related differences in femininity and masculinity in clinical studies may influence an individual's risk for TBI, the symptoms they present with, and the persistence and severity of the symptoms. For example, traditional constructs surrounding masculinity may preclude more masculine males and females from reporting symptomologies for fear of violating gender roles [74]. These messages of masculinity often stigmatize individuals from seeking help, as they do not want to be perceived as weak [74, 82]. Additionally, the constructs of masculinity may alter risk-taking behaviors, promoting risky actions which may increase the actual incidence of TBI [10, 29]. It is important to note, however, that while we dichotomize outcomes as male vs. female or masculine vs. feminine, most variables including social deficits exist on a continuum with a significant degree of overlap. Moreover, preinjury sex differences in social aptitude also contribute to the manifestation of (or resiliency to) TBI-induced social deficits.

10.2.4 Clinical Evidence for TBI-Induced Sex Differences in Social Function

Social deficits generally arise from an inability to interpret the actions of others or from a lack of awareness regarding our own behavior. Given that males and females exhibit different baseline competencies in these areas, it is not surprising that there are sex differences in posttraumatic prevalence rates for social impairments. For example, studies demonstrate that following TBI, men exhibit worse self-awareness of injury-related deficits [55] and reduced emotional insight [54]. However, uninjured men also exhibit significantly higher scores of alexithymia (an inability to access, recognize, label, differentiate, and think about emotions) [54], suggesting that males in general do not have the same baseline aptitude and are therefore unable to compensate when social functioning is disrupted. Conversely, although females have a more accurate appreciation of their injury-induced impairments [55], this insight may lead to increased stress as their need for greater assistance conflicts with their typical caregiving role [15]. In a similar manner, although both men and women with severe TBI exhibit reduced empathy and an inability to recognize emotions based on facial expressions, females maintained their recognition of vocal emotional expression and emotional inferencing [85, 90]. This suggests that the emotional aspects of social processing may exhibit greater redundancy in the female brain, rendering it more resilient to TBI-induced dysfunction.

As described earlier, a history of childhood TBI significantly increases risk for social deficits in adulthood. However, emerging evidence indicates that sex substantially influences the onset and manifestation of these outcomes. An abundance of research demonstrates that females are significantly more likely to develop internalizing disorders such as depression and anxiety, as well as panic attacks and phobias [2, 58]. Women report higher rates of posttraumatic headache and migraine, worse social quality of life, and a loss of confidence and sex drive [15, 39, 81]. Conversely, males are far more likely to develop externalizing disorders, including aggression, antisocial and conduct disorders, substance abuse problems, and criminal behavior [65]. As such, there is an overrepresentation of TBIs in prison inmates which likely results from impairments to social functioning. Although this is true for both males and females, in males, early TBI has been associated with a fourfold increase in risk for the development of mental health disorders concomitant with criminality in adulthood [36, 77]. Conversely, women with a TBI are more likely to become the victim of abuse (both physical and sexual) and, owing in part to their vulnerability, are also less likely to leave violent relationships [79]. Although the exploration of sex differences in social functioning following TBI has been scarce, there is robust evidence demonstrating important sex-dependent distinctions in the socially mediated mechanisms leading to TBI, as well as in the subsequent outcomes, suggesting that further investigation is necessary.

10.2.5 Preclinical Evidence for TBI-Induced Sex Differences in Social Function

Humans typically show sex differences in social behaviors during development, including play behavior for young boys and girls, expansion of peer groups in adolescence, and social intelligence and insight in adulthood [35]. Corresponding sex differences are also evident in the sociality of animals including nonhuman primates, rodents, fruit flies, and zebrafish [16, 71]. For example, male rodents generally engage in more social play and conduct more counterattacks than females, whereas females are more likely to play the defensive role, allowing themselves to be pinned by their partner more often [6].

Importantly, these typical social behaviors can easily be measured—and because they exhibit specific disruptions following experimental TBI, it is possible to examine the mechanistic underpinnings. For example, in a rodent model of adolescent mild TBI, uninjured adolescent rats were less inclined to initiate play with adolescent TBI rats and were more likely to respond to play initiated by a TBI rat with avoidance. This finding of peer rejection and social impairment among adolescent mild TBI rats was significantly more pronounced in females, whereas the males exhibited more resiliency in learning normal play strategies from their uninjured peers [52]. It is possible that these social deficits arise from changes in the ability to regulate one's own behavior. In an analogous study of mild TBI, males were found to be significantly more impulsive and hyperactive whereas females exhibited hypoactivity and reduced motivation for reward [51]. The researchers found that within the mPFC, the males with mild TBI had increased dendritic length, whereas females displayed more substantial increases in spine density [53]. Similarly, while another study found that moderate-to-severe TBI was associated with deficits in social recognition and reduced sociability for male mice, females only exhibited increased social avoidance—while social recognition remained intact for them [67]. In this study, the observed social deficits were associated with decreased dendritic complexity in the hippocampus and mPFC, which was more prominent in males. Taken together, these studies suggest that the sex differences in social impairments that arise in response to TBI are linked to sex differences in both baseline neural connectivity and the neuronal synaptic changes that occur postinjury.

10.3 Knowledge Gaps and Future Research Needs

While advances in medical care have reduced mortality from TBI over the past few decades, morbidity remains high, and we now need strategies to improve quality of life for TBI survivors [45]. To date, social functioning post-TBI remains highly understudied and is rarely incorporated into rehabilitation programs [14]. Given the high rates of reported loneliness in TBI survivors [18], and the strong links between loneliness and mental health issues, there is a dire need to address this gap and improve overall social competency. As there are significant distinctions in the social impairments that arise from differences in TBI severities, age at injury, and sex of the individual, we are unlikely to find a "one-size"-fits-all solution. However, studies have demonstrated that many TBI-related impairments can be attenuated if appropriately addressed and there is no reason to suggest that social deficits would be different.

As described in this chapter, current evidence indicates that both sex and age-at-injury are important determinants of social outcomes after TBI. Considering age-at-injury, while studies support the concept of increased vulnerability during both early brain development and ageing, for distinct biological reasons (i.e., TBI-induced disruption of neurodevelopmental trajectories in the young brain vs. TBI-induced exacerbation of older age-related decline, possibly linked to "inflammaging"), further work is needed to better understand the precise windows of neurodevelopment during which the SBN is most vulnerable to injury and to gain an increased appreciation for how social behavior deficits evolve and change over time after injury relative to ongoing development.

Regarding sex as a variable, it is important to note several caveats to research that compares TBI outcomes between the sexes. For example, factors such as uneven sample sizes (TBI is much more prevalent in men), different study designs, and severity of the TBI often affect whether females are deemed to have fared "better" or "worse" than males. Moreover, female representation in the literature, both clinically and in the laboratory, is a relatively recent addition. As such, we anticipate significant changes to our mechanistic understanding of TBI-induced social dysfunction as we enhance inclusion and delve into the intricacies of this area of research in the future. For example, a recent preclinical study linking the microbiome to social deficits following repeat mild TBI found females to be more resilient, possibly due to greater diversity within the microbiome or increased plasticity within the SBN [70]. In summary, a greater understanding of how age and sex contribute to social outcomes after TBI will substantially modify how we target therapeutics and rehabilitation programs, as well as for the implementation of preventative strategies.

Acknowledgments BDS was supported by an Establishment Grant from Monash University's Central Clinical School and a Near-Miss Grant from Veski. RM was supported by an Investigator Grant from the National Health and Medical Research Council (NHMRC) of Australia.

Disclosures None.

References

1. Achterberg EJM, Vanderschuren L. The neurobiology of social play behaviour: past, present and future. Neurosci Biobehav Rev. 2023;152:105319.
2. Albicini M, McKinlay A. Anxiety disorders in adults with childhood traumatic brain injury: evidence of difficulties more than 10 years postinjury. J Head Trauma Rehabil. 2018;33:191–9.
3. Anderson V, et al. Functional plasticity or vulnerability after early brain injury? Pediatrics. 2005;116:1374–82.
4. Anderson V, et al. Childhood brain insult: can age at insult help us predict outcome? Brain. 2009;132:45–56.
5. Anderson VA, et al. Predicting recovery from head injury in young children: a prospective analysis. J Int Neuropsychol Soc. 1997;3:568–80.
6. Auger A, Olesen K. Brain sex differences and organisation of juvenile social play behaviour. J Neuroendocrinol. 2009;21:519–25.
7. Bajwa NM, et al. Mild concussion, but not moderate traumatic brain injury, is associated with long-term depression-like phenotype in mice. PLoS One. 2016;11:e0146886.
8. Belchev Z, et al. Progressive neurodegeneration across chronic stages of severe traumatic brain injury. J Head Trauma Rehabil. 2022;37:E144–e156.
9. Bellesi G, et al. Pediatric traumatic brain injury and antisocial behavior: are they linked? A systematic review. Brain Inj. 2019;33:1272–92.
10. Berg E, et al. Female football players, the sport ethic and the masculinity-sport nexus. Sport Soc. 2013;17:176–89.
11. Bigler ED, et al. Neuroimaging and social behavior in children after traumatic brain injury: findings from the social outcomes of brain injury in kids (SOBIK) study. NeuroRehabilitation. 2013;32:707–20.
12. Blakemore SJ. The social brain in adolescence. Nat Rev Neurosci. 2008;9:267–77.
13. Boerner K, et al. Conceptual complexity of gender and its relevance to pain. Pain. 2018;159:2137–41.
14. Bozkurt S, et al. Environmental modifications to rehabilitate social behavior deficits after acquired brain injury: what is the evidence? Neurosci Biobehav Rev. 2023;152:105278.
15. Colantonio A, et al. Gender differences in self-reported long term outcomes following moderate to severe traumatic brain injury. BMC Neurol. 2010;10:1–7.
16. Cooke BM, Shukla D. Double helix: reciprocity between juvenile play and brain development. Dev Cogn Neurosci. 2011;1:459–70.
17. Dennis EL, et al. Diverging volumetric trajectories following pediatric traumatic brain injury. Neuroimage Clin. 2017;15:125–35.
18. Di Battista A, et al. Depression and health related quality of life in adolescent survivors of a traumatic brain injury: a pilot study. PLoS One. 2014;9:e101842.

19. Di Battista A, et al. Quality of life in children and adolescents post-TBI: a systematic review and meta-analysis. J Neurotrauma. 2012;29:1717–27.
20. Donders J, Warschausky S. Neurobehavioral outcomes after early versus late childhood traumatic brain injury. J Head Trauma Rehabil. 2007;22:296–302.
21. Ewing-Cobbs L, et al. Longitudinal neuropsychological outcome in infants and preschoolers with traumatic brain injury. J Int Neuropsychol Soc. 1997;3:581–91.
22. Ewing-Cobbs L, et al. Language functions following closed-head injury in children and adolescents. J Clin Exp Neuropsychol. 1987;9:575–92.
23. Faul M, Coronado V. Epidemiology of traumatic brain injury. Handb Clin Neurol. 2015;127:3–13.
24. Faul M, et al. Traumatic brain injury in the United States: emergency department visits, hospitalizations and deaths 2002–2006. Atlanta: CDC, NCIPC; 2010.
25. Finnanger TG, et al. Life after adolescent and adult moderate and severe traumatic brain injury: self-reported executive, emotional, and behavioural function 2–5 years after injury. Behav Neurol. 2015;2015:329241.
26. Fishbein D, et al. Relationships between traumatic brain injury and illicit drug use and their association with aggression in inmates. Int J Offender Ther Comp Criminol. 2016;60:575–97.
27. Fjell AM, Walhovd KB. Structural brain changes in aging: courses, causes and cognitive consequences. Rev Neurosci. 2010;21:187–221.
28. Frith CD. The social brain? Philos Trans R Soc Lond Ser B Biol Sci. 2007;362:671–8.
29. Granie M. Effects of gender, sex-stereotype conformity, age and internalization on risk-taking among adolescent pedestrians. Saf Sci. 2009;47:1277–83.
30. Greco T, et al. The effects of repeat traumatic brain injury on the pituitary in adolescent rats. J Neurotrauma. 2013;30:1983–90.
31. Greco T, et al. Adolescent TBI-induced hypopituitarism causes sexual dysfunction in adult male rats. Dev Neurobiol. 2015;75:193–202.
32. Hamilton WD. The genetical evolution of social behaviour I. J Theor Biol. 1964;7:1–16.
33. Havlicek DF, et al. A single closed head injury in male adult mice induces chronic, progressive white matter atrophy and increased phospho-tau expressing oligodendrocytes. Exp Neurol. 2023;359:114241.
34. Hazari A, et al. Examining changes in rodent temperament following repetitive mild traumatic brain injury in adolescence. Behav Neurosci. 2020;134:384–93.
35. Hines M. Early androgen influences on human neural and behavioural development. Early Hum Dev. 2008;84:805–7.
36. Huw Williams W, et al. Self-reported traumatic brain injury in male young offenders: a risk factor for re-offending, poor mental health and violence? Neuropsychol Rehabil. 2010;20:801–12.
37. Islam M, et al. Differential neuropathology and functional outcome after equivalent traumatic brain injury in aged versus young adult mice. Exp Neurol. 2021;341:113714.
38. Johnson MH, et al. The emergence of the social brain network: evidence from typical and atypical development. Dev Psychopathol. 2005;17:599–619.
39. Jones K, et al. Longitudinal patterns of behavior, cognition, and quality of life after mild traumatic brain injury in children: BIONIC study findings. Brain Inj. 2019;33:887–93.
40. Karver CL, et al. Age at injury and long-term behavior problems after traumatic brain injury in young children. Rehabil Psychol. 2012;57:256–65.
41. Keenan HT, et al. Psychosocial and executive function recovery trajectories one year after pediatric traumatic brain injury: the influence of age and injury severity. J Neurotrauma. 2018;35:286–96.
42. Klemenhagen KC, et al. Repetitive concussive traumatic brain injury interacts with post-injury foot shock stress to worsen social and depression-like behavior in mice. PLoS One. 2013;8:e74510.
43. Koliatsos VE, et al. A mouse model of blast injury to brain: initial pathological, neuropathological, and behavioral characterization. J Neuropathol Exp Neurol. 2011;70:399–416.

44. Loane DJ, et al. Progressive neurodegeneration after experimental brain trauma: association with chronic microglial activation. J Neuropathol Exp Neurol. 2014;73:14–29.
45. Maas AIR, et al. Traumatic brain injury: progress and challenges in prevention, clinical care, and research. Lancet Neurol. 2022;21:1004–60.
46. McKinlay A, et al. Substance abuse and criminal activities following traumatic brain injury in childhood, adolescence, and early adulthood. J Head Trauma Rehabil. 2014a;29:498–506.
47. McKinlay A, et al. Predicting adult offending behavior for individuals who experienced a traumatic brain injury during childhood. J Head Trauma Rehabil. 2014b;29:507–13.
48. Milders M. Relationship between social cognition and social behaviour following traumatic brain injury. Brain Inj. 2019;33:62–8.
49. Mollayeva T, et al. Traumatic brain injury: sex, gender and intersecting vulnerabilities. Nat Rev Neurol. 2018;14:711–22.
50. Mouzon BC, et al. Chronic neuropathological and neurobehavioral changes in a repetitive mild traumatic brain injury model. Ann Neurol. 2014;75:241–54.
51. Mychasiuk R, et al. A mild traumatic brain injury (mTBI) induces secondary attention-deficit hyperactivity disorder-like symptomology in young rats. Behav Brain Res. 2015a;286:285–92.
52. Mychasiuk R, et al. Mean girls: sex differences in the effects of mild traumatic brain injury on the social dynamics of juvenile rat play behaviour. Behav Brain Res. 2014;259:284–91.
53. Mychasiuk R, et al. The development of lasting impairments: a mild pediatric brain injury alters gene expression, dendritic morphology, and synaptic connectivity in the prefrontal cortex of rats. Neuroscience. 2015b;288:145–55.
54. Neumann D, Zupan B. Sex differences in emotional insight after traumatic brain injury. Arch Phys Med Rehabil. 2020;101:1922–8.
55. Niemeier J, et al. Gender differences in awareness and outcomes during acute traumatic brain injury recovery. J Women's Health. 2014;23:573–82.
56. Pandey DK, et al. Depression-like and anxiety-like behavioural aftermaths of impact accelerated traumatic brain injury in rats: a model of comorbid depression and anxiety? Behav Brain Res. 2009;205:436–42.
57. Pischiutta F, et al. Single severe traumatic brain injury produces progressive pathology with ongoing contralateral white matter damage one year after injury. Exp Neurol. 2018;300:167–78.
58. Ponsford J, et al. Epidemiology and natural history of psychiatric disorders after TBI. J Neuropsychiatry Clin Neurosci. 2018;30:262–70.
59. Pullela R, et al. Traumatic injury to the immature brain results in progressive neuronal loss, hyperactivity and delayed cognitive impairments. Dev Neurosci. 2006;28:396–409.
60. Ritzel RM, et al. Old age increases microglial senescence, exacerbates secondary neuroinflammation, and worsens neurological outcomes after acute traumatic brain injury in mice. Neurobiol Aging. 2019;77:194–206.
61. Runyan A, et al. Intranasal administration of oxytocin attenuates social recognition deficits and increases prefrontal cortex inhibitory post-synaptic currents following traumatic brain injury. eNeuro. 2021;8(3):ENEURO.0061-21.2021.
62. Ryan NP, et al. The emergence of age-dependent social cognitive deficits after generalized insult to the developing brain: a longitudinal prospective analysis using susceptibility-weighted imaging. Hum Brain Mapp. 2015;36:1677–91.
63. Ryan NP, et al. Social dysfunction after pediatric traumatic brain injury: a translational perspective. Neurosci Biobehav Rev. 2016;64:196–214.
64. Schofield PW, et al. Does traumatic brain injury lead to criminality? A whole-population retrospective cohort study using linked data. PLoS One. 2015;10:e0132558.
65. Scott C, et al. A comparison of adult outcomes for males compared to females following pediatric traumatic brain injury. Neuropsychology. 2015;29:501–8.
66. Semple BD, et al. Deficits in social behavior emerge during development after pediatric traumatic brain injury in mice. J Neurotrauma. 2012;29:2672–83.
67. Semple BD, et al. Sex-dependent changes in neuronal morphology and psychosocial behaviors after pediatric brain injury. Behav Brain Res. 2017;319:48–62.

68. Semple BD, et al. Sociosexual and communication deficits after traumatic injury to the developing murine brain. PLoS ONE. 2014;9(8):e103386.
69. Sendroy-Terrill M, et al. Aging with traumatic brain injury: cross-sectional follow-up of people receiving inpatient rehabilitation over more than 3 decades. Arch Phys Med Rehabil. 2010;91:489–97.
70. Sgro, M., et al. Under Review. Microbiome depletion prior to repeat mild traumatic brain injury (RmTBI) differentially attenuates social deficits and neuroplasticity within the prefrontal cortex of adolescent and adult rats. iScience.
71. Sgro M, Mychasiuk R. Playful genes: what do we know about the epigenetics of play behaviour? Int J Play. 2020;9:25–38.
72. Shorland J, et al. Insights into social communication following traumatic brain injury sustained in older adulthood. Int J Lang Commun Disord. 2022;59:449.
73. Siracusa ER, et al. Social ageing: exploring the drivers of late-life changes in social behaviour in mammals. Biol Lett. 2022;18:20210643.
74. Steinfeldt J, et al. Gender role conflict and stigma toward help seeking among college football players. Psychol Men Masculinity. 2009;10:261–72.
75. Sun M, et al. Aged rats have an altered immune response and worse outcomes after traumatic brain injury. Brain Behav Immun. 2019;80:536–50.
76. Theadom A, et al. Symptoms and engagement in anti-social behaviour 10 years following mild traumatic brain injury within a community civilian sample: a prospective cohort study with age-sex matched control group. Arch Phys Med Rehabil. 2023;105:295.
77. Timonen M, et al. The association of preceding traumatic brain injury with mental disorders, alcoholism and criminality: the Northern Finland 1966 Birth Cohort Study. Psychiatry Res. 2002;113:217–26.
78. Tuerk C, et al. Altered resting-state functional connectivity within the developing social brain after pediatric traumatic brain injury. Hum Brain Mapp. 2020;41:561–76.
79. Valera E, et al. Understanding traumatic brain injury in females: a state-of-the-art summary and future directions. J Head Trauma Rehabil. 2021;36:1–29.
80. Wade SL, et al. Behavior problems following childhood TBI: the role of sex, age, and time since injury. J Head Trauma Rehabil. 2020;35:E393–e404.
81. Walker W, et al. Prediction of headache severity (density and functional impact) after traumatic brain injury: a longitudinal multicenter study. Cephalagia. 2013;33:998–1008.
82. Wallace J, et al. Sex differences in high school athletes' knowledge of sport-related concussion symptoms and reporting behaviors. J Athl Train. 2017;52:682–8.
83. Wang Y, et al. White matter pathways and social cognition. Neurosci Biobehav Rev. 2018;90:350–70.
84. Wang Y, Olson IR. The original social network: white matter and social cognition. Trends Cogn Sci. 2018;22:504–16.
85. Wearne T, et al. Understanding how others feel: evaluating the relationship between empathy and various aspects of emotion recognition following severe traumatic brain injury. Neuropsychology. 2020;34:288–97.
86. Wilde EA, et al. Diffusion tensor imaging in moderate-to-severe pediatric traumatic brain injury: changes within an 18 month post-injury interval. Brain Imaging Behav. 2012;6:404–16.
87. Yeates KO, et al. Social outcomes in childhood brain disorder: a heuristic integration of social neuroscience and developmental psychology. Psychol Bull. 2007;133:535–56.
88. Yeates KO, et al. Peer relationships of children with traumatic brain injury. J Int Neuropsychol Soc. 2013;19:518–27.
89. Zamani A, et al. Determinants of social behavior deficits and recovery after pediatric traumatic brain injury. Exp Neurol. 2019;314:34–45.
90. Zupan B, et al. Sex differences in emotion recognition and emotional inferencing following severe traumatic brain injury. Brain Impairment. 2016;108:36–48.

Chapter 11
Cognition and Behavior in the Aging Brain Following TBI: Surveying the Preclinical Evidence

Nicholas S. Race, Eleni H. Moschonas, Anthony E. Kline, and Corina O. Bondi

Abstract Elderly individuals (65 years and older) represent the fastest-growing demographic of new clinical traumatic brain injury (TBI) cases, yet there is a paucity of preclinical research in aged animals. The limited preclinical work available aligns with the clinical literature in that there appear to be significant differences in pathophysiology, recovery potential, and response to medications between animals at different points across the age spectrum. The aim of this review is to discuss the limited studies and highlight critical age-related differences in affective, cognitive, and neurobehavioral deficits, to discuss factors that influence functional outcomes, and to identify targets for future research. The consensus is that aged rodents face challenges related to dysregulated inflammation, reduced neuroplasticity, and age-related cellular changes, which hinder their recovery after TBI. The most successful intervention studies in animal models to date, of the limited array available, have explored interventions targeting inflammatory downregulation. Current standards

N. S. Race
Department of Physical Medicine & Rehabilitation and Safar Center for Resuscitation Research, Association of Academic Physiatrists Rehabilitation Medicine Scientist Training Program,University of Pittsburgh, Pittsburgh, PA, USA
e-mail: nsrace.mdphd@gmail.com

E. H. Moschonas
Department of Physical Medicine & Rehabilitation, Center for Neuroscience, and Safar Center for Resuscitation Research, University of Pittsburgh, Pittsburgh, PA, USA
e-mail: ehm18@pitt.edu

A. E. Kline
Departments of Physical Medicine & Rehabilitation, Critical Care Medicine, and Psychology, Center for Neuroscience, Center for the Neural Basis of Cognition, and Safar Center for Resuscitation Research, University of Pittsburgh, Pittsburgh, PA, USA
e-mail: klineae@upmc.edu

C. O. Bondi (✉)
Departments of Physical Medicine & Rehabilitation and Neurobiology, Center for Neuroscience, and Safar Center for Resuscitation Research, University of Pittsburgh, Pittsburgh, PA, USA
e-mail: bondico@upmc.edu

L. J. Noble-Haeusslein, D. M. Schnyer (eds.), *Traumatic Brain Injuries*, Advances in Neurobiology 42, https://doi.org/10.1007/978-3-031-69832-3_11

of neuropsychopharmacology for post-TBI neurocognitive recovery have not been investigated in a significant capacity. In addition, currently available animal models do not sufficiently account for important age-related comorbidities, dual insults, or differences in TBI mechanism of injury in elderly individuals. TBI in the aged population is more likely to lead to additional neurodegenerative diseases that further complicate recovery. The findings underscore the need for tailored therapeutic interventions to improve outcomes in both adult and elderly populations.

Keywords Cognition · Controlled cortical impact · Frontal cortex · Morris water maze · Recovery · Traumatic brain injury

Abbreviations

CCI	Controlled cortical impact
FPI	Fluid percussion injury
MWM	Morris water maze
NSS	Neurologic severity score
NOR	Novel object recognition
RAWM	Radial arm water maze
TBI	Traumatic brain injury

11.1 Introduction

Traumatic brain injury (TBI) is a complex and debilitating condition that affects millions of individuals worldwide annually. Many multiples of the annual incidence live with chronic disabilities related to TBI, effects of which intersect with age-related cognitive/health decline. Elderly individuals account for 20% of emergency department visits, 44% of hospitalizations, and 37% of TBI-related deaths with individuals ≥75 years old comprising the largest proportion of hospitalizations, longest hospital stays, and worst outcomes (including highest death rate) among all age groups [1, 2]. Older individuals who experience TBIs suffer higher rates of persistent adverse neurocognitive and neuropsychiatric outcomes than severity-matched younger adults [3, 4]. TBI in older adults has also been associated with increased risk for and earlier onset of dementia [5–8], a topic reviewed in detail in other chapters. Notably, older adults experienced outsized negative effects on quality of life and rates of permanent and/or severe disability, discharge to nursing facilities [2, 9], new-onset mood disorder [3, 4, 7], and risk of onset of dementia or other neurodegenerative disorders [5–7, 10].

Despite the aging population rapidly increasing and the fact that elderly individuals (65 and older) represent the fastest-growing demographic of new clinical TBI cases [11], there is a relative paucity of preclinical TBI research in aged animals compared to juveniles and young adults. Not only is there little work examining cognitive and behavioral outcomes following TBI in aged animals and differential responses to potential therapeutic interventions, but existing models also do not investigate important issues such as superimposed age-related multicomorbidity states (e.g., hypertension, diabetes, premorbid dementia) [11]. Treatment/intervention studies are also few, marking a significant opportunity landscape for future research. Studies in young adult rodents consistently report better cognitive and motor recovery post-TBI compared to aged rodents [11]. Adult rodents often exhibit faster and more complete recovery, attributed in part to younger animals' more favorable neuroplastic capabilities and more efficient neuronal repair processes [12, 13]. In addition, aged rodents experiencing TBI exhibit dysregulated and prolonged inflammatory responses [14–18], which can exacerbate secondary injury acutely and chronically after TBI. Aged neurons are also prone to cellular senescence and mitochondrial dysfunction, which limit the brain's ability to repair damaged tissue and can lead to persistent functional deficits [19–22].

A fundamental concept to highlight is that there are two major ways in which aging and TBI intersect: (1) aging *after* experiencing TBI in early or midlife and (2) experiencing a TBI for the first time as an older adult. Regarding the former, while some informative early research into chronic degenerative and inflammatory sequelae following TBI in relationship to cognitive/behavioral outcomes has been performed, these studies have not been designed to interrogate long-term cognitive trajectories over time frames representing a typical life span. Clinical studies, however, do report that human patients experiencing TBI earlier in life are at higher risk for neurodegenerative diseases and adverse cognitive/neuropsychiatric outcomes—a topic reviewed in more detail in other chapters. This chapter delves into post-TBI cognitive/behavioral recovery differences between young adult and aged rodents (modeling older adults) experiencing TBIs. In doing so, we aim to highlight critical age-related differences in cognitive/behavioral recovery, discuss factors that influence functional outcomes in aged animals, and identify targets for future research. For consistency, we define "aged" animals as mice or rats at least 15 months old, corresponding to human adults in their mid-50s to mid-60s and onward [23–25]. We hope these insights highlight the importance of age-specific approaches to TBI research and underscore the need for tailored therapeutic interventions to improve outcomes in both adult and elderly populations.

11.2 Overview of Preclinical TBI Cognitive/Behavioral Outcomes in Aged Animals

11.2.1 Mortality and Acute Functional Recovery

Acute neurologic outcome assessments following TBI generally include comparing differences in time duration for recovery of reflexes and basic somatomotor functions between injured and sham-injured animals when recovering from anesthesia given for injury administration. Acute neurological outcome differences between young adult and aged animals have not been studied extensively. One identified study comparing young adult and aged rats utilized a fluid percussion injury (FPI) TBI model and investigated age-related differences in mortality and duration of suppression of reflexes and somatomotor function [26]. At both of two FPI severities tested, mild and moderate, the authors reported higher mortality rates in aged animals [26], which has since been reproduced in other FPI studies [27]. The authors also found nearly all measures had age-related prolongation of acute neurologic recovery in aged as compared to young adult rats [26]. These differences did not correlate with nor could they be attributed to any major divergence in cardiovascular or glycemic responses over the same time period, though the authors did note a tendency toward postinjury bradycardia in aged animals compared to tachycardia in young adult animals [26].

Similar findings of increased mortality and worse acute neurologic recovery measured by the neurologic severity score (NSS) have been reported in aged (21 months) vs. young adult (2 months) mice exposed to a moderate severity right parietal controlled cortical impact (CCI), which was partially but not solely attributable to increased edema in the aged animals and could not be attributed to differences in lesion volume, cardiopulmonary, or glycemic parameters (equivalent between groups) [28]. Other work utilizing a mouse model of moderate CCI overlying the sensorimotor cortex reported prolonged duration of acute edema resolution and greater magnitude with prolonged duration of blood-brain barrier compromise in aged mice, despite comparable lesion sizes [29]. To achieve comparable mortality rates and reflex return times/acute neurologic recovery in the classic Marmarou closed head impact-acceleration model, injury parameters had to be decreased in intensity for aged animals (lower weight and shorter height of weight drop) [30].

In general, findings of heightened mortality and prolonged acute neurologic recovery in aged animals compared to young adult animals across numerous injury models align well with the human experience [1, 2, 31].

11.2.2 Motor Function

Motor recovery following TBI is less commonly studied in both the preclinical and clinical literature than other brain insults such as stroke; however, disability rating scales in the clinical literature are often heavily biased by motor function status. Motor recovery, particularly coordination and balance, is particularly important in the aged/elderly population as the leading cause of TBI, and recurrent TBI in this population is falls [32–34]. Age independently compromises motor function and motor learning, with numerous proposed mechanisms such as decreased striatal dopamine release and loss of cerebellar noradrenergic function [35, 36], among others [37].

Some of the earliest work studying post-TBI motor deficits in young adult vs. aged animals (3 vs. 20 months) found TBI induced by a single midline FPI significantly worsened subacute motor performance in beam balance and beam walk testing in the aged animals from days 1–5 postinjury [38]. Similarly, in a study comparing young adult and aged mice undergoing repeated TBI via a closed-head right lateral impact model, aged mice experienced higher rates of footslips than young adult mice [39]. Moderate right parietal CCI produced worse subacute outcomes in motor recovery in aged (21–24 months) mice as measured by rotarod and grid-walk compared to young adult mice (5–6 months) at time points ranging from 1 to 14 days postinjury [29]. Right-sided, midline-adjacent mild CCI produced more severe motor deficits in aged (20 months) vs. young adult (6 months) rats in the elevated body swing test (EBST), forelimb akinesia, and forelimb grip strength testing at postinjury days 0, 1, 3, and 7 [40]. Aged (18 months) mice receiving moderate left parietal CCI were found to have postinjury impairments in forelimb grip strength, forelimb use asymmetry (cylinder test), and motor coordination (rotarod) which coincided with generally elevated proinflammatory and prosenescence tissue markers as compared to young adult mice (3 months) [22]. In general, motor deficits were worse than in young adult mice; however, interestingly, forelimb grip strength impairments were less severe in the aged animals.

Longer-term motor recovery beyond the first few days to weeks postinjury have not been systematically assessed. Additionally, preclinical studies examining the intersection between diseases of sensorimotor function in the elderly (e.g., motor neuron disease/ALS, Parkinson's disease, peripheral neuropathy) and TBI have not been performed in any significant capacity. A significant opportunity landscape exists to better characterize and investigate potential interventions to address motor recovery after TBI in aged animals to explore putative strategies for improving clinical motor outcomes, reducing fall risk, and preventing adverse outcomes such as recurrent TBI in older adults.

11.2.3 Learning and Memory

Learning and memory in the aging population is a significant concern relevant for TBI research, including in preclinical models. Most commonly in TBI animal studies, rodent learning and memory is assessed via the Morris water maze (MWM). The MWM largely assesses hippocampal-mediated spatial learning and memory and is a robust behavioral task capable of capturing post-TBI cognitive impairments in a wide array of TBI induction models [41]. Aging independently decreases performance in the MWM in naive animals in a strain-specific fashion, with most tested strains exhibiting moderate impairments by 18 months and severe impairments by 24 months [42, 43].

The severity of post-TBI MWM impairments has been found to be worse in aged (20 months) compared to young adult (6 months) rats at 11–15 days postinjury despite comparable FPI at (findings consistent across two severities) and no overt differences in general health/welfare (e.g., animal distress, weight loss) in the aged animals [38]. Similar disproportionately poor radial arm water maze (RAWM) performance was observed in aged rats (20 months) despite comparable lesion volume to young adult rats (6 months) in a right-sided, midline-adjacent mild CCI [40].

Comparable findings were observed in MWM for aged vs. young rats (24 months vs. 2.5 months) following moderate right parietal CCI in the first week postinjury [44]. Aged mice (20–25 months) administered moderate right parietal CCI exhibited significantly worse performance in the RAWM compared to young adult (3–6 months) mice at 1 month postinjury [45]. Notably, associated findings in relevant brain regions, namely, the hippocampus and entorhinal cortex, related to inflammation, neurodegeneration, cellular senescence, and/or dendritic spine density are variably reported across each of these studies.

Neural substrates (including networks, regions, cell types, and neurotransmitters) have not been systematically explored in detail in relation to post-TBI behavioral impairments in aged animals, a distinct contrast from most non-TBI cognitive and behavioral neuroscience work. Notable exceptions to this are found in two studies. At 1 week postinjury, aged rats (20–23 months) exhibited significantly worse impairments in the MWM despite decreasing injury intensity to achieve comparable mortality as young adult rats (2–3 months) in the Marmarou impact-acceleration model [30]. This difference persisted for at least 5 weeks, and the rate of improvement on repeat trials was much worse in the aged compared to the young adult animals [30]. The behavioral differences were accompanied by significantly lower forebrain synaptosomal choline uptake in aged rats, but no differences were observed in entorhinal cortex or hippocampal uptake between the young and aged groups [30]. In a separate study, aged mice (16–18 months) exhibited substantially worse Barnes maze acquisition compared to young adult counterparts (2–4 months), noting correlations between behavioral impairments and hippocampal lesion size [46].

Age-related cognitive decline with or without premorbid dementia prior to TBI as well as increased risk for new onset/accelerated progression of dementia following TBI are significant concerns which would benefit from dedicated research efforts, as detailed extensively in other chapters. We expect synergistically detrimental interaction effects of aging and TBI exist in the learning and memory domains, due to the ability of each to independently negatively impact critical relevant neural substrates. Neurotransmitter-specific, nutritional/metabolic, genetic, inflammatory, senescence, and environmental effects are all hypothesized to contribute to age-related cognitive decline independent of TBI [47, 48]. Modulating cholinergic [49, 50] or dopaminergic [51–54] neural substrates as well as a number of acute neuroprotective strategies have incompletely improved learning and memory following TBI in adolescent to young adult rodents, findings which have yet to translate to clinical use or even demonstrate consistent efficacy in animal models [49, 55, 56].

11.2.4 Attentional and Executive Functions

Frontal/prefrontal-mediated attentional and executive function impairments are common following TBI, and similar impairments have been reported in preclinical TBI studies of adolescent and young adult animals [50, 57–66] (ranging from weeks to 11 months postinjury). Despite this, there is a relative paucity of research assessing neurocognitive outcomes beyond the classical motor, learning, and memory assays described above. Attention and processing speed and executive functions such as working memory, reversal learning, and cognitive flexibility have not been assessed to our knowledge following TBI in aged animals as defined by the criteria at the start of this chapter (15 months and older).

TBI impairs attention [63, 67–69]. Independent of TBI, aging also impairs attentional performance and has been linked to changes in cholinergic function [70–75], which is known to be compromised independently by TBI [76–81]. Notably, lifelong environmental enrichment has been shown to prevent age-related degradation of attentional performance and in parallel has been associated with relative preservation of basal forebrain cholinergic neurons [71]. We are aware of a single study that has examined attention in the context of aging *after* TBI for a period of 12 months (to age 15 months total), which found isolated negative effects of age on attention at the 12-month postinjury time point but no effects of TBI at any severity (graded severity Marmarou close-head impact-acceleration model) [69]. Similarly, borderline-aged rats (14 months) demonstrated worse executive function (working memory) in a bilateral frontal CCI model [82]. There are currently no published studies to our knowledge that have examined attentional or executive functions in aged animals after TBI. This represents a significant opportunity for future research.

11.2.5 Mood, Affect, and Agitation

Anxiety, depression, and mood/affect dysregulation are common problems after TBI [7, 83–85]. Older adults experience adverse neuropsychiatric outcomes at higher rates than other age groups [3, 4]. Anxiety- and depression-like behavioral phenotypes have been regularly observed following experimental TBI and adolescent and young adult animals [86–95] but have not routinely been studied in aged animals. We are aware of one study investigating mood/affect changes in aged mice (18 months). In this study, following moderate left parietal CCI, aged mice exhibited increased anxiety-like behavior, displaying decreased time in the open field center zone compared to young adult (3 months) mice receiving CCI [22]. Mood and affect dysregulation following TBI and interventions to improve/address it mark a significant area for future research.

Agitation is another common neurobehavioral sequela of TBI that is distressing for patients, families, and staff and is associated with worse outcomes [96–101]. Agitation's relevance in the aged population is enhanced due to the potential for comorbid conditions to independently induce agitation such as polypharmacy, dementia, and hospital-acquired delirium. These effects are often worse at night, referred to as "sundowning" [102, 103] and are often managed symptomatically for patient and staff/family safety with antipsychotics [104], which are known to be deleterious to neurologic recovery following TBI even in adolescent and young adult animals [105–107] (to our knowledge, aged animals have not specifically been studied). At present, the authors are not aware of any existing models recapitulating post-TBI agitation in the literature, including aged animals. We also did not identify any studies explicitly investigating the influence of commonly used agents for agitation management on neurologic recovery after TBI in aged animals (e.g., antipsychotics, benzodiazepines, beta-blockers). These represent areas of opportunity for future research.

11.3 Preclinical Post-TBI Intervention Effects on Cognitive/Behavioral Outcomes in Aged Animals

In the clinical realm, arbitrary upper age limits and polypharmacy/comorbid conditions often exclude older individuals from clinical trial inclusion [108–110]. TBI in general has a relative paucity of clinical trials due to practical complexities in conducting them, making sound preclinical investigations even more critical. Very few preclinical TBI treatment studies have included aged animals. Most aged animal treatment investigations have focused on acute neuroprotection targeting improved postinjury tissue preservation with hopes to lead to better neurofunctional outcomes.

Several investigations exploring acute neuroprotective interventions have been performed in aged rats, with the majority focused on progesterone. Aged rats injured via moderate severity bifrontal CCI that received progesterone—in addition to positive tissue endpoints including inflammatory, apoptotic, and edema endpoints—demonstrated dose-dependent improvement in short-term open field spontaneous locomotor outcomes assessed via open field test [111]. It is unclear based on this study design whether these improvements were truly motor recovery in origin or rather related to improved motor initiation in treated animals. Similar positive effects of progesterone on aged animal recovery were observed in subsequent bifrontal CCI studies [112, 113], and coadministration of vitamin D was noted to interfere with positive effects [112]. Subacute improvements in the modified neurological severity score (mNSS), a combined score of motor, sensory, reflex, and balance tests, as well as the MWM have also been noted in aged rats receiving progesterone after lateral CCI [114]. Improvements in post-TBI acute edema in progesterone-treated animals may provide a partial mechanistic basis for improved recovery [115]. Other potential contributors include improved hippocampal long-term potentiation [114], increased endothelial progenitor cells and neovascular proliferation [114], and decreased expression of inflammatory and apoptotic factors (COX-2, IL-6, NF-κB) [111], though notably lesion sizes were not affected [113]. It should be noted that progesterone was trialed for acute neuroprotection in the ProTECT III multicenter, randomized, placebo-controlled double-blind clinical trial in which it failed to provide mortality, neuropsychological outcome, or surrogate biomarker of tissue injury benefits [116–118]. While no specific exclusion criteria existed for older individuals and there were participants up to 94 years of age, the median age was 35. To our knowledge, no published subgroup analysis specific to the older adults included in the study is available.

Beyond progesterone, interventions to improve post-TBI outcomes in aged rats are relatively limited, and only a handful of studies were identified. Aged rats' (20 months) post-TBI performance in MWMMorris water maze (MWM) spatial learning/memory testing was modestly improved by intravenously transplanted human adipose-derived stem cells, but these benefits did not extend to motor recovery nor occur to the same degree as in young adult (6 months) rats, perhaps in part due to decreased stem cell homing to the spleen in aged rats [40]. Treatment of aged mice (16–18 months) with low molecular weight fucoidan, an algae-derived antioxidant and antiinflammatory, resulted in improved neurobehavioral outcomes ranging from one to 3 weeks postinjury, including sensorimotor function (beam walk), and hippocampus-associated spatial learning and memory (MWM, Barnes maze) alongside improved tissue endpoints related to inflammation, lipid peroxidation, and oxidative stress after moderate right parietal CCI [46]. Taurine supplementation in aged animals after lateral CCI over sensorimotor cortex did not improve motor outcomes from days 1 to 14 nor lesion volumes at day 14 postinjury, though comparable treatment paradigms had previously been effective in young adult animals [119].

The currently published subchronic to chronic post-TBI interventions in aged animals have employed immune modulation strategies. Subchronic knockout of CCR2 to inhibit infiltration of peripherally derived monocytes prevented chronic TBI-induced RAWM spatial memory deficits in aged (20–25 months) mice [45]. This was attributed to greater magnitude and longer duration of CCR2-mediated peripherally derived monocyte recruitment in aged brains [120]. In the same injury induction model, novel object recognition (NOR) task memory deficits were observed at 1 month postinjury in aged (20–25 months) mice, which were associated with complement pathway C1q accumulation on hippocampal dendritic spines [15]. Postinjury NOR task memory deficits were prevented in cohorts of animals administered chronic weekly pharmacologic blockade preventing C1q activation [15]. Other recently successful immunomodulating strategies such as depopulating and repopulating microglial populations via CSF1R inhibition [121, 122] have yet to be employed in aged animal models of TBI. Such strategies would be particularly interesting given the reports that microglial proliferation is more significant in aged compared to young adult brains for the first weeks to month following TBI in animal models [15, 45, 123].

Traditional neuropharmacological treatments used clinically to facilitate improved function (not provide acute neuroprotection) in the subacute to chronic post-TBI period including noradrenergic and dopaminergic neurostimulants [51–54, 124–126], cholinergic memory-enhancing agents [50, 55, 56, 58], antianxiety, and antidepressant medications [64, 66, 127] (typically modulating serotonin, norepinephrine, and/or dopamine levels), antipsychotics [105–107], and sleep aides [128, 129] have not been studied to any significant degree in aged animal models of TBI. These represent significant opportunities for future research.

11.4 Potential Biological Contributors to Differences in Aged Post-TBI Recovery and Neurological Outcomes

Rodent models provide invaluable insights into the mechanisms of injury and recovery following TBI, but the literature base in aged animals is far less robust than for adolescents and young adults. Other chapters will extensively delve into the multifaceted, multiscale contributions to worse outcomes for older adults who experience TBI, ranging from molecular mechanisms to environmental factors. However, some high-level salient points merit inclusion in the present chapter.

Studies in adolescent and young adult rodents consistently report faster and more complete cognitive and motor functional recovery post-TBI compared to aged rodents, as detailed in the sections above. Acute edema management, altered cerebrovascular dynamics, and BBB integrity/repair deficiencies may contribute in the early postinjury period [28, 29, 130–132]. Reduced neurorecovery potential is attributed in part to aged animals' inferior neuroplastic capabilities, less efficient neuronal repair processes [27, 133, 134], and decreased cognitive reserve compared to younger animals [135–137]. Potential contributors to reduced neuroplasticity in aged animals include reduced postinjury proliferation of endogenous neural stem cells [138] and altered neurotransmitter homeostatic mechanisms [139, 140]. Aged neurons are also relatively more prone to cellular senescence and mitochondrial dysfunction [20], which can impede TBI recovery and lead to more extensive secondary injury processes postinjury such as oxidative stress and lipid peroxidation [43, 44, 141, 142], neurodegeneration [29, 143], and inflammation [15, 22, 45, 120, 123, 143–145]. These age-related cellular changes limit the brain's ability to repair damaged tissue and can lead to persistent cognitive and motor deficits.

One of the most significant differences in TBI recovery between adolescent/young adult and aged rodents is the altered neuroinflammatory response. Adult rodents generally exhibit a more robust and controlled inflammatory response. In contrast, aged rodents often display a dysregulated and prolonged inflammatory response, which can exacerbate secondary damage and curtail functional neurorecovery [15, 22, 45, 120, 123, 143–145]. The importance of this is exemplified by the success of interventional strategies on neurobehavioral/motor/cognitive recovery which have targeted the immune system [15, 45, 46]. The intertwined nature of hypothalamic-pituitary axis (HPA) dysregulation and excess inflammation following TBI has yet to be explored in detail in the context of aging. TBI and age can independently disrupt HPA function, including levels of glucocorticoids which are intimately linked to inflammatory states and microglial priming [146, 147]. Furthermore, the hippocampus—known to have excess vulnerability to TBI in aged animals [29, 123, 133, 141, 143] and the effects of natural age-related cognitive decline—has been identified as a key component of stress responses related to sensing endogenous glucocorticoid levels and engaging the hypothalamus in a negative feedback loop to maintain homeostasis [148].

Age-related neurotransmitter and receptor density changes may also contribute to reduced recovery potential after TBI. Age-related alterations in acetylcholinesterase expression and function, decreases in biogenic amines (including dopamine and dopamine receptor densities, specifically), and altered neurotransmitter homeostasis have been reported, which are associated with declines in cognitive and motor performance [43, 149]. As detailed above, such changes would likely compromise function across numerous domains, including learning, memory, attention, and executive functions. In addition, motor control/learning systems involving dopaminergic and noradrenergic substrates have been reported to undergo natural age-related declines in function [35, 36], among other proposed mechanisms [37]. Examination of medications modulating neurotransmitter systems is dually important in aged animals, as many centrally acting medications are on the American Geriatric Society's "Beers List" of Potentially Inappropriate Medications in Older Adults (increased potential for adverse events related to the primary mechanism of action or off-target effects). For example, older individuals are known to be more sensitive to off-target anticholinergic effects than younger adults, placing them at increased risk for confusion, falls, urinary retention, and other adverse events. This may spur a need for more selective, novel neuropsychopharmacologic agents with less off-target effects [150].

11.5 Clinical Considerations and Implications for Rodent Models of TBI and Aging

To date, aged animal studies have not substantially contributed to improved clinical care of TBI. Modeling TBI in aged animals to appropriately recapitulate the human condition is more challenging than for adolescent and young adult animals. In addition to mechanism of injury challenges meriting development of new models and species-specific physiologic changes with aging, human TBI patients have a significant array of potential comorbidities likely to have independent interaction effects intertwined with both age and TBI. Estimates suggest 80% of adults over 65 have at least one chronic condition while 50% have at least two conditions, compared with 28% of younger adults with a single medical condition [9, 151].

The most common chronic disease among individuals greater than 65 years of age (prevalence ranging from ~30 to 60%) are reported as hypertension, high cholesterol, cardiovascular disease (which includes coronary and peripheral arterial disease, heart failure, valvular heart disease, and stroke), diabetes, and osteoarthritis [152]. From the clinical literature, we know that some comorbidities such as diabetes increase TBI-related mortality [153]. Diabetes patients are also at increased risk for falls related to hypoglycemic events and peripheral neuropathy, meaning this population likely suffers an outsized incidence of TBI compared to nondiabetic patients [154, 155]. Older patients are also at increased risk for heart attack and associated anoxic injuries, which can copresent as falls with dual traumatic + anoxic brain injury profiles. Other conditions such as chronic substance use (viz., alcohol) and premorbid dementia (e.g., Alzheimer's, Parkinson's, vascular and frontotemporal dementias) in older adults could also be explored, as these (much like DM) can dually cause inherent harm to the baseline neural substrate/cognitive reserve and place individuals at increased risk for falls. Importantly with relevance to dementia, there exist numerous independent pathophysiologic potential mechanistic links between TBI and dementia, some of which have been observed (though examined less extensively) in aged animals including predisposition to synucleinopathy and tauopathy postinjury [156, 157]. Other chapters extensively explore these mechanisms in greater detail, as well as risks of developing dementia after TBI.

From a mechanism of injury standpoint, older adults are more likely to experience closed head injuries and also are more likely to experience intracranial bleeding in tandem (most commonly subdural hematomas). The increased risk in bleeding is in part due to age-related volume loss alongside increased tissue friability, resulting in increased vulnerability of bridging veins to tearing. In addition, the widespread use of antiplatelet and anticoagulant medications in the aged population is associated with increased TBI severity, higher mortality, and worse long-term outcomes [31, 158–160]. We are not aware of any animal models which reliably utilize a closed head injury and produce pathologically relevant degrees of intracranial bleeding. Importantly, intracranial blood product presence and extent of bleeding burden have been demonstrated as independent predictors of postinjury impaired cerebrovascular dynamics [161], which may already be altered at baseline in aged animals [132]. Some dual-insult animal models have been developed in younger animals to incorporate polytrauma, hypoxia, or hypotension in tandem with TBI [162–164], but blood products have been explored to a lesser extent. We are not aware of any published TBI animal studies that have incorporated dual-insult or companion comorbidity modeling in aged animals, representing a significant opportunity for future research. As in other realms of sound fundamentals of animal research, sex differences need to be explored in aged animals exposed to TBI as there are numerous differences reported in both clinical and preclinical research as it relates to the natural animal cognition and behavior, the naturally aging brain, the pathophysiology of comorbidities in the aged population, and the pathophysiologic and neurorecovery responses to TBI [165–171].

11.6 Conclusion

This chapter highlights the importance of age-specific approaches in TBI research and underscores the need for expanded research efforts to better understand the pathophysiology of and improve outcomes related to TBI in aged populations. Traumatic brain injury recovery in rodents varies significantly depending on age, with adolescent and young adult rodents generally demonstrating more favorable neurofunctional recovery across an array of cognitive and motor endpoints. Conversely, aged rodents face challenges related to dysregulated inflammation, reduced neuroplasticity, and age-related cellular changes, which hinder their recovery. The field's characterization of TBI neurofunctional recovery alone across cognitive domains in aged animals and multicomorbidity/dual-insult models remains in its infancy, with much to be explored. Therapies that target neuroinflammation, enhance neuroplasticity, and address age-related cellular changes have had varying degrees of success, with interventions targeting the inflammatory response having particularly promising early results. Neurotransmitter-modulating agents classically used to treat impairments in different cognitive domains after TBI (e.g., learning, memory, attention, executive function, agitation, mood/affect, sleep) have not been explored in a significant capacity in aged animals. Tailored studies, and new animal models to facilitate them, that consider age-specific mechanisms of injury including dual-insults (e.g., hypoxia, intracranial blood products), companion comorbidities, and sex differences will be essential to understand and optimize TBI recovery strategies for the aged population.

Acknowledgments This work was supported, in part, by NIH grants NS110609 (COB), NS084967 and NS121037 (AEK), and a New Investigator Pilot Grant (#36579953) from the Foundation for Physical Medicine & Rehabilitation (NSR). EHM is a PhD candidate and is supported in part by a Brain Injury Association of America Fund as well as a Research Advisory Committee grant supported by the Children's Hospital of Pittsburgh of the UPMC Health System.

Declaration of interest None.

References

1. Taylor CA, Bell JM, Breiding MJ, Xu L. Traumatic brain injury-related emergency department visits, hospitalizations, and deaths—United States, 2007 and 2013. MMWR Surveill Summ. 2017;66(9):1–16.
2. Thompson HJ, McCormick WC, Kagan SH. Traumatic brain injury in older adults: epidemiology, outcomes, and future implications. J Am Geriatr Soc. 2006;54(10):1590–5.
3. Deb S, Burns J. Neuropsychiatric consequences of traumatic brain injury: a comparison between two age groups. Brain Inj. 2007;21(3):301–7.
4. Albrecht JS, Kiptanui Z, Tsang Y, Khokhar B, Smith GS, Zuckerman IH, et al. Patterns of depression treatment in medicare beneficiaries with depression after traumatic brain injury. J Neurotrauma. 2015;32(16):1223–9.
5. Gardner RC, Burke JF, Nettiksimmons J, Kaup A, Barnes DE, Yaffe K. Dementia risk after traumatic brain injury vs nonbrain trauma: the role of age and severity. JAMA Neurol. 2014;71(12):1490–7.
6. Mendez MF. What is the relationship of traumatic brain injury to dementia? J Alzheimers Dis. 2017;57(3):667–81.
7. Perry DC, Sturm VE, Peterson MJ, Pieper CF, Bullock T, Boeve BF, et al. Association of traumatic brain injury with subsequent neurological and psychiatric disease: a meta-analysis. J Neurosurg. 2016;124(2):511–26.
8. Hanrahan JG, Burford C, Nagappan P, Adegboyega G, Rajkumar S, Kolias A, et al. Is dementia more likely following traumatic brain injury? A systematic review. J Neurol. 2023;270(6):3022–51.
9. Papa L, Mendes ME, Braga CF. Mild traumatic brain injury among the geriatric population. Curr Transl Geriatr Exp Gerontol Rep. 2012;1(3):135–42.
10. Barker S, Paul BD, Pieper AA. Increased risk of aging-related neurodegenerative disease after traumatic brain injury. Biomedicines [Internet]. 2023;11(4). Available from: https://doi.org/10.3390/biomedicines11041154.

11. Iboaya A, Harris JL, Arickx AN, Nudo RJ. Models of traumatic brain injury in aged animals: a clinical perspective. Neurorehabil Neural Repair. 2019;33(12):975–88.
12. Inoue R, Nishimune H. Neuronal plasticity and age-related functional decline in the motor cortex. Cells [Internet]. 2023;12(17). Available from: https://doi.org/10.3390/cells12172142.
13. Filogamo G, Vernadakis A, Gremo F, Privat AM, Timiras PS. Brain plasticity: development and aging. Springer Science & Business Media; 2013. p. 332.
14. Delage C, Taib T, Mamma C, Lerouet D, Besson VC. Traumatic brain injury: an age-dependent view of post-traumatic neuroinflammation and its treatment. Pharmaceutics [Internet]. 2021;13(10). Available from: https://doi.org/10.3390/pharmaceutics13101624.
15. Krukowski K, Chou A, Feng X, Tiret B, Paladini MS, Riparip LK, et al. Traumatic brain injury in aged mice induces chronic microglia activation, synapse loss, and complement-dependent memory deficits. Int J Mol Sci [Internet]. 2018;19(12). Available from: https://doi.org/10.3390/ijms19123753.
16. Buenaventura RG, Harvey AC, Burns MP, Main BS. Traumatic brain injury induces an adaptive immune response in the meningeal transcriptome that is amplified by aging. Front Neurosci. 2023;17:1210175.
17. Early AN, Gorman AA, Van Eldik LJ, Bachstetter AD, Morganti JM. Effects of advanced age upon astrocyte-specific responses to acute traumatic brain injury in mice. J Neuroinflammation. 2020;17(1):115.
18. Moro F, Pischiutta F, Portet A, Needham EJ, Norton EJ, Ferdinand JR, et al. Ageing is associated with maladaptive immune response and worse outcome after traumatic brain injury. Brain Commun. 2022;4(2):fcac036.
19. Sun M, Brady RD, Casillas-Espinosa PM, Wright DK, Semple BD, Kim HA, et al. Aged rats have an altered immune response and worse outcomes after traumatic brain injury. Brain Behav Immun. 2019;80:536–50.
20. Gilmer LK, Ansari MA, Roberts KN, Scheff SW. Age-related mitochondrial changes after traumatic brain injury. J Neurotrauma. 2010;27(5):939–50.
21. Ring NAR, Valdivieso K, Grillari J, Redl H, Ogrodnik M. The role of senescence in cellular plasticity: lessons from regeneration and development and implications for age-related diseases. Dev Cell. 2022;57(9):1083–101.
22. Ritzel RM, Doran SJ, Glaser EP, Meadows VE, Faden AI, Stoica BA, et al. Old age increases microglial senescence, exacerbates secondary neuroinflammation, and worsens neurological outcomes after acute traumatic brain injury in mice. Neurobiol Aging. 2019;77:194–206.
23. Flurkey K, M. Currer J, Harrison DE. Chapter 20—Mouse models in aging research. In: Fox JG, Davisson MT, Quimby FW, Barthold SW, Newcomer CE, Smith AL, editors. The mouse in biomedical research. 2nd ed. Burlington: Academic; 2007. p. 637–72.
24. Dutta S, Sengupta P. Men and mice: relating their ages. Life Sci. 2016;152:244–8.
25. Sengupta P. The laboratory rat: relating its age with human's. Int J Prev Med. 2013;4(6):624–30.
26. Hamm RJ, Jenkins LW, Lyeth BG, White-Gbadebo DM, Hayes RL. The effect of age on outcome following traumatic brain injury in rats. J Neurosurg. 1991;75(6):916–21.
27. Shah SA, Prough DS, Garcia JM, DeWitt DS, Hellmich HL. Molecular correlates of age-specific responses to traumatic brain injury in mice. Exp Gerontol. 2006;41(11):1201–5.
28. Timaru-Kast R, Luh C, Gotthardt P, Huang C, Schäfer MK, Engelhard K, et al. Influence of age on brain edema formation, secondary brain damage and inflammatory response after brain trauma in mice. PLoS One. 2012;7(8):e43829.
29. Onyszchuk G, He YY, Berman NEJ, Brooks WM. Detrimental effects of aging on outcome from traumatic brain injury: a behavioral, magnetic resonance imaging, and histological study in mice. J Neurotrauma. 2008;25(2):153–71.
30. Maughan PH, Scholten KJ, Schmidt RH. Recovery of water maze performance in aged versus young rats after brain injury with the impact acceleration model. J Neurotrauma. 2000;17(12):1141–53.
31. Mak CHK, Wong SKH, Wong GK, Ng S, Wang KKW, Lam PK, et al. Traumatic brain injury in the elderly: is it as bad as we think? Curr Transl Geriatr Exp Gerontol Rep. 2012;1(3):171–8.

32. Krishnamoorthy V, Distelhorst JT, Vavilala MS, Thompson H. Traumatic brain injury in the elderly: burden, risk factors, and prevention. J Trauma Nurs. 2015;22(4):204–8; quiz E3–4.
33. Lecky FE, Otesile O, Marincowitz C, Majdan M, Nieboer D, Lingsma HF, et al. The burden of traumatic brain injury from low-energy falls among patients from 18 countries in the CENTER-TBI Registry: a comparative cohort study. PLoS Med. 2021;18(9):e1003761.
34. Fu WW, Fu TS, Jing R, McFaull SR, Cusimano MD. Predictors of falls and mortality among elderly adults with traumatic brain injury: a nationwide, population-based study. PLoS One. 2017;12(4):e0175868.
35. Yurek DM, Hipkens SB, Hebert MA, Gash DM, Gerhardt GA. Age-related decline in striatal dopamine release and motoric function in brown Norway/Fischer 344 hybrid rats. Brain Res. 1998;791(1–2):246–56.
36. Bickford P. Motor learning deficits in aged rats are correlated with loss of cerebellar noradrenergic function. Brain Res. 1993;620(1):133–8.
37. Seidler RD, Bernard JA, Burutolu TB, Fling BW, Gordon MT, Gwin JT, et al. Motor control and aging: links to age-related brain structural, functional, and biochemical effects. Neurosci Biobehav Rev. 2010;34(5):721–33.
38. Hamm RJ, White-Gbadebo DM, Lyeth BG, Jenkins LW, Hayes RL. The effect of age on motor and cognitive deficits after traumatic brain injury in rats. Neurosurgery. 1992;31(6):1072–7. discussion 1078
39. Eyolfson E, Carr T, Fraunberger E, Khan A, Clark I, Mychasiuk R, et al. Repeated mild traumatic brain injuries in mice cause age- and sex-specific alterations in dendritic spine density. Exp Neurol. 2022;357:114172.
40. Tajiri N, Acosta SA, Shahaduzzaman M, Ishikawa H, Shinozuka K, Pabon M, Hernandez-Ontiveros D, Kim DW, Metcalf C, Staples M, Dailey T. Intravenous transplants of human adipose-derived stem cell protect the brain from traumatic brain injury-induced neurodegeneration and motor and cognitive impairments: cell graft biodistribution and soluble factors in young and aged rats. J Neurosci. 2014;34(1):313–26. Available from: https://www.jneurosci.org/content/34/1/313.short
41. Tucker LB, Velosky AG, McCabe JT. Applications of the Morris water maze in translational traumatic brain injury research. Neurosci Biobehav Rev. 2018;88:187–200.
42. Wyss JM, Chambless BD, Kadish I, van Groen T. Age-related decline in water maze learning and memory in rats: strain differences. Neurobiol Aging. 2000;21(5):671–81.
43. Haider S, Saleem S, Perveen T, Tabassum S, Batool Z, Sadir S, et al. Age-related learning and memory deficits in rats: role of altered brain neurotransmitters, acetylcholinesterase activity and changes in antioxidant defense system. Age. 2014;36(3):9653.
44. Itoh T, Imano M, Nishida S, Tsubaki M, Mizuguchi N, Hashimoto S, et al. Increased apoptotic neuronal cell death and cognitive impairment at early phase after traumatic brain injury in aged rats. Brain Struct Funct. 2013;218(1):209–20.
45. Chou A, Krukowski K, Morganti JM, Riparip LK, Rosi S. Persistent infiltration and impaired response of peripherally-derived monocytes after traumatic brain injury in the aged brain. Int J Mol Sci [Internet]. 2018;19(6). Available from: https://doi.org/10.3390/ijms19061616.
46. Wang T, Zhu M, He ZZ. Low-molecular-weight fucoidan attenuates mitochondrial dysfunction and improves neurological outcome after traumatic brain injury in aged mice: involvement of Sirt3. Cell Mol Neurobiol. 2016;36(8):1257–68.
47. Salthouse TA. Neuroanatomical substrates of age-related cognitive decline. Psychol Bull. 2011;137(5):753–84.
48. Hullinger R, Puglielli L. Molecular and cellular aspects of age-related cognitive decline and Alzheimer's disease. Behav Brain Res. 2017;322(Pt B):191–205.
49. Scremin OU, Norman KM, Roch M, Holschneider DP, Scremin AME. Acetylcholinesterase inhibition interacts with training to reverse spatial learning deficits after cortical impact injury. J Neurotrauma. 2012;29(15):2457–64.
50. de la Tremblaye PB, Bondi CO, Lajud N, Cheng JP, Radabaugh HL, Kline AE. Galantamine and environmental enrichment enhance cognitive recovery after experimental trau-

matic brain injury but do not confer additional benefits when combined. J Neurotrauma. 2017;34(8):1610–22.
51. Bleimeister IH, Wolff M, Lam TR, Brooks DM, Patel R, Cheng JP, et al. Environmental enrichment and amantadine confer individual but nonadditive enhancements in motor and spatial learning after controlled cortical impact injury. Brain Res. 2019;1714:227–33.
52. Dixon CE, Kraus MF, Kline AE, Ma X, Yan HQ, Griffith RG, et al. Amantadine improves water maze performance without affecting motor behavior following traumatic brain injury in rats. Restor Neurol Neurosci. 1999;14(4):285–94.
53. Kline AE, Massucci JL, Ma X, Zafonte RD, Dixon CE. Bromocriptine reduces lipid peroxidation and enhances spatial learning and hippocampal neuron survival in a rodent model of focal brain trauma. J Neurotrauma. 2004;21(12):1712–22.
54. Kline AE, Massucci JL, Marion DW, Dixon CE. Attenuation of working memory and spatial acquisition deficits after a delayed and chronic bromocriptine treatment regimen in rats subjected to traumatic brain injury by controlled cortical impact. J Neurotrauma. 2002;19(4):415–25.
55. Shaw KE, Bondi CO, Light SH, Massimino LA, McAloon RL, Monaco CM, et al. Donepezil is ineffective in promoting motor and cognitive benefits after controlled cortical impact injury in male rats. J Neurotrauma. 2013;30(7):557–64.
56. Bondi CO, Yelleswarapu NK, Day-Cooney J, Memarzadeh K, Folweiler KA, Bou-Abboud CE, et al. Systemic administration of donepezil attenuates the efficacy of environmental enrichment on neurobehavioral outcome after experimental traumatic brain injury. Restor Neurol Neurosci. 2018;36(1):45–57.
57. Bondi CO, Cheng JP, Tennant HM, Monaco CM, Kline AE. Old dog, new tricks: the attentional set-shifting test as a novel cognitive behavioral task after controlled cortical impact injury. J Neurotrauma. 2014;31(10):926–37.
58. Njoku I, Radabaugh HL, Nicholas MA, Kutash LA, O'Neil DA, Marshall IP, et al. Chronic treatment with galantamine rescues reversal learning in an attentional set-shifting test after experimental brain trauma. Exp Neurol. 2019;315:32–41.
59. Chou A, Morganti JM, Rosi S. Frontal lobe contusion in mice chronically impairs prefrontal-dependent behavior. PLoS One. 2016;11(3):e0151418.
60. Shaver TK, Ozga JE, Zhu B, Anderson KG, Martens KM, Vonder HC. Long-term deficits in risky decision-making after traumatic brain injury on a rat analog of the Iowa gambling task. Brain Res. 2019;1704:103–13.
61. Vonder Haar C, Martens KM, Riparip LK, Rosi S, Wellington CL, Winstanley CA. Frontal traumatic brain injury increases impulsive decision making in rats: a potential role for the inflammatory cytokine interleukin-12. J Neurotrauma. 2017;34(19):2790–800.
62. Vonder Haar C, Lam FCW, Adams WK, Riparip LK, Kaur S, Muthukrishna M, et al. Frontal traumatic brain injury in rats causes long-lasting impairments in impulse control that are differentially sensitive to pharmacotherapeutics and associated with chronic neuroinflammation. ACS Chem Neurosci. 2016;7(11):1531–42.
63. Kutash LA, Moschonas EH, O'Neil DA, Craine TJ, Iouchmanov AL, Sunleaf CR, et al. Sustained attention performance deficits in the three-choice serial reaction time task in male and female rats after experimental brain trauma. Brain Res. 2023;1808:148336.
64. Craine TJ, Race NS, Kutash LA, Iouchmanov AL, Moschonas EH, O'Neil DA, et al. Milnacipran ameliorates executive function impairments following frontal lobe traumatic brain injury in male rats: a multimodal behavioral assessment. J Neurotrauma. 2023;40(1–2):112–24.
65. Lajud N, Roque A, Cheng JP, Bondi CO, Kline AE. Early life stress preceding mild pediatric traumatic brain injury increases neuroinflammation but does not exacerbate impairment of cognitive flexibility during adolescence. J Neurotrauma [Internet]. 2020; Available from: https://www.liebertpub.com/doi/abs/10.1089/neu.2020.7354?casa_token=sJioKrb-iywAAAAA:CbvgiUMgk9yeCO1N669G8hQlJxYKcvo8PV-thSJ9qAGyOQGtNug9ypfEUtiM0AuyxccHTXqGcNSP

66. Minchew HM, Radabaugh HL, LaPorte ML, Free KE, Cheng JP, Bondi CO. A combined therapeutic regimen of citalopram and environmental enrichment ameliorates attentional set-shifting performance after brain trauma. Eur J Pharmacol. 2021;904:174174.
67. Arciniegas D, Adler L, Topkoff J, Cawthra E, et al. Subject review: attention and memory dysfunction after traumatic brain injury: cholinergic mechanisms, sensory gating, and a hypothesis for further investigation. Brain Inj. 1999;13(1):1–13.
68. Arciniegas DB, Held K, Wagner P. Cognitive impairment following traumatic brain injury. Curr Treat Options Neurol. 2002;4(1):43–57.
69. Arulsamy A, Corrigan F, Collins-Praino LE. Age, but not severity of injury, mediates decline in executive function: validation of the rodent touchscreen paradigm for preclinical models of traumatic brain injury. Behav Brain Res. 2019;368:111912.
70. Yegla B, Parikh V. Effects of sustained proNGF blockade on attentional capacities in aged rats with compromised cholinergic system. Neuroscience. 2014;261:118–32.
71. Harati H, Majchrzak M, Cosquer B, Galani R, Kelche C, Cassel JC, et al. Attention and memory in aged rats: impact of lifelong environmental enrichment. Neurobiol Aging. 2011;32(4):718–36.
72. Jones DN, Barnes JC, Kirkby DL, Higgins GA. Age-associated impairments in a test of attention: evidence for involvement of cholinergic systems. J Neurosci. 1995;15(11):7282–92.
73. McGaughy J, Sarter M. Behavioral vigilance in rats: task validation and effects of age, amphetamine, and benzodiazepine receptor ligands. Psychopharmacology. 1995;117(3):340–57.
74. Muir JL, Fischer W, Björklund A. Decline in visual attention and spatial memory in aged rats. Neurobiol Aging. 1999;20(6):605–15.
75. Sarter M, Turchi J. Age- and dementia-associated impairments in divided attention: psychological constructs, animal models, and underlying neuronal mechanisms. Dement Geriatr Cogn Disord. 2002;13(1):46–58.
76. Gorman LK, Fu K, Hovda DA, Murray M, Traystman RJ. Effects of traumatic brain injury on the cholinergic system in the rat. J Neurotrauma. 1996;13(8):457–63.
77. Östberg A, Virta J, Rinne JO, Oikonen V, Luoto P, Någren K, et al. Cholinergic dysfunction after traumatic brain injury: preliminary findings from a PET study. Neurology. 2011;76(12):1046–50.
78. Leonard JR, Maris DO, Grady MS. Fluid percussion injury causes loss of forebrain choline acetyltransferase and nerve growth factor receptor immunoreactive cells in the rat. J Neurotrauma. 1994;11(4):379–92.
79. Murdoch I, Nicoll JAR, Graham DI, Dewar D. Nucleus basalis of Meynert pathology in the human brain after fatal head injury. J Neurotrauma. 2002;19(2):279–84.
80. Donat CK, Schuhmann MU, Voigt C, Nieber K, Deuther-Conrad W, Brust P. Time-dependent alterations of cholinergic markers after experimental traumatic brain injury. Brain Res. 2008;1246:167–77.
81. Moschonas EH, Leary JB, Memarzadeh K, Bou-Abboud CE, Folweiler KA, Monaco CM, et al. Disruption of basal forebrain cholinergic neurons after traumatic brain injury does not compromise environmental enrichment-mediated cognitive benefits. Brain Res. 2021;1751:147175.
82. Hoane MR, Lasley LA, Akstulewicz SL. Middle age increases tissue vulnerability and impairs sensorimotor and cognitive recovery following traumatic brain injury in the rat. Behav Brain Res. 2004;153(1):189–97.
83. Chun HYY, Newman R, Whiteley WN, Dennis M, Mead GE, Carson AJ. A systematic review of anxiety interventions in stroke and acquired brain injury: efficacy and trial design. J Psychosom Res. 2018;104:65–75.
84. Hicks AJ, Clay FJ, Hopwood M, James AC, Jayaram M, Batty R, et al. Efficacy and harms of pharmacological interventions for neurobehavioral symptoms in post-traumatic amnesia after traumatic brain injury: a systematic review. J Neurotrauma. 2018;35(23):2755–75.
85. Bombardier CH, Fann JR, Temkin NR, Esselman PC, Barber J, Dikmen SS. Rates of major depressive disorder and clinical outcomes following traumatic brain injury. JAMA. 2010;303(19):1938–45.

86. Tucker LB, Burke JF, Fu AH, McCabe JT. Neuropsychiatric symptom modeling in male and female C57BL/6J mice after experimental traumatic brain injury. J Neurotrauma. 2017;34(4):890–905.
87. Klemenhagen KC, O'Brien SP, Brody DL. Repetitive concussive traumatic brain injury interacts with post-injury foot shock stress to worsen social and depression-like behavior in mice. PLoS One. 2013;8(9):e74510.
88. Mao X, Terpolilli NA, Wehn A, Cheng S, Hellal F, Liu B, et al. Progressive histopathological damage occurring up to one year after experimental traumatic brain injury is associated with cognitive decline and depression-like behavior. J Neurotrauma. 2020;37(11):1331–41.
89. Pandey DK, Yadav SK, Mahesh R, Rajkumar R. Depression-like and anxiety-like behavioural aftermaths of impact accelerated traumatic brain injury in rats: a model of comorbid depression and anxiety? Behav Brain Res. 2009;205(2):436–42.
90. Popovitz J, Mysore SP, Adwanikar H. Long-term effects of traumatic brain injury on anxiety-like behaviors in mice: behavioral and neural correlates. Front Behav Neurosci. 2019;13:6.
91. Broussard JI, Acion L, De Jesús-Cortés H, Yin T, Britt JK, Salas R, et al. Repeated mild traumatic brain injury produces neuroinflammation, anxiety-like behaviour and impaired spatial memory in mice. Brain Inj. 2018;32(1):113–22.
92. Beitchman JA, Griffiths DR, Hur Y, Ogle SB, Bromberg CE, Morrison HW, et al. Experimental traumatic brain injury induces chronic glutamatergic dysfunction in amygdala circuitry known to regulate anxiety-like behavior. Front Neurosci. 2019;13:1434.
93. Race NS, Andrews KD, Lungwitz EA, Alvarez SM, Warner TR, Acosta G, Cao J, Lu KH, Liu Z, Dietrich AD, Majumdar S. Psychosocial impairment following mild blast-induced traumatic brain injury in rats. Behav Brain Res. 2021;412:113405.
94. Schwarzbold ML, Rial D, De Bem T, Machado DG, Cunha MP, dos Santos AA, et al. Effects of traumatic brain injury of different severities on emotional, cognitive, and oxidative stress-related parameters in mice. J Neurotrauma. 2010;27(10):1883–93.
95. Jones NC, Cardamone L, Williams JP, Salzberg MR, Myers D, O'Brien TJ. Experimental traumatic brain injury induces a pervasive hyperanxious phenotype in rats. J Neurotrauma. 2008;25(11):1367–74.
96. Bogner JA, Corrigan JD, Fugate L, Mysiw WJ, Clinchot D. Role of agitation in prediction of outcomes after traumatic brain injury. Am J Phys Med Rehabil. 2001;80(9):636–44.
97. Lequerica AH, Rapport LJ, Loeher K, Axelrod BN, Vangel SJ Jr, Hanks RA. Agitation in acquired brain injury: impact on acute rehabilitation therapies. J Head Trauma Rehabil. 2007;22(3):177–83.
98. McNett M, Sarver W, Wilczewski P. The prevalence, treatment and outcomes of agitation among patients with brain injury admitted to acute care units. Brain Inj. 2012;26(9):1155–62.
99. Nott MT, Chapparo C, Baguley IJ. Agitation following traumatic brain injury: an Australian sample. Brain Inj. 2006;20(11):1175–82.
100. Riley GA. Stress and depression in family carers following traumatic brain injury: the influence of beliefs about difficult behaviours. Clin Rehabil. 2007;21(1):82–8.
101. Silva MA, Nakase-Richardson R, Sherer M, Barnett SD, Evans CC, Yablon SA. Posttraumatic confusion predicts patient cooperation during traumatic brain injury rehabilitation. Am J Phys Med Rehabil. 2012;91(10):890–3.
102. Gnanasekaran G. "Sundowning" as a biological phenomenon: current understandings and future directions: an update. Aging Clin Exp Res. 2016;28(3):383–92.
103. Boronat AC, Ferreira-Maia AP, Wang YP. Sundown syndrome in older persons: a scoping review. J Am Med Dir Assoc. 2019;20(6):664–71.e5.
104. Carrier SL, Hicks AJ, Ponsford J, McKay A. Managing agitation during early recovery in adults with traumatic brain injury: an international survey. Ann Phys Rehabil Med. 2021;64(5):101532.
105. Race NS, Moschonas EH, Cheng JP, Bondi CO, Kline AE. Antipsychotic drugs: the antithesis to neurorehabilitation in models of pre-clinical traumatic brain injury. Neurotrauma Rep. 2023;4(1):724–35.

106. Kline AE, Hoffman AN, Cheng JP, Zafonte RD, Massucci JL. Chronic administration of antipsychotics impede behavioral recovery after experimental traumatic brain injury. Neurosci Lett. 2008;448(3):263–7.
107. Phelps TI, Bondi CO, Ahmed RH, Olugbade YT, Kline AE. Divergent long-term consequences of chronic treatment with haloperidol, risperidone, and bromocriptine on traumatic brain injury-induced cognitive deficits. J Neurotrauma. 2015;32(8):590–7.
108. Cusack S, O'Toole PW, ELDERMET Consortium. Challenges and implications for biomedical research and intervention studies in older populations: insights from the ELDERMET study. Gerontology. 2013;59(2):114–21.
109. Mody L, Miller DK, McGloin JM, Freeman M, Marcantonio ER, Magaziner J, et al. Recruitment and retention of older adults in aging research. J Am Geriatr Soc. 2008;56(12):2340–8.
110. Kuchel GA. Inclusion of older adults in research: ensuring relevance, feasibility, and rigor. J Am Geriatr Soc. 2019;67(2):203–4.
111. Cutler SM, Cekic M, Miller DM, Wali B, VanLandingham JW, Stein DG. Progesterone improves acute recovery after traumatic brain injury in the aged rat. J Neurotrauma. 2007;24(9):1475–86.
112. Cekic M, Cutler SM, VanLandingham JW, Stein DG. Vitamin D deficiency reduces the benefits of progesterone treatment after brain injury in aged rats. Neurobiol Aging. 2011;32(5):864–74.
113. Wali B, Sayeed I, Stein DG. Improved behavioral outcomes after progesterone administration in aged male rats with traumatic brain injury. Restor Neurol Neurosci. 2011;29(1):61–71.
114. Li Z, Wang B, Kan Z, Zhang B, Yang Z, Chen J, et al. Progesterone increases circulating endothelial progenitor cells and induces neural regeneration after traumatic brain injury in aged rats. J Neurotrauma. 2012;29(2):343–53.
115. Kasturi BS, Stein DG. Progesterone decreases cortical and sub-cortical edema in young and aged ovariectomized rats with brain injury. Restor Neurol Neurosci. 2009;27(4):265–75.
116. Wright DW, Yeatts SD, Silbergleit R, Palesch YY, Hertzberg VS, Frankel M, et al. Very early administration of progesterone for acute traumatic brain injury. N Engl J Med. 2014;371(26):2457–66.
117. Korley F, Pauls Q, Yeatts SD, Jones CMC, Corbett-Valade E, Silbergleit R, et al. Progesterone treatment does not decrease serum levels of biomarkers of glial and neuronal cell injury in moderate and severe Traumatic Brain Injury subjects: a secondary analysis of the progesterone for Traumatic Brain Injury, Experimental Clinical Treatment (ProTECT) III trial. J Neurotrauma. 2021;38(14):1953–60.
118. Goldstein FC, Caveney AF, Hertzberg VS, Silbergleit R, Yeatts SD, Palesch YY, et al. Very early administration of progesterone does not improve neuropsychological outcomes in subjects with moderate to severe traumatic brain injury. J Neurotrauma. 2017;34(1):115–20.
119. Gupte R, Christian S, Keselman P, Habiger J, Brooks WM, Harris JL. Evaluation of taurine neuroprotection in aged rats with traumatic brain injury. Brain Imaging Behav. 2019;13(2):461–71.
120. Morganti JM, Riparip LK, Chou A, Liu S, Gupta N, Rosi S. Age exacerbates the CCR2/5-mediated neuroinflammatory response to traumatic brain injury. J Neuroinflammation. 2016;13(1):80.
121. Wang Y, Wernersbach I, Strehle J, Li S, Appel D, Klein M, et al. Early posttraumatic CSF1R inhibition via PLX3397 leads to time- and sex-dependent effects on inflammation and neuronal maintenance after traumatic brain injury in mice. Brain Behav Immun. 2022;106:49–66.
122. Henry RJ, Ritzel RM, Barrett JP, Doran SJ, Jiao Y, Leach JB, et al. Microglial depletion with CSF1R inhibitor during chronic phase of experimental traumatic brain injury reduces neurodegeneration and neurological deficits. J Neurosci. 2020;40(14):2960–74.
123. Kumar A, Stoica BA, Sabirzhanov B, Burns MP, Faden AI, Loane DJ. Traumatic brain injury in aged animals increases lesion size and chronically alters microglial/macrophage classical and alternative activation states. Neurobiol Aging. 2013;34(5):1397–411.

124. Okigbo AA, Helkowski MS, Royes BJ, Bleimeister IH, Lam TR, Bao GC, et al. Dose-dependent neurorestorative effects of amantadine after cortical impact injury. Neurosci Lett. 2019;694:69–73.
125. Wagner AK, Drewencki LL, Chen X, Santos FR, Khan AS, Harun R, et al. Chronic methylphenidate treatment enhances striatal dopamine neurotransmission after experimental traumatic brain injury. J Neurochem. 2009;108(4):986–97.
126. Reid WM, Hamm RJ. Post-injury atomoxetine treatment improves cognition following experimental traumatic brain injury. J Neurotrauma. 2008;25(3):248–56.
127. Weaver JL, Eliceiri B, Costantini TW. Fluoxetine reduces organ injury and improves motor function after traumatic brain injury in mice. J Trauma Acute Care Surg. 2022;93(1):38–42.
128. Larson EB, Zollman FS. The effect of sleep medications on cognitive recovery from traumatic brain injury. J Head Trauma Rehabil. 2010;25(1):61–7.
129. Driver S, Stork R. Pharmacological management of sleep after traumatic brain injury. NeuroRehabilitation. 2018;43(3):347–53.
130. Lee P, Kim J, Williams R, Sandhir R, Gregory E, Brooks WM, et al. Effects of aging on blood brain barrier and matrix metalloproteases following controlled cortical impact in mice. Exp Neurol. 2012;234(1):50–61.
131. Hawkins BE, Cowart JC, Parsley MA, Capra BA, Eidson KA, Hellmich HL, et al. Effects of trauma, hemorrhage and resuscitation in aged rats. Brain Res. 2013;1496:28–35.
132. Lartaud I, Bray-des-Boscs L, Chillon JM, Atkinson J, Capdeville-Atkinson C. In vivo cerebrovascular reactivity in Wistar and Fischer 344 rat strains during aging. Am J Phys. 1993;264(3 Pt 2):H851–8.
133. Shimamura M, Garcia JM, Prough DS, Hellmich HL. Laser capture microdissection and analysis of amplified antisense RNA from distinct cell populations of the young and aged rat brain: effect of traumatic brain injury on hippocampal gene expression. Mol Brain Res. 2004;122(1):47–61.
134. Anderson J, Sandhir R, Hamilton ES, Berman NEJ. Impaired expression of neuroprotective molecules in the HIF-1alpha pathway following traumatic brain injury in aged mice. J Neurotrauma. 2009;26(9):1557–66.
135. Sibille E. Molecular aging of the brain, neuroplasticity, and vulnerability to depression and other brain-related disorders. Dialogues Clin Neurosci. 2013;15(1):53–65.
136. Toricelli M, Pereira AAR, Souza Abrao G, Malerba HN, Maia J, Buck HS, et al. Mechanisms of neuroplasticity and brain degeneration: strategies for protection during the aging process. Neural Regeneration Res. 2021;16(1):58–67.
137. Venkatesan UM, Rabinowitz AR, Bernier RA, Hillary FG. Cognitive reserve in individuals aging with traumatic brain injury: independent and interactive effects on cognitive functioning. J Head Trauma Rehabil. 2022;37(3):E196–205.
138. Itoh T, Imano M, Nishida S, Tsubaki M, Nakayama T, Mizuguchi N, et al. Appearance of neural stem cells around the damaged area following traumatic brain injury in aged rats. J Neural Transm. 2013;120(3):361–74.
139. Gupta RK, Prasad S. Early down regulation of the glial Kir4.1 and GLT-1 expression in pericontusional cortex of the old male mice subjected to traumatic brain injury. Biogerontology. 2013;14(5):531–41.
140. Gupta RK, Prasad S. Age-dependent alterations in the interactions of NF-κB and N-myc with GLT-1/EAAT2 promoter in the pericontusional cortex of mice subjected to traumatic brain injury. Mol Neurobiol. 2016;53(5):3377–88.
141. Moor E, Shohami E, Kanevsky E, Grigoriadis N, Symeonidou C, Kohen R. Impairment of the ability of the injured aged brain in elevating urate and ascorbate. Exp Gerontol. 2006;41(3):303–11.
142. Shao C, Roberts KN, Markesbery WR, Scheff SW, Lovell MA. Oxidative stress in head trauma in aging. Free Radic Biol Med. 2006;41(1):77–85.
143. Sun D, McGinn M, Hankins JE, Mays KM, Rolfe A, Colello RJ. Aging- and injury-related differential apoptotic response in the dentate gyrus of the hippocampus in rats following brain trauma. Front Aging Neurosci. 2013;5:95.

144. Sandhir R, Onyszchuk G, Berman NEJ. Exacerbated glial response in the aged mouse hippocampus following controlled cortical impact injury. Exp Neurol. 2008;213(2):372–80.
145. Sandhir R, Berman NEJ. Age-dependent response of CCAAT/enhancer binding proteins following traumatic brain injury in mice. Neurochem Int. 2010;56(1):188–93.
146. Ziebell JM, Rowe RK, Muccigrosso MM, Reddaway JT, Adelson PD, Godbout JP, et al. Aging with a traumatic brain injury: could behavioral morbidities and endocrine symptoms be influenced by microglial priming? Brain Behav Immun. 2017;59:1–7.
147. White-Gbadebo D, Hamm RJ. Chronic corticosterone treatment potentiates deficits following traumatic brain injury in rats: implications for aging. J Neurotrauma. 1993;10(3):297–306.
148. Miller DB, O'Callaghan JP. Aging, stress and the hippocampus. Ageing Res Rev. 2005;4(2):123–40.
149. Bäckman L, Nyberg L, Lindenberger U, Li SC, Farde L. The correlative triad among aging, dopamine, and cognition: current status and future prospects. Neurosci Biobehav Rev. 2006;30(6):791–807.
150. 2023 American Geriatrics Society Beers Criteria® Update Expert Panel. American Geriatrics Society 2023 updated AGS Beers Criteria® for potentially inappropriate medication use in older adults. J Am Geriatr Soc 2023.
151. Mosenthal AC, Livingston DH, Lavery RF, Knudson MM, Lee S, Morabito D, et al. The effect of age on functional outcome in mild traumatic brain injury: 6-month report of a prospective multicenter trial. J Trauma. 2004;56(5):1042–8.
152. Chronic conditions [Internet]. [cited 2023 Oct 31]. Available from: https://www.cms.gov/data-research/statistics-trends-and-reports/chronic-conditions/chronic-conditions
153. Shi J, Dong B, Mao Y, Guan W, Cao J, Zhu R, et al. Review: Traumatic brain injury and hyperglycemia, a potentially modifiable risk factor. Oncotarget. 2016;7(43):71052–61.
154. Kennedy RL, Henry J, Chapman AJ, Nayar R, Grant P, Morris AD. Accidents in patients with insulin-treated diabetes: increased risk of low-impact falls but not motor vehicle crashes—a prospective register-based study. J Trauma Acute Care Surg. 2002;52(4):660.
155. Michael YL, Whitlock EP, Lin JS, Fu R, O'Connor EA, Gold R. Primary care–relevant interventions to prevent falling in older adults: a systematic evidence review for the U.S. Preventive Services Task Force. Ann Intern Med. 2010;153(12):815–25.
156. Uryu K, Giasson BI, Longhi L, Martinez D, Murray I, Conte V, et al. Age-dependent synuclein pathology following traumatic brain injury in mice. Exp Neurol. 2003;184(1):214–24.
157. Ojo JO, Mouzon B, Greenberg MB, Bachmeier C, Mullan M, Crawford F. Repetitive mild traumatic brain injury augments tau pathology and glial activation in aged hTau mice. J Neuropathol Exp Neurol. 2013;72(2):137–51.
158. Li L, Geraghty OC, Mehta Z, Rothwell PM, Oxford Vascular Study. Age-specific risks, severity, time course, and outcome of bleeding on long-term antiplatelet treatment after vascular events: a population-based cohort study. Lancet. 2017;390(10093):490–9.
159. Wong DK, Lurie F, Wong LL. The effects of clopidogrel on elderly traumatic brain injured patients. J Trauma. 2008;65(6):1303–8.
160. Go AS, Hylek EM, Phillips KA, Chang Y, Henault LE, Selby JV, et al. Prevalence of diagnosed atrial fibrillation in adults: national implications for rhythm management and stroke prevention: the AnTicoagulation and Risk Factors in Atrial Fibrillation (ATRIA) Study. JAMA. 2001;285(18):2370–5.
161. Zeiler FA, Donnelly J, Nourallah B, Thelin EP, Calviello L, Smielewski P, et al. Intracranial and extracranial injury burden as drivers of impaired cerebrovascular reactivity in traumatic brain injury. J Neurotrauma. 2018;35(14):1569–77.
162. Xiong Y, Mahmood A, Chopp M. Animal models of traumatic brain injury. Nat Rev Neurosci. 2013;14(2):128–42.
163. Cernak I. Animal models of head trauma. NeuroRx. 2005;2(3):410–22.
164. Animal Models of Acute Neurological Injury. Springer International Publishing; 10 p.
165. Armstrong NM, An Y, Beason-Held L, Doshi J, Erus G, Ferrucci L, et al. Sex differences in brain aging and predictors of neurodegeneration in cognitively healthy older adults. Neurobiol Aging. 2019;81:146–56.

166. Chowen JA, Garcia-Segura LM. Role of glial cells in the generation of sex differences in neurodegenerative diseases and brain aging. Mech Ageing Dev. 2021;196:111473.
167. Díaz A, López-Grueso R, Gambini J, Monleón D, Mas-Bargues C, Abdelaziz KM, et al. Sex differences in age-associated type 2 diabetes in rats—role of estrogens and oxidative stress. Oxid Med Cell Longev [Internet]. 2019 [cited 2023 Oct 31];2019. Available from: https://www.hindawi.com/journals/omcl/2019/6734836/abs/
168. Smith BL, Bollinger JL, De Luca SN, VanRyzin J, Kopec A, Murphy AZ, et al. Sex-specific effects of environment on behavioral outcomes across the lifespan. Frontiers Media SA; 2023. 162 p.
169. Conley YP, Arenth PM, Wagner AK. … pathway gene risk score for cognitive recovery following traumatic brain injury: methodological considerations, preliminary findings, and interactions with sex. J Head [Internet]. 2016; Available from: https://www.ingentaconnect.com/content/wk/htr/2016/00000031/00000005/art00003
170. Rogers E, Wagner AK. Gender, sex steroids, and neuroprotection following traumatic brain injury. J Head Trauma Rehabil. 2006;21(3):279–81.
171. Scholl JL, Afzal A, Fox LC, Watt MJ, Forster GL. Sex differences in anxiety-like behaviors in rats. Physiol Behav. 2019;211:112670.

Chapter 12
The Neuroimmune Interface: Age-Related Responses to Traumatic Brain Injury

John Leonard, Liliana Ladner, Elizabeth A. Harris, Caroline de Jager, and Michelle H. Theus

Abstract Traumatic Brain Injury (TBI) is a significant public health issue, with diverse consequences across the lifespan. This comprehensive review explores the complex interplay between age-related responses and the immune system following TBI. TBI exhibits distinct effects in pediatric, adult, and elderly populations, with profound implications for recovery and long-term outcomes. The immune system, as a key player in the post-TBI inflammatory cascade, exerts age-dependent influences on inflammation, neuroinflammation, and tissue repair. We examine the evolving understanding of age-related neuroinflammatory responses, cytokine profiles, and the role of immune cells, such as microglia and T cells, in the context of TBI. Furthermore, we evaluate the therapeutic implications of age-specific immunomodulation strategies toward mitigating TBI-associated neuropathology. This review consolidates the current knowledge on age-related immune responses in TBI, shedding light on potential avenues for tailored therapeutic interventions across the age spectrum. Understanding these nuanced responses is crucial for optimizing patient care and enhancing recovery outcomes in the aftermath of traumatic brain injury.

Keywords Childhood · Juvenile · Traumatic brain injury · Brain development · Functional outcome · Age dependence

J. Leonard · L. Ladner · E. A. Harris · C. de Jager
Virginia-Maryland Regional College of Veterinary Medicine, Blacksburg, VA, USA
e-mail: leonardj@vt.edu; lilianaladner@vt.edu; eaharris228@vt.edu; cdejager@vt.edu

M. H. Theus (✉)
The Department of Biomedical Sciences and Pathobiology, Virginia-Maryland Regional College of Veterinary Medicine, Blacksburg, VA, USA
e-mail: mtheus@vt.edu

L. J. Noble-Haeusslein, D. M. Schnyer (eds.), *Traumatic Brain Injuries*, Advances in Neurobiology 42, https://doi.org/10.1007/978-3-031-69832-3_12

Abbreviations

CTE	Chronic traumatic encephalopathy
NLR	Neutrophil-lymphocyte ratio
PTE	Post-traumatic epilepsy
TBIs	Traumatic brain injuries

12.1 Introduction

In the scope of neuroscientific medical research, few phenomena command as much attention as traumatic brain injury (TBI). Often referred to as a "silent epidemic" [1, 2] due to a lack of public information on the subject, TBI represents a significant public health concern that transcends geographical and socioeconomic boundaries, with the most common causes being car accidents and falls [3] and environmental factors playing a major role [4]. The US records over two million TBI-related emergency room visits yearly with adults over age 75 and children 4 years or younger having the highest incidence [5]. TBI effects range from subtle neurocognitive changes to severe physical and mental impairments, largely dependent on the injury's severity [6] but increasingly, as this chapter posits, also on the age at which the injury occurs. Thus, while TBI remains a significant issue in itself, the complex overlay of age-related responses adds another layer of intricacy. The complicated interplay between the neuroimmune interface and the age-related responses to TBI is a topic of increasing significance that demands comprehensive exploration and analysis. This chapter delves into the captivating and complex relationship between the immune system, age-related changes, and the outcomes of traumatic brain injury.

12.2 Neuroimmune Response to Brain Injury

The neuroimmune interface, a juncture where the nervous and immune systems converge, stands as a pivotal determinant of both health and disease [7]. A sophisticated network of cells, molecules, and pathways, this interface serves as the guardian of cerebral homeostasis, shielding the brain from harm while orchestrating healing when injury strikes [8]. In the event of a traumatic brain injury, this interface is perturbed, initiating a cascade of intricate immune responses that can influence the injury's aftermath [9]. What renders this interplay even more intricate is its interaction with developmental and aging processes.

Development, an inexorable and multifaceted biological process, exerts a profound impact on the brain's structural and functional integrity. Neurons, the basic functional units of the nervous system, undergo age-related changes that can render them both resilient and vulnerable in the face of injury [10, 11]. Glial cells, often regarded as the brain's caretakers, also experience alterations that influence their reactivity to trauma [12, 13]. Finally, changes in immune system population and function with age can yield protection against some insults or open the door for devastation from others [14]. These age-related modifications can significantly alter the trajectory of neuroinflammatory responses, potentially shaping the course of recovery or decline after TBI.

Understanding how the neuroimmune interface is modulated by age is essential for tailoring effective interventions for TBI in different age groups. The youthful brain possesses remarkable plasticity and regenerative potential [15], allowing it to mount robust responses to injury. However, this vigor diminishes with age [16], necessitating a more nuanced approach to therapeutic strategies, and it has been noted that the plasticity of the young brain may be overstated [17]. Exploring the intricate relationship between age-related changes and the neuroinflammatory response could unveil novel therapeutic targets that promote recovery in both the young and the elderly. As we uncover the intricate tapestry of the neuroimmune interface and its interaction with age-related responses to TBI, several key questions emerge. How do age-related alterations in immune cell populations impact the trajectory of neuroinflammation following injury? What role do neurotrophic factors and immunomodulatory molecules play in shaping the outcomes of TBI in different age groups? Finally, can we leverage our understanding of these mechanisms to develop age-specific therapeutic interventions that enhance recovery and mitigate neurodegenerative processes?

12.3 Development and the Immune System

In the clinical setting, morbidity and mortality from TBI are directly correlated with age-at-injury, with injury grade increasing and favorable outcomes decreasing with every decade of life after reaching adulthood [18, 19] and younger patients generally seeing more favorable outcomes relative to older patients, with the exception of early childhood patients, who exhibit similar outcomes to aged patients [20–22]. Infants and toddlers presenting with traumatic brain injury also exhibit a different pathologic profile relative to adults [23]. Young patients show differences in their immune profile relative to adults and also have ongoing developmental differences in their anatomy, including greater cerebral blood flow, myelination, synaptogenesis, greater head-to-body-size ratio, greater plasticity of the skull, and unfused cranial bones [24, 25]. Key differences in glucose metabolism, calcium regulation, and mitochondrial function have been identified in pediatric populations versus adults which point to specific areas of vulnerability as well as potential neuroprotective mechanisms [26, 27]. All of this highlights the need for an improved understanding of the developing brain and immune system when considering these age-related outcomes.

The human immune system undergoes significant changes during development, from infancy to adulthood. At birth, infants possess an immature immune system with limited defense mechanisms against pathogens. Over the first few years of life, the immune system gradually matures, reaching its full potential during adolescence and early adulthood [14]. Accordingly, infection and injury may be controlled differently in pediatric and adult patients.

One key feature of immune system development is the gradual expansion and differentiation of resident immune cells. Neonates have a limited number of immune cells, such as T cells and B cells, which are responsible for recognizing and responding to pathogens. Indeed, the percentage of T cells (including CD8+ cells) and NK cells is lowest immediately after birth [28]. The number of neutrophils, while initially high, starts decreasing during the teenage years [29]. Innate immune cells of neonates and children also have much less functional diversity and produce fewer cytokines than adults [30]. Pro-inflammatory cytokines in particular are produced in much lower numbers in neonates, while production of IL-8 is higher [31, 32]. Some immune cells, such as monocytes, are also less mobile in neonates compared to adults [33], which can impact their ability to transit to injured tissue. Antigen specificity is still developing, resulting in a failure for neonates to mount responses to certain diseases, such as Haemophilus influenzae b (Hib) [34], although sometimes this becomes advantageous as seen with SARS-CoV-2 infection in children [35]. With age, the diversity and specificity of these immune cells increase with repeated exposure to antigens, enabling a more robust response to infectious challenges [14, 36]. This developmental process is crucial for mounting effective immune responses in the face of TBI-related complications.

As individuals age, their immune system experiences a gradual decline in functionality over time, a phenomenon known as immune senescence. This condition is marked by various changes, including increased susceptibility of cells to apoptosis, reduced ability to proliferate, and an elevated production of inflammatory mediators. These factors collectively contribute to the development of a persistent, low-grade, pro-inflammatory baseline state.

One prominent shift in the aging immune system is decreased thymic production of naive T cells and an enrichment of antigen-experienced memory T cells. Natural killer (NK) cells also transform from a less mature $CD56^{bright}$ $CD16^{dim/-}$ to a more mature $CD56^{dim}$ $CD16^{+}$ subset, which preferentially migrate to acute inflammatory sites, have higher cytotoxic activity [37], and can induce monocytes to produce TNF [38]. Furthermore, monocytes from elderly subjects show higher basal levels of TNF, MCP-1, and reduced TGB-β compared to adult monocytes [39], subset alterations which are prominent [40] and decreased responsiveness to innate immune system ligands, and transcriptional changes in genes associated with defense mechanisms against bacteria and viruses [26, 27]. They also show reduced efferocytosis and impaired amyloid-beta uptake [41, 42]. Overall, this shift in peripheral immune signature suggests that aging may create an imbalance toward a pro-inflammatory environment that may not benefit tissue health in the brain. Likewise, aged resident microglia display altered morphology and reduced capacity for migration, efferocytosis, and limited ability to mount an anti-inflammatory state to regulate tissue repair [43].

12.4 Clinical Age-at-Injury Consequences of TBI

Following TBI, the brain's immune response is triggered, leading to a complex cascade of events involving resident immune cells, such as microglia [44], and the recruitment of peripheral immune cells from circulation to the site of injury. Immediately after mechanical injury, damage-associated molecular patterns (DAMPs) such as high-mobility group box 1 (HMGB1) and adenosine triphosphate (ATP) are released into the extracellular space, where they are picked up by resident microglia-expressing pathogen-associated molecular pattern (PAMP) sensors such as toll-like receptor (TLR) or receptors such as the pattern recognition receptor (PRP) for advanced glycation end-products (RAGE), triggering an inflammatory cascade and recruitment of monocytes and other immune cells from both the brain and the periphery [45–47]. Secretion of cytokines and chemokines by local and infiltrating immune cells induces the inflammatory cascade which recruits even more immune cells from the periphery through the disrupted blood-brain barrier (BBB) [46]. Microglia attempt to rectify barrier breaches in the glial limitans and begin the process of phagocytosis of dead and dying cells around the lesion [48], while continually recruited peripheral derived monocytes, neutrophils, and T cells intensify inflammatory processes via reactive oxygen species (ROS), chemokines, and cytokines. Outcomes are variable, and inflammation can persist for weeks or even years after the initial injury, leading to a milieu of different sequelae [49].

12.4.1 *Age-Related Chronic Neurological Sequelae*

TBI can lead to a range of chronic outcomes, including cognitive deficits, mood disorders, and neurodegenerative conditions. The immune system has been increasingly recognized as a major influencer of these outcomes, and age-related differences in immune function may contribute to divergent trajectories of recovery and risk for long-term complications. In children and adolescents, TBI can disrupt ongoing brain development and lead to long-term cognitive and behavioral deficits [50]. This can be particularly observed in measurements of IQ, executive functioning, and memory [50]. While negligible cognitive developmental effects are observed for mild TBI in pediatric patients [51], severe TBI is directly correlated with lower IQ later in life, with children aged 8–12 experiencing cognitive recovery similar to adults and children aged 3–7 experiencing minimal cognitive recovery [52]. Cognitive outcomes for pediatric moderate TBI fall between the mild and severe groups, with some deficits persisting for up to 2 years [51], highlighting the significant developmental disruptions that can occur due to TBI. Indeed, behavioral outcomes also track injury severity in a linear manner, and age is correlated with how long behavioral problems such as PTSD, internalizing or externalizing feelings, and attention persist [53]. The immune response to TBI in this age group may play a role in shaping these outcomes.

Excessive inflammation in the developing brain can contribute to neurodegenerative processes, which may manifest later in life [54]. In contrast, adults do not have the same developmental disruptions but still develop cognitive and behavioral impairments. Over one-third of adults experience some form of moderate cognitive impairment, and depression, anxiety, and irritability are found in as many as 40% of patients [55]. Adults over age 65 in particular experience drastically higher rates of depression initially after having a TBI [56]. In adults, brain injury severity appears to be correlated to cognitive measures but not emotional ones [57].

TBI is also associated with an increased risk of neurodegenerative diseases, such as Alzheimer's, chronic traumatic encephalopathy (CTE), and Parkinson's disease [58, 59]. Age-related immune system changes can influence the susceptibility and progression of these conditions [60]. For example, older adults may display a more pronounced inflammatory response after TBI, potentially accelerating the onset of neurodegenerative processes [61]. Aβ plaques, a hallmark of Alzheimer's disease, are found in 30% of patients, across all age groups, who die from acute TBI [62], whereas normally they are only found in the elderly [63]. It is likely that Aβ deposits are associated with axonal injury [62], a common feature associated with TBI and the inflammatory immune response [64]. Alpha-Synuclein, a neuronal protein and hallmark of Parkinson's disease [65], has linked expression with major histocompatibility complex II (MHCII), a regulator of the innate immune response [66]. Injection of α-synuclein into the brains of rats showed an eventual increased

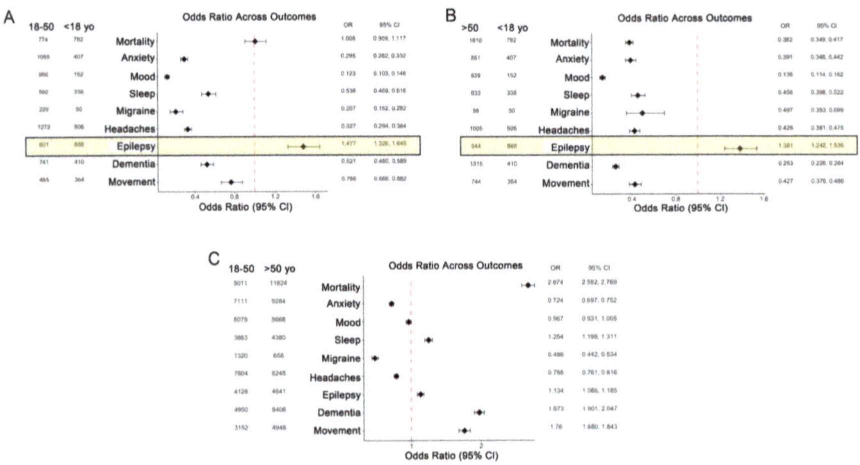

Fig. 12.1 Meta-analysis utilizing the multinational deidentified database TriNetX to characterize outcomes from moderate to severe TBI based on ICD codes. Outcomes were reported for <18-year-old (yo) compared to (**a**) 18–50 yo, (**b**) >50 yo, and (**c**) >50 compared to 18–50 yo patients. Patients were matched using propensity score matching based on sex and race across each comparison group. Patient outcomes were assessed at 1 year following moderate to severe TBI, including mortality, anxiety (F40-F48), mood disorders (F30-F39), sleep disorders (G47), migraines (G43), headaches (R51), epilepsy (G40), dementia or delirium (F01-F09, G30), or motor dysfunction (R26)

expression of MHCII and recruitment of peripheral immune cells in a similar fashion to TBI and immune cell infiltration preceded neurodegeneration [67].

Meta-analysis using the multinational deidentified database TriNetX [68] shows that moderate to severe patients <18 (mean age 4.3 +/− 5.1 years) had increased odds of developing post-traumatic epilepsy (OR = 1.477; 95% Cl [1.326, 1.645]) compared to patients 18–50 years old (yo) (mean age 29.6 +/− 9.5) or patients >50 years old (69.8 +/− 13.3 years) (OR = 1.381; 95% Cl [1.242, 1.536]) (Fig. 12.1a, b, respectively). On the other hand, patients >50 had the greatest odds associated with mortality, dementia, and movement or motor dysfunction (OR = 2.674, 1.973, 1.76; 95% Cl [2.582, 2.769; 1.901, 2.047; 1.680, 1.843] (Fig. 12.1c).

12.5 Post-traumatic Epilepsy

Post-traumatic epilepsy (PTE), a well-known chronic complication of TBI, in the pediatric population is a nuanced and largely understudied area. Our TriNetX meta-analysis shows moderate/severe TBI in patients <18 increase the odds ratio of developing epilepsy at 1 year post-injury (Fig. 12.2). A reported overall incidence of 10% of pediatric patients who have experienced TBI will develop PTE, and those that suffer from a severe TBI are up to ten times more likely to experience seizures than those with a mild injury [69]. PTE is defined as recurrent and unprovoked seizure that occurs at least 1 week after TBI. The development of acquired epilepsies, known as epileptogenesis, is largely resistant to anti-epileptic drugs (AED), and there are limited prognostic indicators of individual risk of developing PTE due to an incomplete understanding of the complex pathophysiological changes involved including those associated with age. However, there is emerging evidence that suggests both innate and peripherally derived immune cells contribute to neuroimmune-mediated epileptogenesis [70], and penetration of inflammatory cells in PTE human tissue has been observed [71]. For example, Ccl2, a ligand for infiltrating monocytes, IL-1β, TNF, IL-6, and HMGB1 are pro-ictogenic. Pro-inflammatory factors including IL-1β, TNFα, IL-6, and HMGB1 have been shown to be upregulated in animal and clinical models following TBI and have also been implicated in ictogenesis through potentiation of the inflammation [72–75]. CCL2, a chemokine constitutively expressed by neurons, is known to increase with TBI and plays a major role in chemoattraction of peripheral monocytes resulting in a more robust leukocyte-associated cytokine response and exacerbation of the inflammatory state [76]. Elevated concentrations of CCL2 have also been shown to increase synaptic activity in the hippocampus, resulting in enhanced neuronal excitability consistent with generation of excitotoxicity and seizure activity [77]. Interestingly, the absence of Ccl2 delays secretion of inflammatory cytokines and mediates the seizure-enhancing effects of systemic inflammation [78]. IL-1β, an inflammatory cytokine that plays a significant role in the recruitment of peripheral monocytes as well as regulation of other inflammatory cytokines, is known to be significantly elevated in response to TBI [79]. IL-1β also plays an important role in temporal lobe neural differentiation

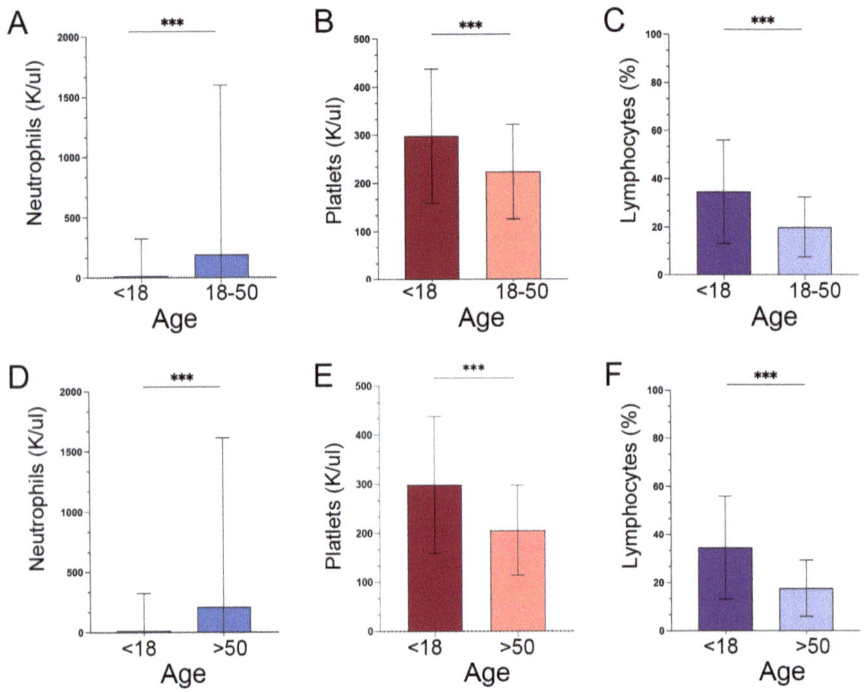

Fig. 12.2 Meta-analysis utilizing the multinational deidentified database TriNetX to characterize lab values from moderate to severe TBI at 24 h based on ICD codes. Patients were split into groups based on age, including less than 18 years (N = 10,156), 18–50 years (N = 68,570), and greater than 50 years (N = 174,439). (**a**) There was a significant decrease in neutrophils in patients <18 years old (yo) compared to those between 18 and 50 yo (**a**) and >50 yo (**d**). The number of platelets and lymphocytes was increased in <18 yo compared to 18–50 (**b**, **c**) and >50 (**e**, **f**). A student's T-test was utilized to compare lab values across age groups ***$P < 0.001$

with the developing brain relying on tight regulation of IL-1β concentrations for development of normal learning and memory [80]. Elevated IL-1β CSF:serum ratios have also been directly associated with increased incidence of PTE in TBI patients [71]. These clinical findings further highlight the potential use of IL-1β as a predictive biomarker as well as a therapeutic target which may be particularly useful in pediatric populations [81].

12.5.1 *Age and the Immune Response*

Pediatric TBI and the immune system's involvement remains an understudied subset of the TBI field. As previously outlined by Giza et al. [82], the response of the developing brain to injury differs from that of the adult brain. Significant physiological distinctions in the developing brain encompass the ongoing processes of myelination and synaptogenesis, higher cerebral blood flow and oxygen consumption, elevated brain water content, and reduced elasticity in the brain parenchyma as reviewed by Fraunberger et al. [24]. Importantly, immune cell prevalence is notable in the pediatric population as lymphocytes increased from 6 months to 18 years, while neutrophils and eosinophils decrease with age [29]. Monocyte chemotaxis is reduced in children under 10 compared to adults, which may shift the immune cell contribution to post-TBI secondary injury. Further this may affect the adaptive arm of the immune response since monocytes and dendritic cells act as antigen-presenting cells that activate T cells, shifting the trajectory of chronic neurological sequelae influenced by adaptive immunity.

Because the homeostatic number and composition of immune cells are altered with age, the acute peripheral innate immune and subsequent chronic adaptive immune response to TBI will undoubtably influence TBI outcome. Additionally, a nonstereotypical response to infectious challenges in the pediatric population may help describe the divergent inflammatory response induced following TBI [83, 84]. The immune system is responsible for both defending against infectious agents and facilitating the development of the CNS. Importantly, pro- and anti-inflammatory messengers, IL-1β, IL-1α, IL-6, IFN-γ, and TNF, in the brain are temporally regulated to coincide with key developmental milestones [85]. Neurophysiological processes and neurocognitive development are contingent upon proper balance and tight regulation of cytokine levels in the maturing CNS. This may explain why the juvenile inflammatory response to TBI in animal models is blunted [86] or shows localized anti-inflammatory activity compared to diffuse neuroinflammation in adults [87]. While a diminished inflammatory response may influence acute outcome from TBI, neuroimmune imbalance may have significant consequences on long-term neurological function which relies on a controlled and coordinated cytokine profile to regulate key epochs of development.

12.6 Clinical Immune Cell Evaluation in TBI

The maximum influx of neutrophils into the injured brain occurs approximately 24 h post-injury, although they infiltrate the site within 5 min of a closed brain injury. Neutrophils release harmful substances, including free radicals, pro-inflammatory cytokines, and proteases, which contribute to secondary damage. Regulatory T cells hinder T cell infiltration into the central nervous system, while promoting reactive astrogliosis and the expression of interferon-γ genes. Therefore, the neutrophil-to-lymphocyte ratio (NLR) has emerged as a low-cost, objective, and available predictor of inflammation as well as a marker of secondary injury associated with neutrophil activation [88, 89]. Inflammation-based scores are being increasingly used for prognosis in TBI. For example, the NLR value on admission might be effective for predicting outcome, and there is evidence of its utility in understanding age-related trajectories [89] or early emergence of traumatic intracerebral hemorrhage [90]. Overall, NLR predicts an increased mortality risk and unfavorable outcome in all ages [91–93]. Moreover, platelet-to-lymphocyte ratio also predicts early mortality in moderate to severe TBI in adults and pediatric patients [94, 95]; both NLR and PLR may be useful in predicting epileptic seizure risk [96]. However, only NLR may have a role in CT decision-making in mild pediatric TBI [97].

12.6.1 Immune Profile Alterations in Pediatric TBI

It has been shown clinically that the cytokines exhibited by pediatric TBI patients are dependent on the severity of injury. Pediatric patients with mild TBI experience greater numbers of IFN-γ immediately after injury, while in severe TBI, it is decreased [98]. IL-6 is greatly elevated in pediatric patients with both mild and severe TBI and is correlated with symptom presentation [99]. IL-8 and IL-10 are elevated in severe TBI and are correlated with worse outcomes [100, 101]. The role that chemotaxis plays in age differences in TBI remains unclear. Monocytes in children display decreased chemotaxis, which could reduce monocyte recruitment to the site of injury [102]. The immature blood-brain barrier and more fragile cerebral vasculature in neonates may allow for increased infiltration of immune cells, potentially altering neuroinflammation [103]. Pediatric patients also have muted responses to oxidative stress and demonstrate iron accumulation following focal TBI [104] which may influence chronic evolution of neurological sequalae as the brain develops. Adults typically mount a balanced immune response to TBI, characterized by the activation of both pro- and anti-inflammatory pathways [27]. This balanced response is thought to promote recovery while limiting excessive inflammation. However, the quality of this response can vary based on individual health and genetics [105].

12.6.2 Immune Cell Shifts in the Elderly After TBI

While more research is needed to understand the outcomes of TBI that worsen with advanced age, the consequence of TBI at old age is equally important due to increased rates of preexisting conditions and normal age-related functional decline. The elderly (>65 yo) are at a greater risk of acquiring TBI due to falls, and they show a divergent immune response following TBI. For example, CD4 T cells are only affected in adults which are driven to effector memory cells; however, in the elderly, both activated CD4 and CD8 T cells are increased and are induced to differentiate into subtypes with low central memory CD4 and memory precursor effector CD8 cells. In addition, TBI reduces production of pro-inflammatory cytokines TNF and IL-6 in monocytes. This is consistent with the fact that during aging, the immune system undergoes a gradual time-dependent functional decline called immune senescence, including reduced cytotoxic activity of T cells [106], and monocytes display diminished responsiveness [106]. Further studies show that IL-6 and IL-10 did not show age-related differences in plasma levels compared to adult patients [106]. While T cells are reduced after TBI in adult and elderly patients, only elderly showed reduced NK cells and altered subsets of CD8 T cells. CD8 T cells are characterized by high levels of granzyme B and TNF in elderly, a common phenotype of aged T cells leading to inflammaging [107]. While adaptive immune cells (T and B cells) are more affected than innate immune cells (NK and monocytes) in elderly TBI patients, a worse outcome is also associated with lower CD86+ monocytes.

Age-dependent neurological outcomes following moderate to severe TBI were correlated with complete blood counts, neutrophils, platelets, and lymphocytes from patients included in Fig. 12.1. TriNetX shows that patients <18 yo display reduced neutrophils (Fig. 12.2a, d) with a concomitant increase in platelets (Fig. 12.2b, e) and lymphocytes (Fig. 12.2c, f) compared to either 18–50 or >50 yo patients. These findings imply that pediatric patients may have lower NLR, which may contribute to the reduced OR for most neurological impairments analyzed in Fig. 12.1, excluding epilepsy.

12.7 Animal Studies and Age-at-Injury Mechanisms of TBI

The influence of age-at-injury is well documented in numerous animal models [108]. Studies in juvenile animal models of TBI have highlighted age-related differences in immune responses. Younger individuals often exhibit less regulated inflammatory responses, potentially leading to increased neuroinflammation after TBI. Additionally, the immature blood-brain barrier in neonates and children may allow for greater infiltration of immune cells into the brain, exacerbating neuroinflammation [109]. Pre-clinical studies in adult animals with TBI have shown a more balanced immune response characterized by both pro- and anti-inflammatory pathways. This balanced response aims to promote recovery while limiting excessive inflammation. However, the quality of the immune response in adults may vary based on individual health and genetic factors. Pre-clinical research in aged animal models has indicated that older adults may experience delayed and less robust immune responses to TBI [110]. Age-related changes in immune function, including decreased immune cell proliferation and impaired cytokine production, can hinder the ability to combat secondary infections and exacerbate neuroinflammation.

Studies have linked age-related differences in immune responses to cognitive and behavioral outcomes following TBI [111]. For juvenile populations, TBI can disrupt ongoing brain development and lead to long-term deficits in memory, attention, and emotional regulation, partly due to excessive neuroinflammation. Animal models also demonstrate that TBI is associated with an increased risk of neurodegenerative diseases, such as Alzheimer's and Parkinson's disease, particularly in the elderly population [112]. The age-related immune response may influence the susceptibility and progression of these conditions. Older adults may experience a more pronounced inflammatory response after TBI, potentially accelerating the onset of neurodegenerative processes.

Studies in rats have shown that N-methyl-D-aspartate receptors (NMDAR) are lost acutely following TBI, leading to dysregulation of calcium. Interestingly, this effect was observed predominantly in the cortex of adult animals; however, juveniles displayed these changes in the hippocampus. This may contribute to regional excitotoxic vulnerability in the temporal lobe, a known contributing factor to seizure disorders [113, 114]. Differences in 45calcium accumulation in the thalamus have also been noted as age-dependent. In a rodent model of mild LFP injury, adult rats exhibited a delayed post-traumatic spike in focal thalamic calcium accumulation that correlated with cell death, which was not observed in P17 rats [27, 115]. It is also postulated that a combination of failed clearance and exaggerated release of glutamate by glial cells during immune activation leads to increased glutamate and synaptic dysfunction [116]. Astrocytes are critical in regulating amino-acid metabolic homeostasis, controlling excessive glutamate clearance through sodium-dependent high-affinity glutamate transporters (EAATs), and preserving the BBB [117]. Astrogliosis is more prominent in juvenile TBI [118, 119]; while this may alleviate acute clearance of excitotoxic amino acids to support injury containment, long-term alterations may impair glutamate regulation and contribute to cognitive impairments and seizure susceptibility in juvenile TBI [120, 121]. Blood monocytes also express cystine/glutamate antiporter and EAATs, and stimulation with TNF leads to functional transport [122]. Given the known differences in immune cell activity and subclass specification during development and aging as well as in response to TBI, their role in the process may be altered with age which may have temporal consequences.

Age has been confirmed to be a determinant of leukocyte infiltration in the juvenile P21-injured cortex compared to injured adults [123]. CD45-positive leukocytes, namely, neutrophils, are elevated in the P21 cortex, which coincide with age-dependent reduction in cortical volume that was exacerbated chronically in P21 mice. The blood-brain barrier is also preserved in P21 mice following cortical impact, which is related to enhanced Tie2 receptor signaling [118]. Further, microglial activation was greater in P17 rats compared with P35 rats following fluid percussion injury [124]. Similarly, aged mice showed greater histological microglial and astrocytic activation than adult mice [125]. Differences in the onset of cerebral inflammation started earlier in old mice, whereas inflammation in young mice was delayed with an expected pattern of T cell migration [126]. These differences highlight the key role that neuroinflammation plays in the age-at-injury acute and chronic histopathological outcomes in several models of TBI and rodent species.

Neuroinflammation following TBI is both a necessity for efficient elimination of dead or damaged cells and significant driving factor of secondary brain injury. Inappropriate persistence of inflammatory cytokines, peripheral immune cells, and altered activity of resident microglia has been strongly associated with an elevated risk of developing post-traumatic epilepsy [127]. Understanding the distinct immune responses of the adolescent and aged brain to TBI is crucial for devising effective therapeutic strategies. As research indicates, the adolescent brain's reaction to injury, particularly from an immune standpoint, differs significantly from other age groups. Potential therapies could include targeted anti-inflammatory agents or modulators of glial activation. Immune-modulating therapies, such as the use of minocycline, an antibiotic with anti-inflammatory properties, have shown promise in pre-clinical models of adolescent TBI [128].

12.8 Conclusions

The age-at-injury spectrum of TBI effects and outcomes is intrinsically linked to age-related variations in immune responses. From the vulnerable pediatric brain to the unique challenges faced by the elderly, TBI-induced pathophysiological changes reveal a compelling interdependence with the immune system. As our understanding of this relationship deepens, it is evident that age dictates not just the initial injury response but also the trajectory of recovery and long-term outcomes. Neuroinflammation, orchestrated by cytokines and immune cells, plays a pivotal role in this nexus. As elucidated in this chapter, therapeutic strategies must be increasingly refined and tailored to address the unique immune profiles associated with different age groups. Such an approach holds the promise of improving patient prognosis, reducing long-term complications, and offering hope to individuals and families affected by TBI. With continued research and a deeper grasp of the molecular and cellular interactions at play, the future of TBI management promises to be more informed, targeted, and effective across all age demographics.

References

1. Coburn K. Traumatic brain injury: the silent epidemic. AACN Clin Issues Crit Care Nurs. 1992;3(1):9–18. Epub 1992/02/01. https://doi.org/10.4037/15597768-1992-1002. PubMed PMID: 1554576.
2. Alkhaibary A, Alshalawi A, Althaqafi RMM, Alghuraybi AA, Basalamah A, Shammaa AM, Altalhy AA, Abdelrahman TM. Traumatic brain injury: a perspective on the silent epidemic. Cureus. 2021;13(5):e15318. Epub 2021/07/06. https://doi.org/10.7759/cureus.15318. PubMed PMID: 34221767; PMCID: PMC8238020.
3. Taylor CA, Bell JM, Breiding MJ, Xu L. Traumatic brain injury-related emergency department visits, hospitalizations, and deaths—United States, 2007 and 2013. MMWR Surveill Summ. 2017;66(9):1–16. Epub 2017/03/17. https://doi.org/10.15585/mmwr.ss6609a1. PubMed PMID: 28301451; PMCID: PMC5829835.
4. Whiteneck GG, Gerhart KA, Cusick CP. Identifying environmental factors that influence the outcomes of people with traumatic brain injury. J Head Trauma Rehabil. 2004;19(3):191–204. Epub 2004/07/13. https://doi.org/10.1097/00001199-200405000-00001. PubMed PMID: 15247842.
5. Capizzi A, Woo J, Verduzco-Gutierrez M. Traumatic brain injury: an overview of epidemiology, pathophysiology, and medical management. Med Clin North Am. 2020;104(2):213–38. Epub 2020/02/10. https://doi.org/10.1016/j.mcna.2019.11.001. PubMed PMID: 32035565.

6. Spikman JM, Timmerman ME, Milders MV, Veenstra WS, van der Naalt J. Social cognition impairments in relation to general cognitive deficits, injury severity, and prefrontal lesions in traumatic brain injury patients. J Neurotrauma. 2012;29(1):101–11. Epub 2011/09/22. https://doi.org/10.1089/neu.2011.2084. PubMed PMID: 21933011.
7. Stock AD, Gelb S, Pasternak O, Ben-Zvi A, Putterman C. The blood brain barrier and neuropsychiatric lupus: new perspectives in light of advances in understanding the neuroimmune interface. Autoimmun Rev. 2017;16(6):612–9. Epub 2017/04/22. https://doi.org/10.1016/j.autrev.2017.04.008. PubMed PMID: 28428121; PMCID: PMC6068398.
8. Zouikr I, Hasegawa-Ishii S, Shimada A. Editorial: Neuroimmune interface in health and diseases. Front Immunol. 2017;8:1315. Epub 2017/11/08. https://doi.org/10.3389/fimmu.2017.01315. PubMed PMID: 29109722; PMCID: PMC5660279.
9. Bouras M, Asehnoune K, Roquilly A. Immune modulation after traumatic brain injury. Front Med (Lausanne). 2022;9:995044. Epub 2022/12/20. https://doi.org/10.3389/fmed.2022.995044. PubMed PMID: 36530909; PMCID: PMC9751027.
10. Finch CE. Neurons, glia, and plasticity in normal brain aging. Neurobiol Aging. 2003;24 Suppl 1:S123–7; discussion S31. Epub 2003/06/28. https://doi.org/10.1016/s0197-4580(03)00051-4. PubMed PMID: 12829120.
11. Wirak GS, Florman J, Alkema MJ, Connor CW, Gabel CV. Age-associated changes to neuronal dynamics involve a disruption of excitatory/inhibitory balance in C. elegans. Elife. 2022;11. Epub 2022/06/16. https://doi.org/10.7554/eLife.72135. PubMed PMID: 35703498; PMCID: PMC9273219.
12. Loane DJ, Kumar A. Microglia in the TBI brain: the good, the bad, and the dysregulated. Exp Neurol. 2016;275 Pt 3(0 3):316–27. Epub 2015/09/08. https://doi.org/10.1016/j.expneurol.2015.08.018. PubMed PMID: 26342753; PMCID: PMC4689601.
13. Salas IH, Burgado J, Allen NJ. Glia: victims or villains of the aging brain? Neurobiol Dis. 2020;143:105008. Epub 2020/07/06. https://doi.org/10.1016/j.nbd.2020.105008. PubMed PMID: 32622920.
14. Simon AK, Hollander GA, McMichael A. Evolution of the immune system in humans from infancy to old age. Proc Biol Sci. 2015;282(1821):20143085. Epub 2015/12/25. https://doi.org/10.1098/rspb.2014.3085. PubMed PMID: 26702035; PMCID: PMC4707740.
15. Johnston MV. Clinical disorders of brain plasticity. Brain Dev. 2004;26(2):73–80. Epub 2004/03/24. https://doi.org/10.1016/s0387-7604(03)00102-5. PubMed PMID: 15036425.
16. Park DC, Bischof GN. The aging mind: neuroplasticity in response to cognitive training. Dialogues Clin Neurosci. 2013;15(1):109–19. Epub 2013/04/12. https://doi.org/10.31887/DCNS.2013.15.1/dpark. PubMed PMID: 23576894; PMCID: PMC3622463.
17. Dennis M, Spiegler BJ, Juranek JJ, Bigler ED, Snead OC, Fletcher JM. Age, plasticity, and homeostasis in childhood brain disorders. Neurosci Biobehav Rev. 2013;37(10 Pt 2):2760–73. Epub 2013/10/08. https://doi.org/10.1016/j.neubiorev.2013.09.010. PubMed PMID: 24096190; PMCID: PMC3859812.
18. Hukkelhoven CW, Steyerberg EW, Rampen AJ, Farace E, Habbema JD, Marshall LF, Murray GD, Maas AI. Patient age and outcome following severe traumatic brain injury: an analysis of 5600 patients. J Neurosurg. 2003;99(4):666–73. Epub 2003/10/22. https://doi.org/10.3171/jns.2003.99.4.0666. PubMed PMID: 14567601.
19. Dhandapani S, Manju D, Sharma B, Mahapatra A. Prognostic significance of age in traumatic brain injury. J Neurosci Rural Pract. 2012;3(2):131–5. Epub 2012/08/07. https://doi.org/10.4103/0976-3147.98208. PubMed PMID: 22865961; PMCID: PMC3409980.
20. Erlebach R, Pagnamenta A, Klinzing S, Stretti F, Cottini S, Schüpbach R, Steiger P, Brandi G. Age-related outcome of patients after traumatic brain injury: a single-center observation. Minerva Anestesiol. 2017;83(11):1169–77. Epub 2017/06/24. https://doi.org/10.23736/s0375-9393.17.11837-7. PubMed PMID: 28643996.
21. Gardner RC, Dams-O'Connor K, Morrissey MR, Manley GT. Geriatric traumatic brain injury: epidemiology, outcomes, knowledge gaps, and future directions. J Neurotrauma.

2018;35(7):889–906. Epub 2017/12/08. https://doi.org/10.1089/neu.2017.5371. PubMed PMID: 29212411; PMCID: PMC5865621.

22. Skaansar O, Tverdal C, Rønning PA, Skogen K, Brommeland T, Røise O, Aarhus M, Andelic N, Helseth E. Traumatic brain injury—the effects of patient age on treatment intensity and mortality. BMC Neurol. 2020;20(1):376. https://doi.org/10.1186/s12883-020-01943-6.
23. Ciurea AV, Gorgan MR, Tascu A, Sandu AM, Rizea RE. Traumatic brain injury in infants and toddlers, 0–3 years old. J Med Life. 2011;4(3):234–43. Epub 2012/05/09. PubMed PMID: 22567045; PMCID: PMC3168813.
24. Fraunberger E, Esser MJ. Neuro-inflammation in pediatric traumatic brain injury-from mechanisms to inflammatory networks. Brain Sci. 2019;9(11). Epub 2019/11/14. https://doi.org/10.3390/brainsci9110319. PubMed PMID: 31717597; PMCID: PMC6895990.
25. Araki T, Yokota H, Morita A. Pediatric traumatic brain injury: characteristic features, diagnosis, and management. Neurol Med Chir (Tokyo). 2017;57(2):82–93. Epub 2017/01/24. https://doi.org/10.2176/nmc.ra.2016-0191. PubMed PMID: 28111406; PMCID: PMC5341344 of interest.
26. Anderson V, Catroppa C, Morse S, Haritou F, Rosenfeld J. Functional plasticity or vulnerability after early brain injury? Pediatrics. 2005;116(6):1374–82. Epub 2005/12/03. https://doi.org/10.1542/peds.2004-1728. PubMed PMID: 16322161.
27. Serpa RO, Ferguson L, Larson C, Bailard J, Cooke S, Greco T, Prins ML. Pathophysiology of pediatric traumatic brain injury. Front Neurol. 2021;12:696510. Epub 2021/08/03. https://doi.org/10.3389/fneur.2021.696510. PubMed PMID: 34335452; PMCID: PMC8319243.
28. Hannet I, Erkeller-Yuksel F, Lydyard P, Deneys V, DeBruyère M. Developmental and maturational changes in human blood lymphocyte subpopulations. Immunol Today. 1992;13(6):215, 8. Epub 1992/06/01. https://doi.org/10.1016/0167-5699(92)90157-3. PubMed PMID: 1627249.
29. Aldrimer M, Ridefelt P, Rödöö P, Niklasson F, Gustafsson J, Hellberg D. Population-based pediatric reference intervals for hematology, iron and transferrin. Scand J Clin Lab Invest. 2013;73(3):253–61. Epub 2013/03/02. https://doi.org/10.3109/00365513.2013.769625. PubMed PMID: 23448533.
30. Ygberg S, Nilsson A. The developing immune system—from foetus to toddler. Acta Paediatr. 2012;101(2):120–7. Epub 2011/10/19. https://doi.org/10.1111/j.1651-2227.2011.02494.x. PubMed PMID: 22003882.
31. Strunk T, Temming P, Gembruch U, Reiss I, Bucsky P, Schultz C. Differential maturation of the innate immune response in human fetuses. Pediatr Res. 2004;56(2):219–26. Epub 2004/06/08. https://doi.org/10.1203/01.Pdr.0000132664.66975.79. PubMed PMID: 15181184.
32. Kollmann TR, Crabtree J, Rein-Weston A, Blimkie D, Thommai F, Wang XY, Lavoie PM, Furlong J, Fortuno ES, 3rd, Hajjar AM, Hawkins NR, Self SG, Wilson CB. Neonatal innate TLR-mediated responses are distinct from those of adults. J Immunol. 2009;183(11):7150–60. Epub 2009/11/18. https://doi.org/10.4049/jimmunol.0901481. PubMed PMID: 19917677; PMCID: PMC4556237.
33. Wilson CB. Immunologic basis for increased susceptibility of the neonate to infection. J Pediatr. 1986;108(1):1–12. Epub 1986/01/01. https://doi.org/10.1016/s0022-3476(86)80761-2. PubMed PMID: 3511202.
34. Cadoz M. Potential and limitations of polysaccharide vaccines in infancy. Vaccine. 1998;16(14–15):1391–5. Epub 1998/08/26. https://doi.org/10.1016/s0264-410x(98)00097-8. PubMed PMID: 9711777.
35. Carsetti R, Quintarelli C, Quinti I, Piano Mortari E, Zumla A, Ippolito G, Locatelli F. The immune system of children: the key to understanding SARS-CoV-2 susceptibility? Lancet Child Adolesc Health. 2020;4(6):414–6. Epub 2020/05/28. https://doi.org/10.1016/s2352-4642(20)30135-8. PubMed PMID: 32458804; PMCID: PMC7202830.

36. West LJ. Defining critical windows in the development of the human immune system. Hum Exp Toxicol. 2002;21(9–10):499–505. Epub 2002/12/03. https://doi.org/10.1191/0960327102ht288oa. PubMed PMID: 12458907.
37. Cooper MA, Fehniger TA, Caligiuri MA. The biology of human natural killer-cell subsets. Trends Immunol. 2001;22(11):633–40. https://doi.org/10.1016/s1471-4906(01)02060-9. PubMed PMID: 11698225.
38. Dalbeth N, Gundle R, Davies RJ, Lee YC, McMichael AJ, Callan MF. CD56bright NK cells are enriched at inflammatory sites and can engage with monocytes in a reciprocal program of activation. J Immunol. 2004;173(10):6418–26. https://doi.org/10.4049/jimmunol.173.10.6418. PubMed PMID: 15528382.
39. Pinke KH, Calzavara B, Faria PF, do Nascimento MP, Venturini J, Lara VS. Proinflammatory profile of in vitro monocytes in the ageing is affected by lymphocytes presence. Immun Ageing. 2013;10(1):22. Epub 20130608. https://doi.org/10.1186/1742-4933-10-22. PubMed PMID: 23758797; PMCID: PMC3707736.
40. Nyugen J, Agrawal S, Gollapudi S, Gupta S. Impaired functions of peripheral blood monocyte subpopulations in aged humans. J Clin Immunol. 2010;30(6):806–13. Epub 20100812. https://doi.org/10.1007/s10875-010-9448-8. PubMed PMID: 20703784; PMCID: PMC2970801.
41. Chen SH, Tian DY, Shen YY, Cheng Y, Fan DY, Sun HL, He CY, Sun PY, Bu XL, Zeng F, Liu J, Deng J, Xu ZQ, Chen Y, Wang YJ. Amyloid-beta uptake by blood monocytes is reduced with ageing and Alzheimer's disease. Transl Psychiatry. 2020;10(1):423. Epub 20201208. https://doi.org/10.1038/s41398-020-01113-9. PubMed PMID: 33293506; PMCID: PMC7722845.
42. De Maeyer RPH, Chambers ES. The impact of ageing on monocytes and macrophages. Immunol Lett. 2021;230:1–10. Epub 20201210. https://doi.org/10.1016/j.imlet.2020.12.003. PubMed PMID: 33309673.
43. Harry GJ. Microglia during development and aging. Pharmacol Ther. 2013;139(3):313–26. Epub 20130430. https://doi.org/10.1016/j.pharmthera.2013.04.013. PubMed PMID: 23644076; PMCID: PMC3737416.
44. Bachstetter AD, Rowe RK, Kaneko M, Goulding D, Lifshitz J, Van Eldik LJ. The p38α MAPK regulates microglial responsiveness to diffuse traumatic brain injury. J Neurosci. 2013;33(14):6143–53. Epub 2013/04/05. https://doi.org/10.1523/jneurosci.5399-12.2013. PubMed PMID: 23554495; PMCID: PMC3667712.
45. Hsieh CL, Niemi EC, Wang SH, Lee CC, Bingham D, Zhang J, Cozen ML, Charo I, Huang EJ, Liu J, Nakamura MC. CCR2 deficiency impairs macrophage infiltration and improves cognitive function after traumatic brain injury. J Neurotrauma. 2014;31(20):1677–88. Epub 2014/05/09. https://doi.org/10.1089/neu.2013.3252. PubMed PMID: 24806994; PMCID: PMC4545982.
46. Shi K, Zhang J, Dong JF, Shi FD. Dissemination of brain inflammation in traumatic brain injury. Cell Mol Immunol. 2019;16(6):523–30. Epub 2019/03/09. https://doi.org/10.1038/s41423-019-0213-5. PubMed PMID: 30846842; PMCID: PMC6804599.
47. Balança B, Desmurs L, Grelier J, Perret-Liaudet A, Lukaszewicz AC. DAMPs and RAGE pathophysiology at the acute phase of brain injury: an overview. Int J Mol Sci. 2021;22(5). Epub 2021/03/07. https://doi.org/10.3390/ijms22052439. PubMed PMID: 33670976; PMCID: PMC7957733.
48. Roth TL, Nayak D, Atanasijevic T, Koretsky AP, Latour LL, McGavern DB. Transcranial amelioration of inflammation and cell death after brain injury. Nature. 2014;505(7482):223–8. Epub 2013/12/10. https://doi.org/10.1038/nature12808. PubMed PMID: 24317693; PMCID: PMC3930079.
49. Postolache TT, Wadhawan A, Can A, Lowry CA, Woodbury M, Makkar H, Hoisington AJ, Scott AJ, Potocki E, Benros ME, Stiller JW. Inflammation in Traumatic Brain Injury. J Alzheimers Dis. 2020;74(1):1–28. Epub 2020/03/17. https://doi.org/10.3233/jad-191150. PubMed PMID: 32176646; PMCID: PMC8190673.

50. Giza CC, Prins ML. Is being plastic fantastic? Mechanisms of altered plasticity after developmental traumatic brain injury. Dev Neurosci. 2006;28(4–5):364–79. Epub 2006/09/01. https://doi.org/10.1159/000094163. PubMed PMID: 16943660; PMCID: PMC4297630.
51. Babikian T, Asarnow R. Neurocognitive outcomes and recovery after pediatric TBI: meta-analytic review of the literature. Neuropsychology. 2009;23(3):283–96. Epub 2009/05/06. https://doi.org/10.1037/a0015268. PubMed PMID: 19413443; PMCID: PMC4064005.
52. Anderson V, Catroppa C, Morse S, Haritou F, Rosenfeld J. Recovery of intellectual ability following traumatic brain injury in childhood: impact of injury severity and age at injury. Pediatr Neurosurg. 2000;32(6):282–90. Epub 2000/09/06. https://doi.org/10.1159/000028956. PubMed PMID: 10971189.
53. Wade SL, Kaizar EE, Narad ME, Zang H, Kurowski BG, Miley AE, Moscato EL, Aguilar JM, Yeates KO, Taylor HG, Zhang N. Behavior problems following childhood TBI: the role of sex, age, and time since injury. J Head Trauma Rehabil. 2020;35(5):E393–e404. Epub 2020/02/29. https://doi.org/10.1097/htr.0000000000000567. PubMed PMID: 32108717; PMCID: PMC7483173.
54. Bach AM, Xie W, Piazzoli L, Jensen SKG, Afreen S, Haque R, Petri WA, Nelson CA. Systemic inflammation during the first year of life is associated with brain functional connectivity and future cognitive outcomes. Dev Cogn Neurosci. 2022;53:101041. Epub 2022/01/02. https://doi.org/10.1016/j.dcn.2021.101041. PubMed PMID: 34973509; PMCID: PMC8728426.
55. Devi Y, Khan S, Rana P, Dhandapani M, Ghai S, Gopichandran L, Dhandapani S. Cognitive, behavioral, and functional impairments among traumatic brain injury survivors: impact on caregiver burden. J Neurosci Rural Pract. 2020;11(4):629–35. Epub 2020/11/05. https://doi.org/10.1055/s-0040-1716777. PubMed PMID: 33144802; PMCID: PMC7595773.
56. Albrecht JS, Kiptanui Z, Tsang Y, Khokhar B, Liu X, Simoni-Wastila L, Zuckerman IH. Depression among older adults after traumatic brain injury: a national analysis. Am J Geriatr Psychiatry. 2015;23(6):607–14. Epub 2014/08/27. https://doi.org/10.1016/j.jagp.2014.07.006. PubMed PMID: 25154547; PMCID: PMC4306647.
57. Dikmen SS, Machamer JE, Powell JM, Temkin NR. Outcome 3 to 5 years after moderate to severe traumatic brain injury. Arch Phys Med Rehabil. 2003;84(10):1449–57. Epub 2003/10/31. https://doi.org/10.1016/s0003-9993(03)00287-9. PubMed PMID: 14586911.
58. Simon DW, McGeachy MJ, Bayır H, Clark RS, Loane DJ, Kochanek PM. The far-reaching scope of neuroinflammation after traumatic brain injury. Nat Rev Neurol. 2017;13(3):171–91. Epub 2017/02/12. https://doi.org/10.1038/nrneurol.2017.13. PubMed PMID: 28186177; PMCID: PMC5675525.
59. Padmakumar S, Kulkarni P, Ferris CF, Bleier BS, Amiji MM. Traumatic brain injury and the development of parkinsonism: understanding pathophysiology, animal models, and therapeutic targets. Biomed Pharmacother. 2022;149:112812. Epub 2022/03/16. https://doi.org/10.1016/j.biopha.2022.112812. PubMed PMID: 35290887; PMCID: PMC9050934.
60. Heavener KS, Bradshaw EM. The aging immune system in Alzheimer's and Parkinson's diseases. Semin Immunopathol. 2022;44(5):649–57. Epub 2022/05/04. https://doi.org/10.1007/s00281-022-00944-6. PubMed PMID: 35505128; PMCID: PMC9519729.
61. Houle S, Kokiko-Cochran ON. A Levee to the flood: pre-injury neuroinflammation and immune stress influence traumatic brain injury outcome. Front Aging Neurosci. 2021;13:788055. Epub 2022/02/01. https://doi.org/10.3389/fnagi.2021.788055. PubMed PMID: 35095471; PMCID: PMC8790486.
62. Johnson VE, Stewart W, Smith DH. Traumatic brain injury and amyloid-β pathology: a link to Alzheimer's disease? Nat Rev Neurosci. 2010;11(5):361–70. Epub 2010/03/11. https://doi.org/10.1038/nrn2808. PubMed PMID: 20216546; PMCID: PMC3979339.
63. Roberts GW, Gentleman SM, Lynch A, Murray L, Landon M, Graham DI. Beta amyloid protein deposition in the brain after severe head injury: implications for the pathogenesis of Alzheimer's disease. J Neurol Neurosurg Psychiatry. 1994;57(4):419–25. Epub 1994/04/01. https://doi.org/10.1136/jnnp.57.4.419. PubMed PMID: 8163989; PMCID: PMC1072869.

64. Lin Y, Wen L. Inflammatory response following diffuse axonal injury. Int J Med Sci. 2013;10(5):515–21. Epub 2013/03/28. https://doi.org/10.7150/ijms.5423. PubMed PMID: 23532682; PMCID: PMC3607236.
65. Stefanis L. α-Synuclein in Parkinson's disease. Cold Spring Harb Perspect Med. 2012;2(2):a009399. Epub 2012/02/23. https://doi.org/10.1101/cshperspect.a009399. PubMed PMID: 22355802; PMCID: PMC3281589.
66. Delic V, Beck KD, Pang KCH, Citron BA. Biological links between traumatic brain injury and Parkinson's disease. Acta Neuropathol Commun. 2020;8(1):45. Epub 2020/04/09. https://doi.org/10.1186/s40478-020-00924-7. PubMed PMID: 32264976; PMCID: PMC7137235 of the Department of Veterans Affairs or the United States Government and the opinions, interpretations, conclusions and recommendations are those of the authors and are not necessarily endorsed by the Department of Defense.
67. Harms AS, Delic V, Thome AD, Bryant N, Liu Z, Chandra S, Jurkuvenaite A, West AB. α-Synuclein fibrils recruit peripheral immune cells in the rat brain prior to neurodegeneration. Acta Neuropathol Commun. 2017;5(1):85. Epub 2017/11/23. https://doi.org/10.1186/s40478-017-0494-9. PubMed PMID: 29162163; PMCID: PMC5698965.
68. Palchuk MB, London JW, Perez-Rey D, Drebert ZJ, Winer-Jones JP, Thompson CN, Esposito J, Claerhout B. A global federated real-world data and analytics platform for research. JAMIA Open. 2023;6(2):ooad035. Epub 20230513. https://doi.org/10.1093/jamiaopen/ooad035. PubMed PMID: 37193038; PMCID: PMC10182857.
69. Arndt DH, Goodkin HP, Giza CC. Early posttraumatic seizures in the pediatric population. J Child Neurol. 2016;31(1):46–56. Epub 20150106. https://doi.org/10.1177/0883073814562249. PubMed PMID: 25564481.
70. Rana A, Musto AE. The role of inflammation in the development of epilepsy. J Neuroinflammation. 2018;15(1):144. Epub 20180515. https://doi.org/10.1186/s12974-018-1192-7. PubMed PMID: 29764485; PMCID: PMC5952578.
71. Klein P, Dingledine R, Aronica E, Bernard C, Blumcke I, Boison D, Brodie MJ, Brooks-Kayal AR, Engel J, Jr., Forcelli PA, Hirsch LJ, Kaminski RM, Klitgaard H, Kobow K, Lowenstein DH, Pearl PL, Pitkanen A, Puhakka N, Rogawski MA, Schmidt D, Sillanpaa M, Sloviter RS, Steinhauser C, Vezzani A, Walker MC, Loscher W. Commonalities in epileptogenic processes from different acute brain insults: do they translate? Epilepsia. 2018;59(1):37–66. Epub 20171215. https://doi.org/10.1111/epi.13965. PubMed PMID: 29247482; PMCID: PMC5993212.
72. Webster KM, Sun M, Crack PJ, O'Brien TJ, Shultz SR, Semple BD. Age-dependent release of high-mobility group box protein-1 and cellular neuroinflammation after traumatic brain injury in mice. J Comp Neurol. 2019;527(6):1102–17. Epub 20181230. https://doi.org/10.1002/cne.24589. PubMed PMID: 30499129.
73. Morganti-Kossmann MC, Rancan M, Stahel PF, Kossmann T. Inflammatory response in acute traumatic brain injury: a double-edged sword. Curr Opin Crit Care. 2002;8(2):101–5. https://doi.org/10.1097/00075198-200204000-00002. PubMed PMID: 12386508.
74. Shohami E, Bass R, Wallach D, Yamin A, Gallily R. Inhibition of tumor necrosis factor alpha (TNFalpha) activity in rat brain is associated with cerebroprotection after closed head injury. J Cereb Blood Flow Metab. 1996;16(3):378–84. https://doi.org/10.1097/00004647-199605000-00004. PubMed PMID: 8621742.
75. Maroso M, Balosso S, Ravizza T, Liu J, Aronica E, Iyer AM, Rossetti C, Molteni M, Casalgrandi M, Manfredi AA, Bianchi ME, Vezzani A. Toll-like receptor 4 and high-mobility group box-1 are involved in ictogenesis and can be targeted to reduce seizures. Nat Med. 2010;16(4):413–9. Epub 20100328. https://doi.org/10.1038/nm.2127. PubMed PMID: 20348922.
76. Woodcock T, Morganti-Kossmann MC. The role of markers of inflammation in traumatic brain injury. Front Neurol. 2013;4:18. Epub 20130304. https://doi.org/10.3389/fneur.2013.00018. PubMed PMID: 23459929; PMCID: PMC3586682.

77. Zhou Y, Tang H, Liu J, Dong J, Xiong H. Chemokine CCL2 modulation of neuronal excitability and synaptic transmission in rat hippocampal slices. J Neurochem. 2011;116(3):406–14. Epub 20101213. https://doi.org/10.1111/j.1471-4159.2010.07121.x. PubMed PMID: 21105875; PMCID: PMC3018532.
78. Cerri C, Genovesi S, Allegra M, Pistillo F, Puntener U, Guglielmotti A, Perry VH, Bozzi Y, Caleo M. The chemokine CCL2 mediates the seizure-enhancing effects of systemic inflammation. J Neurosci. 2016;36(13):3777–88. https://doi.org/10.1523/JNEUROSCI.0451-15.2016. PubMed PMID: 27030762; PMCID: PMC6601741.
79. Alam A, Thelin EP, Tajsic T, Khan DZ, Khellaf A, Patani R, Helmy A. Cellular infiltration in traumatic brain injury. J Neuroinflammation. 2020;17(1):328. Epub 20201103. https://doi.org/10.1186/s12974-020-02005-x. PubMed PMID: 33143727; PMCID: PMC7640704.
80. Araujo DM, Cotman CW. Differential effects of interleukin-1 beta and interleukin-2 on glia and hippocampal neurons in culture. Int J Dev Neurosci. 1995;13(3–4):201–12. https://doi.org/10.1016/0736-5748(94)00072-b. PubMed PMID: 7572276.
81. Diamond ML, Ritter AC, Failla MD, Boles JA, Conley YP, Kochanek PM, Wagner AK. IL-1beta associations with posttraumatic epilepsy development: a genetics and biomarker cohort study. Epilepsia. 2015;56(7):991–1001. https://doi.org/10.1111/epi.13100. PubMed PMID: 26149793.
82. Giza CC, Mink RB, Madikians A. Pediatric traumatic brain injury: not just little adults. Curr Opin Crit Care. 2007;13(2):143–52. Epub 2007/03/01. https://doi.org/10.1097/MCC.0b013e32808255dc. PubMed PMID: 17327734.
83. Farrington L, Vance H, Rek J, Prahl M, Jagannathan P, Katureebe A, Arinaitwe E, Kamya MR, Dorsey G, Feeney ME. Both inflammatory and regulatory cytokine responses to malaria are blunted with increasing age in highly exposed children. Malar J. 2017;16(1):499. Epub 2017/12/30. https://doi.org/10.1186/s12936-017-2148-6. PubMed PMID: 29284469; PMCID: PMC5747142.
84. Lilic D, Cant AJ, Abinun M, Calvert JE, Spickett GP. Cytokine production differs in children and adults. Pediatr Res. 1997;42(2):237–40. Epub 1997/08/01. https://doi.org/10.1203/00006450-199708000-00018. PubMed PMID: 9262229.
85. Garay PA, Hsiao EY, Patterson PH, McAllister AK. Maternal immune activation causes age- and region-specific changes in brain cytokines in offspring throughout development. Brain Behav Immun. 2013;31:54–68. Epub 20120725. https://doi.org/10.1016/j.bbi.2012.07.008. PubMed PMID: 22841693; PMCID: PMC3529133.
86. Chhor V, Moretti R, Le Charpentier T, Sigaut S, Lebon S, Schwendimann L, Ore MV, Zuiani C, Milan V, Josserand J, Vontell R, Pansiot J, Degos V, Ikonomidou C, Titomanlio L, Hagberg H, Gressens P, Fleiss B. Role of microglia in a mouse model of paediatric traumatic brain injury. Brain Behav Immun. 2017;63:197–209. Epub 20161104. https://doi.org/10.1016/j.bbi.2016.11.001. PubMed PMID: 27818218; PMCID: PMC5441571.
87. Tajiri N, Hernandez D, Acosta S, Shinozuka K, Ishikawa H, Ehrhart J, Diamandis T, Gonzales-Portillo C, Borlongan MC, Tan J, Kaneko Y, Borlongan CV. Suppressed cytokine expression immediately following traumatic brain injury in neonatal rats indicates an expeditious endogenous anti-inflammatory response. Brain Res. 2014;1559:65–71. Epub 20140303. https://doi.org/10.1016/j.brainres.2014.02.041. PubMed PMID: 24602693; PMCID: PMC4021707.
88. Nguyen A, Nguyen A, Hsu TI, Lew HD, Gupta N, Nguyen B, Mandavalli A, Diaz MJ, Lucke-Wold B. Neutrophil to lymphocyte ratio as a predictor of postoperative outcomes in traumatic brain injury: a systematic review and meta-analysis. diseases. 2023;11(1). Epub 2023/03/29. https://doi.org/10.3390/diseases11010051. PubMed PMID: 36975600; PMCID: PMC10047119.
89. Ghozy S, El-Qushayri AE, Varney J, Kacimi SEO, Bahbah EI, Morra ME, Shah J, Kallmes KM, Abbas AS, Elfil M, Alghamdi BS, Ashraf G, Alhabbab R, Dmytriw AA. The prognostic value of neutrophil-to-lymphocyte ratio in patients with traumatic brain injury: a systematic review. Front Neurol. 2022;13:1021877. Epub 2022/11/11. https://doi.org/10.3389/fneur.2022.1021877. PubMed PMID: 36353130; PMCID: PMC9638118.

90. Zhuang D, Sheng J, Peng G, Li T, Cai S, Din F, Li L, Huang M, Tian F, Li K, Wang S, Chen W. Neutrophil to lymphocyte ratio predicts early growth of traumatic intracerebral haemorrhage. Ann Clin Transl Neurol. 2021;8(8):1601–9. Epub 2021/06/25. https://doi.org/10.1002/acn3.51409. PubMed PMID: 34165245; PMCID: PMC8351393.
91. Mishra RK, Galwankar S, Gerber J, Jain A, Yunus M, Cincu R, Moscote-Salazar LR, Quiñones-Ossa GA, Agrawal A. Neutrophil-lymphocyte ratio as a predictor of outcome following traumatic brain injury: systematic review and meta-analysis. J Neurosci Rural Pract. 2022;13(4):618–35. Epub 2023/02/07. https://doi.org/10.25259/jnrp-2022-4-21. PubMed PMID: 36743744; PMCID: PMC9893942.
92. Marchese P, Lardone C, Canepele A, Biondi S, Roggi C, Massart F, Bonuccelli A, Peroni D, Giotta Lucifero A, Luzzi S, Foiadelli T, Orsini A. Pediatric traumatic brain injury: a new relation between outcome and neutrophil-to-lymphocyte ratio. Acta Biomed. 2022;92(S4):e2021417. Epub 2022/04/21. https://doi.org/10.23750/abm.v92iS4.12666. PubMed PMID: 35441607; PMCID: PMC9179059 consultancies, stock ownership, equity interest, patent/licensing arrangement etc.) that might pose a conflict of interest in connection with the submitted article.
93. Mukherjee S, Sivakumar G, Goodden JR, Tyagi AK, Chumas PD. Prognostic value of leukocytosis in pediatric traumatic brain injury. J Neurosurg Pediatr. 2020;27(3):335–45. Epub 2020/12/29. https://doi.org/10.3171/2020.7.Peds19627. PubMed PMID: 33361484.
94. Li W, Deng W. Platelet-to-lymphocyte ratio predicts short-term mortality in patients with moderate to severe traumatic brain injury. Sci Rep. 2022;12(1):13976. Epub 2022/08/18. https://doi.org/10.1038/s41598-022-18242-4. PubMed PMID: 35978006; PMCID: PMC9385644.
95. Alimohammadi E, Foroushani AZ, Moradi F, Ebrahimzadeh K, Nadersepahi MJ, Asadzadeh S, Amiri A, Hosseini S, Eden SV, Bagheri SR. Dynamics of neutrophil-to-lymphocyte ratio can be associated with clinical outcomes of children with moderate to severe traumatic brain injury: a retrospective observational study. Injury. 2022;53(3):999–1004. Epub 2021/10/10. https://doi.org/10.1016/j.injury.2021.09.052. PubMed PMID: 34625239.
96. Gunes M, Buyukgol H. Relationship between generalized epileptic seizure and neutrophil/lymphocyte ratio, platelet/lymphocyte ratio, and neutrophil mediated inflammation. Int J Neurosci. 2020;130(11):1095–100. Epub 20200206. https://doi.org/10.1080/00207454.2020.1722662. PubMed PMID: 31983256.
97. Siempis T, Georgalis PA, Lianos G, Kafritsas G, Metaxas D, Alexiou ES, Zika J, Sotiropoulos A, Alexiou GA, Voulgaris S. Blood Biomarkers for Prediction of Positive CT Findings in Mild Traumatic Brain Injury in Paediatric Population. J Integr Neurosci. 2023;22(4):91. Epub 2023/07/31. https://doi.org/10.31083/j.jin2204091. PubMed PMID: 37519178.
98. Ryan E, Kelly L, Stacey C, Huggard D, Duff E, McCollum D, Leonard A, Boran G, Doherty DR, Bolger T, Molloy EJ. Mild-to-severe traumatic brain injury in children: altered cytokines reflect severity. J Neuroinflammation. 2022;19(1):36. Epub 2022/02/09. https://doi.org/10.1186/s12974-022-02390-5. PubMed PMID: 35130911; PMCID: PMC8822689.
99. Nitta ME, Savitz J, Nelson LD, Teague TK, Hoelzle JB, McCrea MA, Meier TB. Acute elevation of serum inflammatory markers predicts symptom recovery after concussion. Neurology. 2019;93(5):e497-e507. Epub 2019/07/05. https://doi.org/10.1212/wnl.0000000000007864. PubMed PMID: 31270219; PMCID: PMC6693429.
100. Lo TY, Jones PA, Minns RA. Combining coma score and serum biomarker levels to predict unfavorable outcome following childhood brain trauma. J Neurotrauma. 2010;27(12):2139–45. Epub 2010/09/23. https://doi.org/10.1089/neu.2010.1387. PubMed PMID: 20858121.
101. O'Hare FM, Watson RW, O'Neill A, Segurado R, Sweetman D, Downey P, Mooney E, Murphy J, Donoghue V, Molloy EJ. Serial cytokine alterations and abnormal neuroimaging in newborn infants with encephalopathy. Acta Paediatr. 2017;106(4):561–7. Epub 2017/01/18. https://doi.org/10.1111/apa.13745. PubMed PMID: 28097694.
102. Yegin O. Chemotaxis in childhood. Pediatr Res. 1983;17(3):183–7. Epub 1983/03/01. https://doi.org/10.1203/00006450-198303000-00002. PubMed PMID: 6835722.

103. Saunders NR, Liddelow SA, Dziegielewska KM. Barrier mechanisms in the developing brain. Front Pharmacol. 2012;3:46. Epub 2012/04/06. https://doi.org/10.3389/fphar.2012.00046. PubMed PMID: 22479246; PMCID: PMC3314990.
104. Potts MB, Koh SE, Whetstone WD, Walker BA, Yoneyama T, Claus CP, Manvelyan HM, Noble-Haeusslein LJ. Traumatic injury to the immature brain: inflammation, oxidative injury, and iron-mediated damage as potential therapeutic targets. NeuroRx. 2006;3(2):143–53. Epub 2006/03/24. https://doi.org/10.1016/j.nurx.2006.01.006. PubMed PMID: 16554253; PMCID: PMC3593438.
105. McAllister TW. Genetic factors modulating outcome after neurotrauma. PM R 2010;2(12 Suppl 2):S241–52. Epub 2011/02/09. https://doi.org/10.1016/j.pmrj.2010.10.005. PubMed PMID: 21172686.
106. Magatti M, Pischiutta F, Ortolano F, Pasotti A, Caruso E, Cargnoni A, Papait A, Capuzzi F, Zoerle T, Carbonara M, Stocchetti N, Borsa S, Locatelli M, Erba E, Prati D, Silini AR, Zanier ER, Parolini O. Systemic immune response in young and elderly patients after traumatic brain injury. Immun Ageing. 2023;20(1):41. Epub 2023/08/13. https://doi.org/10.1186/s12979-023-00369-1. PubMed PMID: 37573338; PMCID: PMC10422735.
107. Akbar AN, Henson SM, Lanna A. Senescence of T lymphocytes: implications for enhancing human immunity. Trends Immunol. 2016;37(12):866–76. Epub 2016/10/11. https://doi.org/10.1016/j.it.2016.09.002. PubMed PMID: 27720177.
108. Doust YV, Rowe RK, Adelson PD, Lifshitz J, Ziebell JM. Age-at-injury determines the extent of long-term neuropathology and microgliosis after a diffuse brain injury in male rats. Front Neurol. 2021;12:722526. Epub 2021/09/28. https://doi.org/10.3389/fneur.2021.722526. PubMed PMID: 34566867; PMCID: PMC8455817.
109. Semple BD, Blomgren K, Gimlin K, Ferriero DM, Noble-Haeusslein LJ. Brain development in rodents and humans: identifying benchmarks of maturation and vulnerability to injury across species. Prog Neurobiol. 2013;106–107:1–16. Epub 2013/04/16. https://doi.org/10.1016/j.pneurobio.2013.04.001. PubMed PMID: 23583307; PMCID: PMC3737272.
110. Thompson HJ, Rivara F, Becker KJ, Maier R, Temkin N. Impact of aging on the immune response to traumatic brain injury (AIm:TBI) study protocol. Inj Prev. 2020;26(5):471–7. Epub 20190903. https://doi.org/10.1136/injuryprev-2019-043325. PubMed PMID: 31481600; PMCID: PMC8026101.
111. Moro F, Pischiutta F, Portet A, Needham EJ, Norton EJ, Ferdinand JR, Vegliante G, Sammali E, Pascente R, Caruso E, Micotti E, Tolomeo D, di Marco Barros R, Fraunberger E, Wang KKW, Esser MJ, Menon DK, Clatworthy MR, Zanier ER. Ageing is associated with maladaptive immune response and worse outcome after traumatic brain injury. Brain Commun. 2022;4(2):fcac036. Epub 20220216. https://doi.org/10.1093/braincomms/fcac036. PubMed PMID: 35350551; PMCID: PMC8947244.
112. Barker S, Paul BD, Pieper AA. Increased risk of aging-related neurodegenerative disease after traumatic brain injury. Biomedicines. 2023;11(4). Epub 20230411. https://doi.org/10.3390/biomedicines11041154. PubMed PMID: 37189772; PMCID: PMC10135798.
113. Babikian T, Merkley T, Savage RC, Giza CC, Levin H. Chronic aspects of pediatric traumatic brain injury: review of the literature. J Neurotrauma. 2015;32(23):1849–60. Epub 2015/09/29. https://doi.org/10.1089/neu.2015.3971. PubMed PMID: 26414654.
114. Kumar A, Zou L, Yuan X, Long Y, Yang K. N-methyl-D-aspartate receptors: transient loss of NR1/NR2A/NR2B subunits after traumatic brain injury in a rodent model. J Neurosci Res. 2002;67(6):781–6. Epub 2002/03/14. https://doi.org/10.1002/jnr.10181. PubMed PMID: 11891792.
115. Osteen CL, Moore AH, Prins ML, Hovda DA. Age-dependency of 45calcium accumulation following lateral fluid percussion: acute and delayed patterns. J Neurotrauma. 2001;18(2):141–62. Epub 2001/03/07. https://doi.org/10.1089/08977150150502587. PubMed PMID: 11229708.
116. Haroon E, Miller AH, Sanacora G. Inflammation, glutamate, and glia: a trio of trouble in mood disorders. Neuropsychopharmacology. 2017;42(1):193–215. Epub 20160915. https://doi.org/10.1038/npp.2016.199. PubMed PMID: 27629368; PMCID: PMC5143501.

117. Mahmoud S, Gharagozloo M, Simard C, Gris D. Astrocytes maintain glutamate homeostasis in the CNS by controlling the balance between glutamate uptake and release. Cells. 2019;8(2). Epub 20190220. https://doi.org/10.3390/cells8020184. PubMed PMID: 30791579; PMCID: PMC6406900.
118. Brickler TR, Hazy A, Guilhaume Correa F, Dai R, Kowalski EJA, Dickerson R, Chen J, Wang X, Morton PD, Whittington A, Ahmed A, Theus MH. Angiopoietin/Tie2 axis regulates the age-at-injury cerebrovascular response to traumatic brain injury. J Neurosci. 2018;38(45):9618–34. Epub 20180921. https://doi.org/10.1523/JNEUROSCI.0914-18.2018. PubMed PMID: 30242049; PMCID: PMC6222064.
119. Panchenko PE, Hippauf L, Konsman JP, Badaut J. Do astrocytes act as immune cells after pediatric TBI? Neurobiol Dis. 2023;185:106231. Epub 20230717. https://doi.org/10.1016/j.nbd.2023.106231. PubMed PMID: 37468048; PMCID: PMC10530000.
120. Shandra O, Winemiller AR, Heithoff BP, Munoz-Ballester C, George KK, Benko MJ, Zuidhoek IA, Besser MN, Curley DE, Edwards GF, 3rd, Mey A, Harrington AN, Kitchen JP, Robel S. Repetitive diffuse mild traumatic brain injury causes an atypical astrocyte response and spontaneous recurrent seizures. J Neurosci. 2019;39(10):1944–63. Epub 20190121. https://doi.org/10.1523/JNEUROSCI.1067-18.2018. PubMed PMID: 30665946; PMCID: PMC6407295.
121. Gudenschwager-Basso EK, Shandra O, Volanth T, Patel DC, Kelly C, Browning JL, Wei X, Harris EA, Mahmutovic D, Kaloss AM, Correa FG, Decker J, Maharathi B, Robel S, Sontheimer H, VandeVord PJ, Olsen ML, Theus MH. Atypical neurogenesis, astrogliosis, and excessive hilar interneuron loss are associated with the development of post-traumatic epilepsy. Cells. 2023;12(9). Epub 20230425. https://doi.org/10.3390/cells12091248. PubMed PMID: 37174647; PMCID: PMC10177146.
122. Rimaniol AC, Haik S, Martin M, Le Grand R, Boussin FD, Dereuddre-Bosquet N, Gras G, Dormont D. Na+-dependent high-affinity glutamate transport in macrophages. J Immunol. 2000;164(10):5430–8. https://doi.org/10.4049/jimmunol.164.10.5430. PubMed PMID: 10799909.
123. Claus CP, Tsuru-Aoyagi K, Adwanikar H, Walker B, Manvelyan H, Whetstone W, Noble-Haeusslein LJ. Age is a determinant of leukocyte infiltration and loss of cortical volume after traumatic brain injury. Dev Neurosci. 2010;32(5–6):454–65. Epub 2010/09/18. doi: 10.1159/000316805. PubMed PMID: 20847543; PMCID: PMC3073760.
124. Green TRF, Murphy SM, Ortiz JB, Rowe RK. Age-at-injury influences the glial response to traumatic brain injury in the cortex of male juvenile rats. Front Neurol. 2021;12:804139. Epub 2022/02/04. https://doi.org/10.3389/fneur.2021.804139. PubMed PMID: 35111130; PMCID: PMC8802670.
125. Wangler LM, Bray CE, Packer JM, Tapp ZM, Davis AC, O'Neil SM, Baetz K, Ouviña M, Witzel M, Godbout JP. Amplified gliosis and interferon-associated inflammation in the aging brain following diffuse traumatic brain injury. J Neurosci. 2022;42(48):9082–96. Epub 2022/10/19. https://doi.org/10.1523/jneurosci.1377-22.2022. PubMed PMID: 36257689; PMCID: PMC9732830.
126. Timaru-Kast R, Luh C, Gotthardt P, Huang C, Schäfer MK, Engelhard K, Thal SC. Influence of age on brain edema formation, secondary brain damage and inflammatory response after brain trauma in mice. PLoS One. 2012;7(8):e43829. Epub 2012/09/07. https://doi.org/10.1371/journal.pone.0043829. PubMed PMID: 22952778; PMCID: PMC3431406.
127. Bosco DB, Tian DS, Wu LJ. Neuroimmune interaction in seizures and epilepsy: focusing on monocyte infiltration. Febs J. 2020;287(22):4822–37. Epub 2020/05/31. https://doi.org/10.1111/febs.15428. PubMed PMID: 32473609; PMCID: PMC7817085.
128. Celorrio M, Shumilov K, Payne C, Vadivelu S, Friess SH. Acute minocycline administration reduces brain injury and improves long-term functional outcomes after delayed hypoxemia following traumatic brain injury. Acta Neuropathol Commun. Epub 2022/01/30. 2022;10(1):10. https://doi.org/10.1186/s40478-022-01310-1. PubMed PMID: 35090569; PMCID: PMC8796448.

Chapter 13
Age at Injury as a Modifier of Preclinical TBI Behavioral, Neuropathological, and Inflammatory Outcomes

Wai Hang Cheng, Mehwish Anwer, Jianjia Fan, Honor Cheung, Kevin Zhang, and Cheryl Wellington

Abstract Traumatic brain injury (TBI) is a leading cause of injury-related death and disability. In high-income countries, TBI is most prevalent among the older population (≥65 years), commonly caused by falls. Though age at injury is associated with increased risk of mortality and poor outcome, the underlying mechanisms are unclear. Studies in animal models may yield insights into the intersection of TBI with age. Here we review recent studies in animal models where TBI induced in aged animals is associated with exacerbated behavioral deficits (e.g., mortality, thigmotaxis, and cognitive deficits), neuropathology (microgliosis and astrogliosis), neuroinflammation (e.g., cytokines and iNOS), microglial alterations (e.g., more cellular vesicles and adopting a damage-associated microglia gene signature), and cell signaling and pathway changes (e.g., complement, phagocytosis, autophagy, trophic factor signaling). As relatively few preclinical studies focus on aged animals, more research is needed to fully understand the pathophysiology of TBI in the aged population. Particularly, we recommend that (1) more aged animals should be used, (2) closed-head TBI models should be considered, and (3) animal models of comorbidity or polytrauma should be considered.

Keywords Traumatic brain injury · Ageing · Age at injury · Age-dependent · Neuroinflammation · Behavior

---

W. H. Cheng · M. Anwer · J. Fan · H. Cheung · K. Zhang
University of British Columbia, Vancouver, BC, Canada
e-mail: chengwh@mail.ubc.ca; mehwish.anwer@ubc.ca; jianjia.fan@ubc.ca; honor.cheung@ubc.ca; kzhang17@mail.ubc.ca

C. Wellington (✉)
Department of Pathology and Laboratory Medicine, Djavad Mowafaghian Centre for Brain Health, International Collaboration on Repair Discoveries, School of Biomedical Engineering, University of British Columbia, Vancouver, BC, Canada
e-mail: cheryl.wellington@ubc.ca

L. J. Noble-Haeusslein, D. M. Schnyer (eds.), *Traumatic Brain Injuries*, Advances in Neurobiology 42, https://doi.org/10.1007/978-3-031-69832-3_13

Abbreviations

CHIMERA	Closed Head Impact Model of Engineered Rotational Acceleration
FPI	Fluid percussion injury
OH-CCI	Open-head Controlled Cortical Impact
PND	Postnatal day
scRNA-seq	Single cell RNA sequencing
TBI	Traumatic brain injury
WD	Weight drop

13.1 Introduction

Traumatic brain injury (TBI) occurs when the head experiences a physical force that leads to disrupted brain function. TBI has the highest incidence of all neurological disorders and is a leading cause of injury-related death and disability worldwide [1]. In high-income countries, TBI is common among older individuals (≥65 years), who are estimated to account for >30% of TBI-related hospital admissions. For example, in the USA, rates of TBI-related hospitalizations were 313 per 100,000 for adults ≥75 years and 105 per 100,000 for adults aged 65–74 years. Falls are the most common injury mechanism for these patients [2]. In contrast, motor vehicle accidents are the most common cause of TBI-related hospitalizations in low-income and middle-income countries such as India, where motor vehicle accidents account for up to 50% of moderate-severe TBI patients from 21 to 40 years [3]. These statistics highlight the fact that TBI affects people across the lifespan. However, as the majority of TBI studies typically focus on young to middle-aged cohorts, there is a paucity of data on TBI outcomes in older patients and a need to determine how TBI pathophysiology is modified by age at injury..

Since ageing is associated with several other health comorbidities such as metabolic disease, cerebrovascular disease, and frailty, it is not surprising that an older age at TBI is associated with increased risk of death, increased hospitalization, poorer recovery, and worse functional outcomes [4–9]. A meta-analysis involving 5600 severe TBI patients showed that every 10 years of age increased the odds ratio for death or unfavorable outcome by 1.4 [10]. Another clinical study involving 244 TBI patients (mild, moderate, and severe) showed that age above 30, 40, or 50 years is associated with unfavorable outcome, even after adjusting for Glasgow coma score at admission [11]. These studies clearly demonstrate that TBI outcome is age-dependent. However, the underlying mechanisms that exacerbate unfavorable TBI outcomes are unknown.

In preclinical studies, reports have shown that ageing may induce alterations in multiple pathways that may predispose to worse TBI outcomes. Examples of ageing-related alterations include neuronal and synaptic loss [12–14], upregulation of inflammation-related genes (e.g., transcriptional profiles of microglia and astrocytes, cytokine response, chemotaxis) [15], and mitochondrial dysfunction (e.g., release of reactive oxygen species and oxidation of cellular macromolecules) [16–19]. It is postulated that these factors exacerbate TBI pathophysiology through enhanced oxidative stress, inflammatory responses, perturbations in blood-brain barrier integrity, and compromised bioenergetic functionality [20–24]. This paper thus aims to review recent preclinical research on how age modifies post-TBI behavior, neuropathology, and inflammation.

13.2 Literature Review Method

A literature search using keywords was performed on PMC and PubMed. A total of 785 recent papers (<10 years) were returned, and a short list of 41 papers was generated after excluding articles that are redundant or irrelevant (Table 13.1). All of the 41 articles are rodent studies. We then established further selection criteria: the study must include wild-type animals or non disease-related transgenes (e.g., Thy1-YFP); the study must contain at least two age groups; the results must include aspects of behavior, pathology, or inflammation. A final list of 16 articles was selected for this review (Table 13.2).

Table 13.1 Literature review: search criteria and short-listed articles

Search #	Date	Search engine	Keyword	Total hits	First short list
1	2023-07-24	PMC	((((“TBI” OR “traumatic brain injury”) AND ((“aging“ OR “ageing“) OR ((“young” OR juvenile OR adolescent OR infant OR infantile) AND (“old” OR “mature” OR “aged” OR elderly OR senior OR (middle-age OR middle-aged OR middle age))) AND (“rodent” OR “mice” OR “rat” OR “murine” OR “mouse”) AND (preclinical OR pre-clinical) AND (“CHIMERA” OR “Blast” OR (“CHI” OR “closed head impact”) OR “OHI” OR (“fluid percussion” OR “FP”) OR (“controlled cortical impact“ OR “CCI”) OR “weight drop” OR “impact acceleration” OR (“mechanical shaking” OR “mechanical shake”)))) AND “author manuscript”[Filter] AND “last 10 years”[PDat]	631	16
2	2023-09-17	PubMed	(“TBI” OR “traumatic brain injury”) AND (aging OR aged) AND (rodent OR murine) AND ((“CHIMERA” OR “Blast” OR (“CHI” OR “closed head impact”) OR “OHI” OR (“fluid percussion” OR “FP”) OR (“controlled cortical impact” OR “CCI”) OR “weight drop” OR “impact acceleration” OR (“mechanical shaking” OR “mechanical shake”)) & publish date <10 years	154	25
Subtotal				785	41
Final list	16				

Table 13.2 A summary of articles reviewed

Article information			Animal information				TBI setting				
Authors	Year	DOI	Species	Strain/genotype	Sex	Age at injury	TBI device	Injury parameter	Severity (according to authors)	Anesthesia	Treatment
Tajiri et al. [25]	2014	https://doi.org/10.1523/JNEUROSCI.2425-13.2014	Rat	Fisher344	M	6 months vs 20 months	OH-CCI (right frontoparietal)	Craniotomy (4 mm radius); 6.0 m/s impact, 150 msec duration; 0.5 mm depth	Mild	Isoflurane	Human adipose-derived stem cell or conditioned media
Timaru-Kast et al. [26]	2015	https://doi.org/10.1089/neu.2014.3784	Mouse	C57BL/6 N	M	2 months vs 21 months	OH-CCI (right parietal)	Craniotomy (4 mm × 4 mm); 3 mm diameter impactor tip; 8 m/s impact; 150 msec duration; 1 mm brain penetration	Not specified	Sevoflurane	–
Glushakov et al. [27]	2016	https://doi.org/10.3389/fmolb.2016.00034	Mouse	C57BL/6	M	2–4 months vs 7–8 months	OH-CCI	Craniotomy; 3 mm impactor tip diameter; 3 m/s; 100 ms duration; 1 mm depth	Mild-moderate	Isoflurane	–
Gupta et al. [28]	2016	https://doi.org/10.1007/s12035-015-9287-y	Mouse	AVR	M	20 weeks vs 70 weeks	Weight-drop closed head injury	impactor 310 g, 3 mm end diameter, 3 cm drop height on the skull	Not specified	Not specified	–
Rowe et al. [29]	2016	https://doi.org/10.3171/2018.1.PEDS17449	Rat	Sprague Dawley Rat	M	PND17, PND35, 2 months, 4 months, 6 months	FPI (midline)	Craniotomy; age-adjusted pressure: 1.5, 1.9, 2.1, 2.2, 2.7 atm	Moderate	Isoflurane	–

Cheng et al. [30]	2018	https://doi.org/10.1016/j.expneurol.2017.12.007	Mouse	APP/PS1 (C3HxB6)	M	6 months vs 13 months	CHIMERA (direct)	2 × 0.5 J, 24 h apart	Mild	Isoflurane	–
			Mouse	WT (C3HxC57BL/6)	M	6 months vs 13 months	CHIMERA (direct)	2 × 0.5 J, 24 h apart	Mild	Isoflurane	–
Rowe et al. [31]	2018	https://doi.org/10.3171/2018.1.PEDS17449	Rat	Sprague Dawley	Not specified	PND17–PND35	FPI (midline)	Craniotomy; 1.3–1.5 atm (PND17); 1.8–2.0 atm (PND35); 1.9–2.2 atm (2-mo)	Mild	Isoflurane	–
Collins et al. [32]	2019	https://doi.org/10.1089/neu.2018.5982	Mouse	APP/PS1	M	3 months vs 6 months	FPI (lateral)	Craniotomy (3 mm); 1.251.35 atm	Mild-to-moderate	Isoflurane	–
			Mouse	WT (C57BL/6)	M	6 months	FPI (lateral)	Craniotomy (3 mm); 1.25–1.35 atm	Mild-to-moderate	Isoflurane	–
Ritzel et al. [33]	2019	https://doi.org/10.1016/j.neurobiolaging.2019.02.010	Mouse	C57BL/6	M	3 months vs 18 months	OH-CCI (left parietal)	Craniotomy (5 mm); 6 m/s; 1 mm depth; 50 ms dwell time	Moderate	Isoflurane	–
Sun et al. [34]	2019	https://doi.org/10.1016/j.bbi.2019.04.038	Rat	Wistar	M	10 weeks vs 1 year	FPI (lateral)	Craniotomy (5 mm); 2 atm	Moderate	Isoflurane	–
Timaru-Kast et al. [35]	2019	https://doi.org/10.3389/fnagi.2019.00063	Mouse	C57BL/6 N	M	2 months vs 21 months	OH-CCI (right parietal)	Craniotomy (4 mm × 4 mm); impactor tip 3 mm diameter, 8 m/s impact, 150 ms duration, 1 mm depth	Not specified	Midazolam	Candesartan (AT1 inhibitor)

(continued)

Table 13.2 (continued)

Article information			Animal information				TBI setting				
Authors	Year	DOI	Species	Strain/genotype	Sex	Age at injury	TBI device	Injury parameter	Severity (according to authors)	Anesthesia	Treatment
Zhu et al. [36]	2020	https://doi.org/10.1016/j.neurobiolaging.2020.03.023	Mouse	Thy1-YFPH (C57BL/6)	M	1.5 months vs 3 months vs 12 months	FPI (lateral)	1.4 atm	Mild	Isoflurane	Epothilone D (microtubule-stabilizing agent)
Islam et al. [37]	2021	https://doi.org/10.1016/j.expneurol.2021.113714	Mouse	C57BL/6	M	14 weeks vs 80 weeks	OH-CCI (left parietal)	Craniotomy (5 mm); 2.5 m/s; 2 mm depth; 0.1 s dwell time	Not specified	Ketamine + Xylazine	–
Ritzel et al. [38]	2022	https://doi.org/10.1007/s11357-022-00562-y	Mouse	C57BL/6	M	10–12 weeks vs 18 months	OH-CCI (left parietal)	Craniotomy (4 mm); 3.5 mm diameter tip; vel 3.9 m/s; 1.2 mm depth	Moderate	Isoflurane	Trehalose (autophagic enhancer)
Wangler et al. [39]	2022	https://doi.org/10.1523/JNEUROSCI.1377-22.2022	Mouse	C57BL/6	M	2 months vs 16–18 months	FPI (midline)	Craniotomy (2 mm); 1.2 atm	Moderate	Isoflurane	STING agonist DMXAA (IFN signaling enhancer)
Chen et al. [40]	2023	https://doi.org/10.1097/SHK.0000000000002027	Mouse	C57BL/6	M	14 weeks vs 80 weeks	OH-CCI (left parietal)	Craniotomy (5 mm); 3 mm impactor tip, 2.5 m/s, 0.1 s dwell time	Not specified	Ketamine + Xylazine	–

13.3 Age-Dependent TBI Outcomes: Behavioral Outcomes

In preclinical TBI studies in rodent models, common behavioral assessments include time of righting reflex, neurological severity score, motor performance (rotarod, grip strength), exploration (open field, zero maze), spatial learning and memory (Morris water maze, Barnes maze), associative fear memory (contextual, cued, or passive avoidance), and novelty recognition (novel object or place). The following section reviews recent publications that studied post-TBI behavioral deficits in non-transgenic mice by comparing young adult mice (young-TBI) to older adult mice (aged-TBI).

Our group previously published an age comparison study (6 months vs 13 months) using male mice on a C57Bl/6xC3H WT background [30]. We used the Closed Head Impact Model of Engineered Rotational Acceleration (CHIMERA) to induce two mild TBI (2x0.5 J, 24 h apart) at 6 or 13 months. We showed spatial learning (Barnes Maze at D8 to D12 post-TBI) was impaired only in the aged-TBI group but not in the young-TBI group. However, other post-TBI behaviors were not affected by age of the mice. For example, both aged-TBI and young-TBI mice showed similar loss of righting reflex (immediate post-TBI), neurological deficits (D1–D7 post-TBI), and motor performance (rotarod) (D1–D14 post-TBI).

Glushakov et al. [27] induced open-head Controlled Cortical Impact (OH-CCI) to male C57BL/6 mice (2–4 mo vs 7–8 mo). They showed that mice injured at 7–8 months had a higher neurological deficit score at 48 h than mice injured at 2–4 months.

Ritzel et al. [33] investigated the effect of age on TBI outcomes in male C57BL/6 mice (3 months vs 18 months), using a moderate level of OH-CCI. They showed that while the young-TBI group had only transient loss of % body weight at 1 day (D1) post-OH-CCI, the aged-TBI group showed sustained body weight loss (up to D3 post-injury). Using rotarod, Ritzel et al. found that aged mice had poorer motor performance and grip strength at baseline. After OH-CCI, both the aged-TBI and young-TBI groups showed impaired motor function and grip. Intriguingly, aged-TBI mice showed early recovery of motor performance (by D3 post-OH-CCI), an effect not observed in the young-TBI group. Nevertheless, the absolute grip strength of the aged-TBI group was poorest of all groups at D3 post-OH-CCI. Ritzel et al. tested locomotor activity using open field. Compared to young-TBI mice, aged-TBI mice showed significantly less distance travelled, slower average speed, more time spent immobile, and more thigmotactic behavior, suggesting age-related worsening of locomotor impairment and anxiety-like behavior.

Islam et al. [37] performed OH-CCI in male C57BL/6 mice (14-wk vs 80-wk). Using the zero maze (D31 post-OH-CCI), they showed that young-TBI mice spent more time in the open regions, suggesting disinhibition of anxiety-like behavior, which was absent in aged-TBI mice. In the open field task (D67 post-OH-CCI), thigmotactic behavior was observed in aged-TBI but not in young-TBI mice. During the Morris Water Maze task (D32 post-OH-CCI), both young-TBI and aged-TBI groups required a longer distance traveled to locate the hidden platform compared to sham mice, suggesting impaired spatial learning. In addition, the aged-TBI group showed greater impairment than young-TBI mice, as they had less proximity to the platform. In contextual and cued fear conditioning tasks (D93 post-OH-CCI), both young-TBI and aged-TBI showed decreased freezing behavior compared to sham controls. Specifically, in contextual learning, aged-TBI mice showed greater exacerbated deficits (less freezing) than young-TBI mice. In contrast, in the cued fear task, young-TBI mice showed a greater deficit than aged-TBI mice.

Wangler et al. [39] performed midline fluid percussion injury (FPI) in male C57BL/6 mice (2 months vs 16–18 months). They showed that aged-TBI mice had longer time of righting than young-TBI mice. In the novel object recognition (D6 post-FPI) and the novel object location (D7 post-FPI) tasks, both injury and age were found to be significant factors that reduced novelty exploration. However, post-hoc analyses did not show significant differences between aged-TBI and young-TBI groups.

Rowe et al. [29] induced midline fluid percussion injury (FPI) to male Sprague Dawley (SD) rats at various ages. The pressure was adjusted to achieve moderate severity: 1.5 atm (postnatal day (PND) 17), 1.9 atm (PND 35), 2.1 atm (2 months), 2.2 atm (4 months), and 2.7 atm (6 months). The behavioral performance of all groups was then compared at multiple ages. Compared to naïve rats, rats injured at PND 17 showed significantly more foot faults (beam walk) until 3 months of age (2.5 months post-FPI). Rats injured at PND 35 showed significantly more motor deficits until 5 months of age (4 months post-FPI). Rats injured at 2 months showed deficits until 3 months of age (1 month post-FPI). At 8 months of age, Rowe et al. assessed anxiety-like behavior and spatial memory deficits. Only rats injured at 2 months, 4 months, and 6 months showed thigmotactic behavior. Only rats injured at 17 days, 35 days, and 6 months showed deficits in the novel location task. No spatial memory deficits were observed at 10 months of age. Rowe et al. thus concluded that age at injury and time since injury both significantly modified post-injury behavior. TBI in juvenile/adolescent rats tended to induce motor and cognitive deficits, whereas TBI in adult rats tended to induce anxiety-like behavior. In a follow-up study [31], Rowe et al. induced midline FPI to SD rats at PND 17 (1.3–1.5 atm), PND 35 (1.8–2.0 atm), and 2 months (1.9–2.2 atm). Compared to other adult-TBI groups, juvenile-TBI rats (PND 17) induced a longer duration of apnea and a higher prevalence of seizure. The juvenile-TBI rats (PND 17 and PND 35) also had visible hematoma despite an intact dura, indicating that TBI at a juvenile/adolescent age may induce more complications than TBI in adulthood.

Sun et al. [34] induced moderate lateral FPI to male Wistar rats (10 weeks vs 1 year). Compared to young-TBI, aged-TBI rats showed greater motor deficits (composite neuromotor score at D2 and D4 post-FPI, beam testing at D5 post-FPI). Aged-TBI rats also had poorer locomotion (open field at D4). In the Morris water maze task, age and injury were both significant factors that contributed to longer search time.

In summary, many post-TBI deficits are exacerbated by older age at injury. Particularly, anxiety-like behavior (thigmotaxis) and cognitive deficits (spatial learning/memory) can be more pronounced in aged-TBI animals. Higher mortality was observed in aged-TBI groups in multiple OH-CCI studies [26, 35, 38] (Table 13.3).

Table 13.3 A summary of TBI-induced outcomes that are modified by age at injury

Assessment	TBI outcomes exacerbated by age at injury
Behavior	Spatial learning deficit (CHIMERA) [30] Neurological deficit score (OH-CCI) [27] Thigmotactic behavior and loss of righting reflex (midline FPI) [29, 31] Sustained body weight loss, deficits in grip strength, mobility, and thigmotactic behavior (OH-CCI) [33] Motor and locomotor deficits and thigmotactic behavior (OH-CCI) [34] Thigmotactic behavior and deficits in spatial learning and contextual fear learning (OH-CCI) [37] Loss of righting reflex (midline FPI) [39] Mortality (OH-CCI) [26, 35, 38]
Neuropathology	White matter microgliosis (CHIMERA) [30] Cortical cavitation (OH-CCI) [27] Shorter telomere length (lateral FPI) [34] Number of activated microglia in cortex (lateral FPI) [36] GFAP and Iba1 immunoactivity in cortex and hippocampus (midline FPI) [39]
Inflammation, transcriptomics, and gene pathway analyses	Earlier Il-6 response (OH-CCI) [26] Decreased angiotensin II receptor type 1a (At1a) by 40% (D3) in aged-TBI; higher expression of iNOS and eNOS; different temporal profile of Ym1 expression in microglia (OH-CCI) [35] Delayed interaction of transcription factors (NF-κB and N-myc) with GLT1/EAAT2 promoter; higher expression of Gfap and Aqp4 (WD) [28] Increased level of microglia and brain-infiltrating leukocytes; impaired microglial phagocytic activity; increased level of the senescence marker ($p16^{ink4a}$) in microglia; lymphocyte infiltration; ROS production in microglia; increased level of CD68, autophagosome, and lysosomes in microglia (OH-CCI) [33] Activation of pathways involved in oligodendrocyte function and neurons/neurotransmission (cortex); activation of pathways involved in insulin signaling and carbohydrate metabolism (hippocampus); pathways related to astrocyte function, innate immune response, epigenetic response, microglia function, angiogenesis, and autophagy (OH-CCI) [33] Activation of pathways related to neuroinflammation signaling, pathogen induced cytokine storm signaling, FAK signaling, phagosome formation, NO and ROS production; activation of master regulators involved in complement, cytokine/chemokines, NF-κ, and interferon, etc.; inhibition of master regulators involved in apolipoprotein E (apoE), transforming growth factor (TGF) receptor 1, corticosterone, etc. (midline FPI) [39] Weaker activation of inflammatory mediator genes; lack of activation of histone genes; activation of pathways related to leukocyte migration, chemotaxis, and BDNF signaling (OH-CCI) [37]

13.4 Age-Dependent TBI Outcomes: Neuropathology

Common neuropathologies examined in preclinical TBI studies include microgliosis, astrogliosis, axonal injury, neuronal loss, tissue loss, disruption of the blood-brain barrier (BBB), and cerebrovascular injury.

In our report using CHIMERA to induce two mild TBI [30] (6 months vs 13 months), we showed that TBI induced axonal injury at the optic tract (NeuroSilver at D14 post-TBI) in both aged-TBI and young-TBI WT mice. However, white matter inflammation (Iba1 at D7 post-TBI) was only significantly increased in aged-TBI but not young-TBI mice. In contrast, axonal bulb injury (neurofilament at D2 post-TBI) was only significantly increased in young-TBI but not aged-TBI mice.

In the OH-CCI study by Glushakov et al. [27] (2–4 months vs 7–8 months), they showed mice injured at 7–8 months had greater cortical cavitation than mice injured at 2–4 months.

Zhu et al. [36] performed mild lateral FPI in male Thy1-YFP-H mice on a C57BL/6 background (1.5 months (juvenile) vs 3 months (young adult) vs 12 months (middle-age adult)). This mouse line expresses high level of yellow fluorescent protein (YFP) in motor and sensory neurons, which acts as an endogenous Golgi-like label. Zhu et al. analyzed axonal degeneration by quantifying the percent area of circular axonal enlargements at the internal capsule. FPI induced significant axonal degeneration with different age-specific profiles: When compared to respective sham, 1.5-month TBI mice showed significant increase of axonal degeneration at 1 week post-FPI, which was resolved by 4 weeks post-FPI. In contrast, in the 3-month TBI and 12-month TBI groups, axonal degeneration was significantly increased at 4 weeks post-FPI but not at 1 week post-FPI. Overall Zhu et al. showed that juvenile mice developed axonopathy early (but also resolved earlier), whereas adult mice had a delayed manifestation of axonopathy.

Compared to respective sham controls, cortical astrogliosis (GFAP % area) was significantly increased in 3 month TBI and 12-month TBI but not in 1.5-mo-TBI groups, all measured at 1 week post-FPI. Astrogliosis in all age groups were resolved by 4 week post-FPI. Regarding cortical microglia, the 12-month group (both 12-month sham and 12-month TBI) had higher baseline Iba1-immunopositivity than the 1.5-month and 3-month groups. However, compared to respective sham, total Iba1 area was significantly increased in the 3-month TBI group but not in the 1.5-month TBI or 12-month TBI (1 week post-FPI) groups. Interestingly, when counting the number of activated microglia (hypertrophic morphology) only, there was a significant increase in the 12-month TBI but not in the 1.5-month TBI or 3-month TBI groups (1 week and 4 weeks post-FPI). Overall, Zhu et al. showed that TBI induced astrogliosis in adult mice only (resolved by 4 weeks). In addition, TBI induced an altered microglia profile in middle-aged adult mice.

In the OH-CCI study by Islam et al. [37] (14 weeks vs 80 weeks), they showed that at D30 post-OH-CCI, young-TBI mice had more hippocampal edema (H&E) and neuronal loss (NeuN) than aged-TBI mice. MRI evaluation (D30 post-OH-CCI) showed that TBI increased ventricle size and decreased fractional anisotropy in both age groups. However, young-TBI mice had greater ventricular enlargement and decreased FA than the aged-TBI group.

In the FPI study by Wangler et al. [39] (2 months vs 16–18 months), they performed immunohistological analysis at D7 post-FPI. In somatosensory cortex, both TBI groups displayed GFAP-immunopositive astrocytes and Iba1-immunopositive microglia. However, aged-TBI mice had greater GFAP and Iba1 immunoactivity than young-TBI mice. In the hippocampus, increased GFAP and Iba1 immunoactivity were observed in aged-TBI but not young-TBI mice.

In the rat FPI study by Sun et al. [34] (10 weeks vs 1 year), they showed that age and TBI were both significant factors that contributed to smaller cortical volumes (MRI at D7 post-FPI). In addition, TBI led to lower fractional anisotropy in both aged and young rats. Sun et al. also extracted genomic DNA from ear samples and performed telomere length analysis (24 h and 1 week). They showed that age, injury, and age * injury were both significant factors for telomere length. Specifically, young-TBI did not show shortened telomeres. However, aged-TBI mice had shorter telomere length when compared to young-TBI mice.

In summary, TBI neuropathologies such as microgliosis and astrogliosis tend to be exacerbated in aged-TBI. However, some TBI-induced axonal pathologies (neurofilament or YFP-filled axonal degeneration) may have different temporal profiles in young vs old mice, with young mice having earlier manifestation and resolution of axonal bulbs. Therefore, it will be important to compare TBI pathological outcomes at multiple time points to better understand how age modifies neuropathological responses to TBI (Table 13.3).

13.5 Age-Dependent TBI Outcomes: Inflammation, Transcriptomics, and Gene Pathway Analyses

Timar-Kast et al. [26] performed OH-CCI in male C57BL/6 N mice at 2 months and 21 months of age. They investigated the absolute copy number of eight housekeeping genes and showed that expressions of β2-microglobulin (B2m), β-actin (Actb), and porphobilinogen deaminase (Pbgd) were significantly increased in young-TBI and aged-TBI mice, whereas the expression of cyclophilin A (Ppia), glyceraldehyde-3-phosphate dehydrogenase (Gapdh), hypoxanthine ribosyltransferase (Hprt), S100b, and 18srRNA remained stable. In addition, B2m expression was age-dependent. Timar-Kast et al. also showed that while the expression of interleukin 6 (Il-6) increased in both TBI groups, the increase was earlier in aged-TBI (15 min post-TBI) compared to young-TBI (24 h post-TBI) mice.

In a follow-up study [35], Timaru-Kast et al. further showed that OH-CCI decreased the expression level of angiotensin II receptor type 1a (At1a) by 40% (D3) in aged-TBI but not in young-TBI mice. In contrast, At1b and At2 were induced in young-TBI but not in aged-TBI mice. Interestingly, aged-TBI mice had higher expression of inducible nitric oxide synthase (iNOS) and endothelial NOS (eNOS). In addition, OH-CCI induced gene expression of Il-1β, Il-6, and tumor necrosis factor (Tnfα), in both young-TBI and aged-TBI mice. However, at D1-post-OH-CCI, Tnfα and Il-6 levels were higher in young-TBI mice, but at D5-post-OH-CCI, Il-1β and Il-6 were higher in aged-TBI mice. At D5-post-OH-CCI, aged-TBI mice had 6–8 times higher T-lymphocyte infiltration ($CD3^+$) than young-TBI mice. Microglia from aged-TBI mice also had a different temporal profile of Ym1 (a marker of "M2a" microglia or microglia with a neuroregenerative phenotype). In young-TBI mice, Ym1 levels were highest at D1 post-OH-CCI; however, in aged-TBI mice, Ym1 levels were highest at D5-post-OH-CCI.

Gupta et al. [28] performed weight drop closed head TBI (WD) in male AVR mice (20 weeks vs 70 weeks). They performed electrophoretic mobility shift assays to study the interaction of NF-κB and N-myc with the promoter of astrocyte-associated glutamate transporter 1/excitatory amino acid transporter 2 (GLT1/EAAT2). They found that NF-κB and N-myc bound to the GLT1/EAAT2 promoter in both young-TBI and aged-TBI groups; however, the interactions occurred earlier in aged-TBI (1 h post-WD) than young-TBI (4 h post-WD) mice. This temporal difference may explain differences in temporal expression of GLT1/EAA2. In addition, using semi-quantitative RT-PCR, they showed that WD induced glial fibrillary acidic protein (Gfap) and aquaporin 4 (Aqp4) expression in both young-TBI and aged-TBI mice. However, aged-TBI mice had higher expression of Gfap (72 h post-WD) and Aqp4 (1 h) than young-TBI mice.

Ritzel et al. [33] induced OH-CCI in male C57BL/6 mice (3 months vs 18 months). Using flow cytometry, they showed that, at baseline, aged mice had less $CD45^{int}$ $CD11b^+$ $Ly6C^-$ microglia. OH-CCI acutely (D3) induced the number of microglia and infiltrating $CD45^{hi}$ leukocytes in both aged-TBI and young-TBI mice. However, the absolute number of microglia in aged-TBI mice was lower than that of young-TBI mice. Interestingly, when expressed as fold change over respective sham levels, the microglia count in aged-TBI mice was higher than that of young-TBI mice, suggesting a very dramatic change from baseline. Aged-TBI mice also had more brain-infiltrating leukocytes than young-TBI mice. Using a bead-engulfment assay, they showed that TBI-induced microglial phagocytic activity. However, aged-TBI mice showed impaired bead engulfment in both microglia and infiltrating myeloid cells ($CD45^{int}$ $CD11b^-$). Ritzel et al. then compared cytokine production in microglia. They found that age and injury were both significant factors that increased TNFα and IL-1β levels. However, infiltrating myeloid cells from aged-TBI mice had lower production of TNFα and IL-1β. They also showed that microglia from both young-TBI and aged-TBI had increased size and granularity, with those in young-TBI mice showing a greater fold change over respective sham. OH-CCI increased mRNA and protein levels of the senescence marker ($p16^{ink4a}$) in microglia, with aged-TBI mice expressing higher levels than young-TBI mice. Ritzel et al. also showed that OH-CCI-induced splenomegaly in young-TBI but not in aged-TBI mice.

In a further follow-up study, Ritzel et al. [38] reported the chronic (16 weeks) microglia profile after OH-CCI. They showed that at 48 h and 16 weeks post-injury, infiltration of lymphocyte ($CD45^{hi}$ $CD11b^{-}$) was greater in aged-TBI than young-TBI mice. Comparing absolute reactive oxygen species (ROS) level, microglia from aged mice produced higher ROS, both at baseline and 16 weeks post-OH-CCI. However, when compared to respective shams, microglia from aged-TBI mice had less fold change in increased ROS production. Ritzel et al. then investigated microglial phagocytosis. They found that ageing and TBI were significant factors that contributed to increased % of phagocytic microglia ($NeuN^{+}$, vesicular glutamate transport 1 ($vGlu1^{+}$), or $CD68^{+}$), as well as significantly contributing to lysosome (LysoTracker or Lamp1) and autophagosome (Cyto-ID or $p62^{+}$) content in microglia. Post-hoc analyses showed that, compared to young-TBI mice, microglia from aged-TBI mice had higher levels of CD68 (16 weeks post-OH-CCI), Cyto-ID autophagosome (48 h), and lysosomes (48 h).

Using NanoString analysis, Ritzel et al. showed that in both cortex and hippocampus, aged-TBI mice had a more distinct gene expression pattern than young-TBI, aged-sham, and young-sham mice. Specifically, in cortex, genes that facilitate oligodendrocyte function and neurons/neurotransmission were more likely to be activated in aged-TBI mice than in young-TBI mice. In hippocampus, genes related to insulin signaling and carbohydrate metabolism were more likely to be activated in aged-TBI mice. Other pathways modified by age include astrocyte function, innate immune response, epigenetic response, microglia function, angiogenesis, and autophagy. Interestingly, genes that were strongly upregulated by both age and chronic injury (Axl, CD14, Cd68, CD74, Ctss, Clec7a, Trem2, Tyrobp) are consistent with a disease-activated microglia (DAM) signature.

Wangler et al. [39] performed FPI in male C57BL/6 (2 months vs 16 ~ 18 months) mice. Using NanoString (D7 post-FPI), they first compared young-TBI vs young-sham and identified ten differentially expressed genes (DEG). They then compared aged-TBI vs aged-sham and identified 113 DEG. The following ten genes were found to be upregulated by TBI, regardless of age: Gfap, Clec7a, C1qa, C1qb, C1qc, Grn, Hexb, Aqp4, Mapk3, and Sirt2. Pathway analyses of these DEG show that both young-TBI and aged-TBI mice activated pattern recognition receptor signaling. Wangler et al. then compared aged-TBI vs young-TBI and identified 103 DEG, including Cxc10, Stat1, Ccl5, Ccl12, Proteasome 20S subunit β8 (Psmb8), etc. Pathway analyses of these DEG show that aged-TBI further activated genes involved in neuroinflammation signaling, pathogen-induced cytokine storm signaling, FAK signaling, phagosome formation, NO and ROS production, etc. These DEG pointed to a list of master regulators that were specifically activated by aged-TBI (complement, cytokine/chemokines, NF-κ, and interferon, etc.), as well as a list of master regulators that were specifically inhibited by aged-TBI (apolipoprotein E (apoE), transforming growth factor (TGF) receptor 1, corticosterone, etc.).

In a follow-up study, [37] Chen et al. [40] induced OH-CCI in male C57BL/6 (14 weeks vs 80 weeks) mice. By performing flow cytometry at a chronic time point (4 months post-OH-CCI), they showed that aged-TBI mice had greater infiltration of $CD3^{+}$, $CD4^{+}$, or $CD8^{+}$ T-cells compared to young-TBI mice. Using single cell RNA sequencing (scRNA-seq), they analyzed the transcriptome of microglia and found that microglia from aged and young mice map into distinct clusters. A subset of the aged microglia cluster showed the DAM gene signature and had high expression level of immune-inflammatory genes, including galectin (Lgals3), lysozyme2 (Lyz2), CD63, and macrophage migration inhibitory factor (Mif). Chen et al. suggested that this subset may indicate senescent and dysfunctional microglia, which ultimately lead to poor TBI outcomes. They then compared DEG from aged-TBI vs aged-sham to DEG from young-TBI vs young-sham groups and found that most genes upregulated in aged-TBI mice were also expressed in young-TBI mice. However, many inflammatory mediator genes (e.g., S100a8/a9 and CDC-like kinase 1 (Clk1)) in aged-TBI mice were weakly upregulated compared to the response in young-TBI mice. In addition, young-TBI mice specifically showed activated histone genes (Hist1he, Hist1h1c, Hist1h2bc). Pathway analysis showed that aged-TBI mice specifically induced genes related to leukocyte migration and chemotaxis, as well as BDNF signaling. Chen et al. then investigated BDNF levels in hippocampus and prefrontal cortex and found that BDNF levels were similar in young-TBI vs young sham groups but markedly decreased in aged-TBI mice.

In the rat FPI study by Sun et al. (10 weeks vs 1 year) [34], TBI induced total leukocytes ($CD45^{+}$), all myeloid cells ($CD45^{+}$ $OX82^{+}$)(both infiltrating myeloid cells ($CD45^{hi+}$ $OX82^{+}$) and microglia ($CD45^{lo+}$ $OX82^{+}$)) in brains of both aged-TBI and young-TBI rats. Compared to aged-TBI rats, young-TBI rats had more leukocytes and total and infiltrating myeloid cells.

In summary, recent technical advancements allow gene expression and pathway analysis to be studied using various technical approaches (e.g., qPCR, flow cytometry, NanoString, scRNA-seq, etc.). It is difficult to compare results across these studies that use different methods. However, in general, TBI in aged mice is typically associated with higher expression of proinflammatory genes (e.g., cytokines and iNOS). In addition, when compared to microglia from young-TBI mice, microglia from aged-TBI mice have a higher content of cellular vesicles (including phagosomes, autophagosomes, and lysosomes) while being functionally impaired (as shown in bead engulfment assays). Furthermore, ageing and TBI are both significant factors that contribute to converting microglia to adopt a DAM gene signature. These findings suggest that microglia from aged-TBI animals are more prone to induce oxidative stress and manifest deficits in phagocytosis and autophagy, thus leading to accumulation of cellular vesicles. Pathway analyses show that aged-TBI mice, when compared to young-TBI mice, tend to overactivate genes related to inflammation (e.g., complement, cytokine, chemokines, and phagocytosis). Aged-TBI mice also have potential impairments in activating pathways related to neuroregeneration (e.g., BDNF and TGF signaling) and epigenetic regulation of microglia. These changes are aligned with previous studies on gene expression profiles of ageing in the mouse brain (5 months vs 30 months) [41] and strengthen the hypothesis [42, 43] that TBI in aged mice further exacerbates the senescence status of the brain (Table 13.3).

13.6 Age-Dependent TBI Outcomes: Intervention Studies

The OH-CCI study by Timaru-Kast et al. [35] included an intervention arm where 0.1 mg/kg candesartan (angiotensin II receptor 1 blocker) or vehicle (Na_2CO_3) solution was delivered by subcutaneous injection at 30 min post-OH-CCI, followed by daily injection for 4 days. At D5 post-OH-CCI, candesartan treatment reduced lesion volume by 20% and reduced perilesional Iba1-immunopositive cells, in both young-TBI and aged-TBI. In aged-TBI, but not young-TBI mice, candesartan reduced the number of brain-infiltrating neutrophil granulocytes. Candesartan treatment also increased the expression level of Ym1 (a marker of "M2a" microglia) in both young-TBI (D1 and D5) and aged-TBI (D5).

In the FPI study by Zhu et al. [36], they administered 2 mg/kg Epothilone D (EpoD, a microtubule-stabilizing agent) at 24 h post-FPI to 1.5-month and 3-month mice, followed by weekly injection of EpoD for 4 weeks. EpoD treatment was not performed in 12-month mice. DMSO served as the vehicle control. Compared to vehicle-treated mice, EpoD exacerbated axonal degeneration in 3-month-TBI but not 1.5-month TBI (4 weeks post-FPI). Cortical GFAP and Iba1 levels were not altered by EpoD treatment.

In the OH-CCI study by Ritzel et al. [38], 5% trehalose (autophagic enhancer) or sucrose (control) was injected subcutaneously in aged mice (18 months), immediately post-OH-CCI. Five percent trehalose or sucrose was then administered into drinking water for 7 days, followed by continuous administration of 2.5% trehalose or sucrose. Treatment was not performed in young mice (10–12 weeks). While sucrose-treated mice showed significant deficits in motor functions and locomotion (rotarod 1 week to 8 weeks post-OH-CCI; open field distance and speed at 1 week post-OH-CCI), no significant deficits were observed in trehalose-treated mice. Sucrose-treated mice showed thigmotactic behavior and impairment in novel object recognition and social interaction tasks. These deficits were prevented by trehalose treatment (8 weeks post-OH-CCI). Trehalose treatment also reduced microglia proliferation, prevented delayed recruitment of peripheral immune cells, and reduced the level of Vglu1 and NeuN engulfed by microglia.

In the FPI study by Wangler et al. [39], DMXAA (25 mg/kg, a murine stimulator of interferon agonist) was injected intraperitoneally to young mice (2 months), at 1 h before FPI and then every day up to D6 post-injury. DMXAA treatment was not performed in aged mice. Compared to vehicle treatment, DMXAA prolonged time of righting reflex post-FPI. At D7, DMXAA further increased cortical immunopositivity of Iba1, when compared to vehicle-treated young-TBI mice. DMXAA also enhanced the expression of inflammatory genes, including IFN-associated genes (Irf7, Stat1), complement (C3, C1qa), chemokine (Ccl2, Ccl12, Ccl5, Cxcl10, Cxcl16), pattern recognition receptors (Clec7a, Tlr2), and injury-associated genes (Tyrobp, Atf3) at D7 post-FPI.

Tajiri et al. [25] induced OH-CCI to male Fisher344 rats (6 months vs 20 months). At 3 h post-OH-CCI, human adipose-derived stem cells (hADSC, 4×10^6 cells) or conditioned media (CM) was administered to the jugular vein of the rats. In vehicle (unconditioned media)-treated rats, both young-TBI and aged-TBI rats manifested motor deficits (asymmetric body swing, forelimb akinesia, paw grasp, D0-D7 post-OH-CCI), cognitive deficits (reversal learning of radial arm water maze, D16), and cortical tissue loss and hippocampal neuronal loss (D11). In young-TBI rats, hADSC or CM treatment ameliorated motor and cognitive deficits, reduced cortical lesion area, and hippocampal neuronal loss. None of these treatment effects were observed in aged-TBI rats. The treatment effects were abolished when CM were harvested from hADSC with knockdown of NEAT1 (nuclear enriched abundant transcript 1) or MALAT1 (metastasis associated lung adenocarcinoma transcript 1), which are long non-coding RNA that regulate cellular differentiation and gene expression. Since knockdown of NEAT or MALAT1 partially reduced the presence of vascular endothelial growth factor (VEGF) and stem cell factor (SCF) while increasing the expression of tissue inhibitor of metalloproteinase 3 (TIMP3), Tajiri et al. postulated that hADSC and CM may exert their effects through these factors.

In summary, many preclinical experiments have been performed in the quest for TBI interventions. Some current possible beneficial targets include trophic factors, angiotensin II receptor 1 inhibitor, and autophagy enhancer, whereas a microtubule stabilizing agent and a stimulator of interferons agonist may exacerbated TBI outcomes. However, existing evidence suggests that young and aged animals may react to TBI treatments differently at different time points. Therefore, it will be important to consider both age at injury and time of assessment as important factors when designing future preclinical intervention studies.

13.7 Challenges and Future Directions

TBI is a recognized public health concern. Though it is an active research area, clinical features of TBI are insufficiently incorporated into preclinical study designs. For example, most preclinical TBI research is largely focused on young animals, even though TBI incidence is now highest among the aged population, particularly in high-income countries. In addition, most TBI-ageing studies are performed using open-head TBI models (e.g., OH-CCI and FPI). However, the most frequent cause of TBI in the aged population is unintentional falls [2], which often happens during walking and slipping [44]. These same-level falls are not frequently associated with skull fracture [45]. Furthermore, the older population has a higher chance of having underlying comorbidities and developing polytrauma during falls. Finally, the vast majority of TBI research is based on male animals, even though the incidence of TBI-related hospitalization and TBI-related deaths is comparable between men and women [2]. Therefore, preclinical TBI researchers should consider testing older animals, employing closed-head TBI mechanisms, using animal models with comorbidities (e.g., transgenic disease models) or polytrauma, and balancing male and female animals in their studies.

There are specific animal considerations when designing ageing experiments, for example, mouse age and body weight. The maturation and life stages in mice are not linearly correlated with human maturation and ageing [46]. For example, 3–6 months in mice corresponds to 20–30 years in humans (mice mature 45 times faster), whereas 18–24 months in mice corresponds to 56–69 years in humans (mice mature 25 times faster). However, aged mice have higher research costs (due to housing and mortality). In addition, the body weight of aged mice is different from that of young mice and is often more variable [47]. Special considerations may be needed during experimental design (e.g., anesthesia, treatment dosage, or outcome measurements that are influenced by body weight, for example, behavior that involves balance tasks).

The central role of neuroinflammatory pathways induced by TBI deserves particular attention. One often-overlooked area of reporting in most preclinical studies relates to the health status of animals across different animal facilities, as factors such as conventional or specific pathogen-free housing and diet can have marked effects on baseline inflammatory tone. For example, mice of exactly same genetic background may respond entirely different to the same TBI protocol at different facilities. One suggestion for the preclinical research community to consider would be to include microbiome status in their TBI studies as a potential anchor point for harmonization of baseline inflammatory status.

Another point of consideration when designing TBI-ageing experiments is statistical analysis. Many preclinical TBI studies often adopt a 2x2 grouping design (injury, sham vs TBI; age, young vs aged). Thus, when linear regression models (e.g., two-way ANOVA) are employed during statistical analysis, it is important to report the statistical significance of the injury factor, the age factor, and the injury * age interaction. Post-hoc tests should be performed where appropriate. If the effects of injury, age, and post-hoc comparisons are all significant, an insignificant interaction effect may suggest that the effects of injury and age are additive, whereas a significant interaction effect may suggest that the effects of injury more than proportionally exacerbate the effects of ageing (or cancel the effects of one another, depending on the direction of change). It is not preferable to report an injury effect in the young cohort and an injury effect in the aged cohort separately, as it may be susceptible to Type I error due to multiple comparison. Moreover, the baseline performance of a young mouse could be very different from that of aged mouse (e.g., behavior performance or gene expression), and the researchers may choose to report the outcome measurements in terms of absolute change in performance or fold change over respective sham. If the outcome is a measurement of functional deficit (e.g., motor or cognitive performance), the absolute change is preferred. However, if the outcome is a biological correlate that does not have a predefined upper bound or lower bound (e.g., gene expression level), a relative change over respective sham may be more convenient for interpretation.

Finally, the challenge of calibrating a constant injury severity level across different age groups should not be overlooked, as brain volume, the mechanical properties of brain tissue, and skull thickness and rigidity may all play a role in the biological response to the same mechanical insult [48, 49].

In the previous sections, we have discussed several challenges faced by preclinical TBI researchers who are working to define how age at injury modifies TBI outcomes. The development of neurological blood biomarkers may potentially help to calibrate injury severity across different models and ages. Blood biomarkers are objectively measurable substances in serum or plasma that can be used as a measure of health/disease status. They can be used to guide clinical diagnosis (diagnostic biomarker), to evaluate disease risk or prognosis (prognostic biomarker), or to monitor therapeutic interventions (pharmacodynamics biomarker). Impressive advances have been made in the discovery of blood biomarkers for various neurological diseases and conditions including dementia (e.g., Alzheimer's disease [50], frontotemporal dementia, Parkinson dementia), neurotrauma (TBI [51], spinal cord injury), neuroinflammatory disease (multiple sclerosis, COVID-19), and cerebrovascular disease (hypoxic ischemic brain injury, stroke). Blood biomarkers have several advantages over neuroimageing and cerebrospinal fluid biomarkers as they are more high-throughput, less invasive, and less expensive [52]. In 2018, the US Food and Drug Administration (FDA) approved the use of two blood biomarkers for concussion (as prescreening tools to decide if a patient requires a computed tomography (CT) scan): GFAP and ubiquitin C- terminal hydrolase 1 (UCH-L1) [53]. GFAP is of particular interest as it is also implicated in several other neurological diseases and conditions such as Alzheimer's disease [54]. More recently, neurofilament light chain (NF-L) has been revealed to be another promising diagnostic and prognostic blood biomarker for mild, moderate, and severe TBI [55]. NF-L and GFAP are among the most pursued blood biomarkers in clinical neurotrauma studies.

However, despite tremendous advances in assay validation for human studies, the preclinical TBI field lacks comparable advances for assays that can be consistently applied in preclinical settings, as blood volume, cross-species antibody reactivity, and expense of biomarker assays are major considerations that have precluded wide adoption in preclinical TBI studies. Further, many preclinical TBI studies that do use biomarkers (reviewed in Zanier et al., under review) have inadequate reporting of assay manufacturer, catalogue, and lot number, and some fail to show evidence of cross-species assay performance. To begin to fill the gap of robust assays of blood-based TBI biomarkers that work in mice, our group recently developed and validated a novel sensitive immunoassay to quantify murine GFAP in blood on the MesoScale Discovery (MSD) platform [56]. We also validated the use of MSD human NF-L antibody set (cat# F217X) on murine samples as these antibodies do cross-react with murine NF-L. These blood biomarker assays may facilitate the translational value of preclinical research on TBI-ageing and provide insights into the mechanistic pathophysiology.

Acknowledgments This research was supported by the Canadian Institutes of Health Research (PJT186157), Weston Brain Institute (TR192003), and US Department of Defense (AZ190011).

Disclosures None.

References

1. Maas AIR, et al. Traumatic brain injury: progress and challenges in prevention, clinical care, and research. Lancet Neurol. 2022;21:1004–60. https://doi.org/10.1016/S1474-4422(22)00309-X.
2. Centers for Disease Control and Prevention. Surveillance report of traumatic brain injury-related hospitalizations and deaths by age group, sex, and mechanism of injury—United States, 2016 and 2017. (2021).
3. Kamal VK, Agrawal D, Pandey RM. Prognostic models for prediction of outcomes after traumatic brain injury based on patients admission characteristics. Brain Inj. 2016;30:393–406. https://doi.org/10.3109/02699052.2015.1113568.
4. Skaansar O, et al. Traumatic brain injury-the effects of patient age on treatment intensity and mortality. BMC Neurol. 2020;20:376. https://doi.org/10.1186/s12883-020-01943-6.
5. Faul, M., Xu, L., Wald, M., Coronado, V. (ed Centers for Disease Control and Prevention US Department of Health and Human Services, National Center for Injury Prevention and Control) (2010).
6. Mosenthal AC, et al. The effect of age on functional outcome in mild traumatic brain injury: 6-month report of a prospective multicenter trial. J Trauma. 2004;56:1042–8.
7. Willemse-van Son AH, Ribbers GM, Verhagen AP, Stam HJ. Prognostic factors of long-term functioning and productivity after traumatic brain injury: a systematic review of prospective cohort studies. Clin Rehabil. 2007;21:1024–37. https://doi.org/10.1177/0269215507077603.
8. Marquez de la Plata CD, et al. Impact of age on long-term recovery from traumatic brain injury. Arch Phys Med Rehabil. 2008;89:896–903. https://doi.org/10.1016/j.apmr.2007.12.030.
9. Brickler, T., Morton, P. D., Hazy, A. & Theus, M. In: Traumatic brain injury—pathobiology, advanced diagnostics and acute management (eds N. Gorbunov & J. Long) Ch. 3, (IntechOpen, 2018).
10. Hukkelhoven CW, et al. Patient age and outcome following severe traumatic brain injury: an analysis of 5600 patients. J Neurosurg. 2003;99:666–73. https://doi.org/10.3171/jns.2003.99.4.0666.
11. Dhandapani S, Manju D, Sharma B, Mahapatra A. Prognostic significance of age in traumatic brain injury. J Neurosci Rural Pract. 2012;3:131–5. https://doi.org/10.4103/0976-3147.98208.
12. Noda S, Sato S, Fukuda T, Tada N, Hattori N. Aging-related motor function and dopaminergic neuronal loss in C57BL/6 mice. Mol Brain. 2020;13:46. https://doi.org/10.1186/s13041-020-00585-6.
13. Gao J, et al. Influence of aging on the dopaminergic neurons in the substantia nigra pars compacta of rats. Curr Aging Sci. 2011;4:19–24. https://doi.org/10.2174/1874609811104010019.
14. Gu J, et al. Plasmalogens eliminate aging-associated synaptic defects and microglia-mediated neuroinflammation in mice. Front Mol Biosci. 2022;9:815320. https://doi.org/10.3389/fmolb.2022.815320.
15. Pan J, Ma N, Yu B, Zhang W, Wan J. Transcriptomic profiling of microglia and astrocytes throughout aging. J Neuroinflammation. 2020;17:97. https://doi.org/10.1186/s12974-020-01774-9.
16. Payne BA, Chinnery PF. Mitochondrial dysfunction in aging: much progress but many unresolved questions. Biochim Biophys Acta. 1847:1347, 2015–53. https://doi.org/10.1016/j.bbabio.2015.05.022.
17. West AP, Shadel GS, Ghosh S. Mitochondria in innate immune responses. Nat Rev Immunol. 2011;11:389–402. https://doi.org/10.1038/nri2975.
18. Giorgi C, et al. Mitochondria and reactive oxygen species in aging and age-related diseases. Int Rev Cell Mol Biol. 2018;340:209–344. https://doi.org/10.1016/bs.ircmb.2018.05.006.
19. Navarro A, Boveris A. Rat brain and liver mitochondria develop oxidative stress and lose enzymatic activities on aging. Am J Physiol Regul Integr Comp Physiol. 2004;287:R1244–9. https://doi.org/10.1152/ajpregu.00226.2004.
20. Gilmer LK, Ansari MA, Roberts KN, Scheff SW. Age-related mitochondrial changes after traumatic brain injury. J Neurotrauma. 2010;27:939–50. https://doi.org/10.1089/neu.2009.1181.

21. Onyszchuk G, He YY, Berman NE, Brooks WM. Detrimental effects of aging on outcome from traumatic brain injury: a behavioral, magnetic resonance imaging, and histological study in mice. J Neurotrauma. 2008;25:153–71. https://doi.org/10.1089/neu.2007.0430.
22. Sandhir R, Onyszchuk G, Berman NE. Exacerbated glial response in the aged mouse hippocampus following controlled cortical impact injury. Exp Neurol. 2008;213:372–80. https://doi.org/10.1016/j.expneurol.2008.06.013.
23. Timaru-Kast R, et al. Influence of age on brain edema formation, secondary brain damage and inflammatory response after brain trauma in mice. PLoS One. 2012;7:e43829. https://doi.org/10.1371/journal.pone.0043829.
24. Unterberg A, Schneider GH, Gottschalk J, Lanksch WR. Development of traumatic brain edema in old versus young rats. Acta Neurochir Suppl (Wien). 1994;60:431–3. https://doi.org/10.1007/978-3-7091-9334-1_117.
25. Tajiri N, et al. Stem cell-paved biobridge facilitates neural repair in traumatic brain injury. Front Syst Neurosci. 2014;8:116. https://doi.org/10.3389/fnsys.2014.00116.
26. Timaru-Kast R, Herbig EL, Luh C, Engelhard K, Thal SC. Influence of age on cerebral housekeeping gene expression for normalization of quantitative polymerase chain reaction after acute brain injury in mice. J Neurotrauma. 2015;32:1777–88. https://doi.org/10.1089/neu.2014.3784.
27. Glushakov AV, Arias RA, Tolosano E, Dore S. Age-dependent effects of haptoglobin deletion in neurobehavioral and anatomical outcomes following traumatic brain injury. Front Mol Biosci. 2016;3:34. https://doi.org/10.3389/fmolb.2016.00034.
28. Gupta RK, Prasad S. Age-dependent alterations in the interactions of NF-kappaB and N-myc with GLT-1/EAAT2 promoter in the pericontusional cortex of mice subjected to traumatic brain injury. Mol Neurobiol. 2016;53:3377–88. https://doi.org/10.1007/s12035-015-9287-y.
29. Rowe RK, Griffiths DR, Lifshitz J. Midline (central) fluid percussion model of traumatic brain injury. Methods Mol Biol. 2016;1462:211–30. https://doi.org/10.1007/978-1-4939-3816-2_13.
30. Cheng WH, et al. Age at injury and genotype modify acute inflammatory and neurofilament-light responses to mild CHIMERA traumatic brain injury in wild-type and APP/PS1 mice. Exp Neurol. 2018;301:26–38. https://doi.org/10.1016/j.expneurol.2017.12.007.
31. Rowe RK, Harrison JL, Ellis TW, Adelson PD, Lifshitz J. Midline (central) fluid percussion model of traumatic brain injury in pediatric and adolescent rats. J Neurosurg Pediatr. 2018;22:22–30. https://doi.org/10.3171/2018.1.PEDS17449.
32. Collins JM, King AE, Woodhouse A, Kirkcaldie MTK, Vickers JC. Age moderates the effects of traumatic brain injury on beta-amyloid plaque load in APP/PS1 Mice. J Neurotrauma. 2019;36:1876–89. https://doi.org/10.1089/neu.2018.5982.
33. Ritzel RM, et al. Old age increases microglial senescence, exacerbates secondary neuroinflammation, and worsens neurological outcomes after acute traumatic brain injury in mice. Neurobiol Aging. 2019;77:194–206. https://doi.org/10.1016/j.neurobiolaging.2019.02.010.
34. Sun M, et al. Aged rats have an altered immune response and worse outcomes after traumatic brain injury. Brain Behav Immun. 2019;80:536–50. https://doi.org/10.1016/j.bbi.2019.04.038.
35. Timaru-Kast R, et al. Angiotensin II receptor 1 blockage limits brain damage and improves functional outcome after brain injury in aged animals despite age-dependent reduction in AT1 expression. Front Aging Neurosci. 2019;11:63. https://doi.org/10.3389/fnagi.2019.00063.
36. Zhu Z, Chuckowree JA, Musgrove R, Dickson TC, Blizzard CA. The pathologic outcomes and efficacy of epothilone treatment following traumatic brain injury is determined by age. Neurobiol Aging. 2020;93:85–96. https://doi.org/10.1016/j.neurobiolaging.2020.03.023.
37. Islam M, et al. Differential neuropathology and functional outcome after equivalent traumatic brain injury in aged versus young adult mice. Exp Neurol. 2021;341:113714. https://doi.org/10.1016/j.expneurol.2021.113714.
38. Ritzel RM, et al. Functional and transcriptional profiling of microglial activation during the chronic phase of TBI identifies an age-related driver of poor outcome in old mice. Geroscience. 2022;44:1407–40. https://doi.org/10.1007/s11357-022-00562-y.

39. Wangler LM, et al. Amplified gliosis and interferon-associated inflammation in the aging brain following diffuse traumatic brain injury. J Neurosci. 2022;42:9082–96. https://doi.org/10.1523/JNEUROSCI.1377-22.2022.
40. Chen Z, et al. Microglia and infiltrating T-Cells adopt long-term, age-specific, transcriptional changes after traumatic brain injury in mice. Shock. 2023;59:267–76. https://doi.org/10.1097/SHK.0000000000002027.
41. Lee CK, Weindruch R, Prolla TA. Gene-expression profile of the ageing brain in mice. Nat Genet. 2000;25:294–7. https://doi.org/10.1038/77046.
42. Boone DR, et al. Traumatic brain injury induces long-lasting changes in immune and regenerative signaling. PLoS One. 2019;14:e0214741. https://doi.org/10.1371/journal.pone.0214741.
43. Ritzel RM, et al. Brain injury accelerates the onset of a reversible age-related microglial phenotype associated with inflammatory neurodegeneration. Sci Adv. 2023;9:eadd1101. https://doi.org/10.1126/sciadv.add1101.
44. Unguryanu TN, Grjibovski AM, Trovik TA, Ytterstad B, Kudryavtsev AV. Mechanisms of accidental fall injuries and involved injury factors: a registry-based study. Inj Epidemiol. 2020;7:8. https://doi.org/10.1186/s40621-020-0234-7.
45. Liew TYS, et al. Changing demographic profiles of patients with traumatic brain injury: an aging concern. Front Surg. 2019;6:37. https://doi.org/10.3389/fsurg.2019.00037.
46. Flurkey K, Currer JM, Harrison DE. In: The mouse in biomedical research. history, wild mice, and genetics (eds Fox JG et al.) Ch. 20, (Academic, 2007).
47. The Jackson Laboratory. Body weight information for aged C57BL/6J.https://www.jax.org/jax-mice-and-services/strain-data-sheet-pages/body-weight-chart-aged-b6
48. Namjoshi DR, et al. Towards clinical management of traumatic brain injury: a review of models and mechanisms from a biomechanical perspective. Dis Model Mech. 2013;6:1325–38. https://doi.org/10.1242/dmm.011320.
49. Meaney DF, Morrison B, Dale Bass C. The mechanics of traumatic brain injury: a review of what we know and what we need to know for reducing its societal burden. J Biomech Eng. 2014;136:021008. https://doi.org/10.1115/1.4026364.
50. Blennow K, Hampel H, Weiner M, Zetterberg H. Cerebrospinal fluid and plasma biomarkers in Alzheimer disease. Nat Rev Neurol. 2010;6:131–44. https://doi.org/10.1038/nrneurol.2010.4.
51. Zetterberg H, Blennow K. Fluid biomarkers for mild traumatic brain injury and related conditions. Nat Rev Neurol. 2016;12:563–74. https://doi.org/10.1038/nrneurol.2016.127.
52. Hampel H, et al. Blood-based biomarkers for Alzheimer disease: mapping the road to the clinic. Nat Rev Neurol. 2018;14:639–52. https://doi.org/10.1038/s41582-018-0079-7.
53. U.S. Food and Drug Administration. FDA authorizes marketing of first blood test to aid in the evaluation of concussion in adults. https://www.fda.gov/news-events/press-announcements/fda-authorizes-marketing-first-blood-test-aid-evaluation-concussion-adults (2018).
54. Middeldorp J, Hol EM. GFAP in health and disease. Prog Neurobiol. 2011;93:421–43. https://doi.org/10.1016/j.pneurobio.2011.01.005.
55. Shahim P, et al. Time course and diagnostic utility of NfL, tau, GFAP, and UCH-L1 in subacute and chronic TBI. Neurology. 2020;95:e623–36. https://doi.org/10.1212/WNL.0000000000009985.
56. Button EB, et al. Development of a novel, sensitive translational immunoassay to detect plasma glial fibrillary acidic protein (GFAP) after murine traumatic brain injury. Alzheimers Res Ther. 2021;13:58. https://doi.org/10.1186/s13195-021-00793-9.

Index

L. J. Noble-Haeusslein, D. M. Schnyer (eds.), *Traumatic Brain Injuries*, Advances in Neurobiology 42, https://doi.org/10.1007/978-3-031-69832-3

MIX
Papier aus verantwortungsvollen Quellen
Paper from responsible sources
FSC® C105338

If you have any concerns about our products,
you can contact us on
ProductSafety@springernature.com

In case Publisher is established outside the EU,
the EU authorized representative is:
Springer Nature Customer Service Center GmbH
Europaplatz 3, 69115 Heidelberg, Germany

Printed by Libri Plureos GmbH
in Hamburg, Germany